Level 3

INSTALLING ELECTROTECHNICAL SYSTEMS & EQUIPMENT

Book A
NVQ/SVQ AND DIPLOMA

JTL
DELIVERING SKILLS
FOR THE FUTURE

www.pearsonschoolsandfecolleges.co.uk

✓ Free online support
✓ Useful weblinks
✓ 24 hour online ordering

0845 630 44 44

Heinemann

Part of Pearson

Heinemann is an imprint of Pearson Education Limited, Edinburgh Gate, Harlow, Essex, CM20 2JE.

www.pearsonschoolsandfecolleges.co.uk

Heinemann is a registered trademark of Pearson Education Limited

Text © JTL 2011
Typeset by Saxon Graphics Ltd
Original illustrations © Pearson Education Ltd 2011
Illustrated by Oxford Designers
Cover design by Wooden Ark
Cover photo/illustration © Construction Photography: Xavier de Canto

The right of JTL to be identified as author of this work has been asserted by them in accordance with the Copyright, Designs and Patents Act 1988

First published 2011

14 13 12 11
10 9 8 7 6 5 4 3 2 1

British Library Cataloguing in Publication Data
A catalogue record for this book is available from the British Library

ISBN 978 0 435 031268

Printed and bound in Spain by Grafos

Websites
Pearson Education Limited is not responsible for the content of any external internet sites. It is essential for tutors to preview each website before using it in class so as to ensure that the URL is still accurate, relevant and appropriate. We suggest that tutors bookmark useful websites and consider enabling students to access them through the school/college intranet.

The information and activities in this book have been prepared according to the standards reasonably to be expected of a competent trainer in the relevant subject matter. However, you should be aware that errors and omissions can be made and that different employers may adopt different standards and practices over time. Before doing any practical activity, you should always carry out your own risk assessment and make your own enquiries and investigations into appropriate standards and practices to be observed.

Acknowledgements

Every effort has been made to contact copyright holders of material reproduced in this book. Any omissions will be rectified in subsequent printings if notice is given to the publishers.

The author and publisher would like to thank the following individuals and organisations for permission to reproduce photographs:

(Key: b-bottom; c-centre; l-left; r-right; t-top)
Alamy Images: APIX 110tr, Cernan Elias 299tr, Charles Stirling 49bc, Geoff du Feu 311br, Jean Schweitzer 150, maxstock 50cr, Paul Thompson Images 110cr; Stephen Barnes / Living History 190tl; **Andrew Wilde:** 372br; **Construction Photography:** Anthony Weller 159, buildpix 84, Paul McMullin 108, Sally-Ann Norman 112; **Corbis:** Image100 237; **CSCS:** 201, 202; **Sid Frisby:** 311cr; **Getty Images:** StockByte 119; **Graham Hare:** 338; **Robert Harding World Imagery:** age fotostock 240; **iStockphoto:** Nancy Honeycutt 301; **Pearson Education Ltd:** Clark Wiseman, Studio 8 49bl, 50tl, 51br, 52bl, 52br, 53tr, David Sanderson 51tr, 53bl, 99, Gareth Boden 31, 57, 58, 59, 73, 75, 83, 88, 123, 171cr, 171br, 214, 255tr, 255cr, 255br, 256tl, 264tl, 265tr, 266tl, 266bl, 267, 268cl, 268bl, 318tl, 318cl, 318bl, 319t, 319b, 320tl, 320tr, 320cl, 320bl, 321tr, 321cr, 321br, 322, 322tl, 322tc, 322tr, 322cl, 322bl, 323cl, 323cr, 323br, 324tl, 324cl/1, 324cl/2, 324bl, 324bc, 331, 344tl, 344b, 345, 346, 347, 348t, 348cl ,349, 350, 352tl, 352tr, 352b, 353, 354, 355, 357, 358, 360tl, 360tr, 361cl, 361cr, 367, 369, 375tr, 375cr, 375br, 417, 420bl, 423, 424cl, 424bl, 425tr, 425cr, 425br, 426tl, 426cl, 426bl, 428, 430tl, 430tr, 430cl, 432, 433tl, 433tr, 450cl, Naki Photography 25, 33, 256, 257tr, 257br, 258cr, 258bl, 259, 264cl, 264cr, 265br, 372bc, 431tl, 431tr; **Wayne Howes Photography:** 304bl; **Science Photo Library Ltd:** GIPhotostock 260tl, 260cl, Tek Image 415; **Shutterstock.com:** Alaettin YILDIRIM 310, Alex Hinds 300, Alexey Nikolaev 49br, auremar 53br, 232, Claude Beaubien 295, Crstrbrt 191, DavidXu 411, Diego Cervo 96, Dmitry Kalinovsky 343, Edd Westmacott 51cr, ekipai 288, Elnur 53bc, Frances A. Miller 111, jennyt 299cr, Justin Kral 100, Kodda 121, LoopAll 70, Nightman1965 305, Olinchuk 50cl, Skripko Ievgen 202/INSET, Skyline 311tr, ssuaphotos 140, Tom Mc Nemar 1, trekandshoot 304tl, ULKASTUDIO 262, WDG Photo 146, Yuri Arcurs 201/INSET, Zakharoff 190; **Warren A Stallard:** 372bl

All other images © Pearson Education

Every effort has been made to trace the copyright holders and we apologise in advance for any unintentional omissions. We would be pleased to insert the appropriate acknowledgement in any subsequent edition of this publication.

Contents

Introduction

This book is designed to support the new NVQ Level 3 Diploma in Installing Electrotechnical Systems and Equipment. This new Diploma has been prepared by SummitSkills (the SSC) in consultation with employers, the main exam boards and training providers. This means that all exam boards offering this qualification will have the same unit structure and assessment strategy. Exam boards will then design their own assessment content.

This book is designed to support the following qualifications:

- 2357 City and Guilds Level 3 NVQ Diploma in Installing Electrotechnical Systems and Equipment
- 501/1605/8 EAL Level 3 NVQ Diploma in Installing Electrotechnical Systems and Equipment (Buildings, Structures and the Environment)

These qualifications are approved on the Qualifications and Credit Framework (QCF). The QCF is a new government framework which regulates all vocational qualifications to ensure they are structured and titled consistently and quality assured. SummitSkills have developed the new Diploma qualification with the awarding bodies.

Who the qualification is aimed at

The new diploma qualification is aimed at both new entrants (such as apprentices or adult career changers) as well as the existing workforce (so those looking to upskill). It is intended to train and assess candidates so that they can be recognized as occupationally competent by the industry in installing electrotechnical systems. Learners should gain the skills to:

- work as a competent electrician
- achieve a qualification recognized by the Joint Industry Board (JIB) for professional grading to the industry
- complete an essential part of the SummitSkills Advanced Apprenticeship.

About this book

This book supports the first five units of the Level 3 Diploma. In combination with Book B, which covers the remaining four units (04a, 06, 07 and 08), it is designed to cover all the information you will need to attain your Level 3 qualification in Installing Electrotechncial Systems and Equipment.

Each chapter of this book relates to a particular unit of the Diploma and provides the information needed to gain the required knowledge and understanding of that area.

This book has been prepared by expert JTL trainers, who have many years of experience of training learners and delivering electrical qualifications. The content of each unit will underpin the various topics which you will be assessed on by your exam board.

Each unit has knowledge checks throughout, as well as a set of multiple choice questions at its conclusion, to allow you to measure your knowledge and understanding.

This book will also be a useful reference tool for you in your professional life once you have gained your qualifications and are a practicing electrician.

Using this book

It is important to note that this book is intended to be used for training. It should not be regarded as being relevant to an actual installation. You should always make specific reference to the British Standards or manufacturer's data when designing electrical installations.

Features of this book

This book has been fully illustrated with artworks and photographs. These will help to give you more information about a concept or a procedure, as well helping you to follow a step-by-step procedure or identify a particular tool or material.

This book also contains a number of different features to help your learning and development.

Key term

These are new or difficult words. They are picked out in **bold** in the text and then defined in the margin.

Remember

This highlights key facts or concepts, sometimes from earlier in the text, to remind you of important things you will need to think about.

Safety tip

This feature gives you guidance for working safely on the tasks in this book.

Did you know?

This feature gives you interesting facts about the building trade.

Find out

These are short activities and research opportunities, designed to help you gain further information about, and understanding of, a topic area.

Working life

This feature gives you a chance to read about and debate a real life work scenario or problem. Why has the situation occurred? What would you do?

Progress check

These are a series of short questions, usually appearing at the end of each learning outcome, which gives you the opportunity to check and revise your knowledge.

Getting ready for assessment

This feature provides guidance for preparing for the practical assessment. It will give you advice on using the theory you have learnt about in a practical way.

CHECK YOUR KNOWLEDGE

This is a series of multiple choice questions at the end of each unit, in the style of the GOLA end of unit tests.

Acknowledgements

JTL would like to express its appreciation to all those members of staff who contributed to thie development of this book, ensuring that the professional standards expected were delivered and generally overseeing the high quality of the final product. Without their commitment this project would not have been seen through successfully.

Particular thanks to Dave Allan, who revised the content of the previous editions of this book and prepared extensive new material to cover the diploma specifications.

The Working Life, Check Your Knowledge and Getting Ready for Assessment features were prepared by Bill Brady of EAL (EMTA Awards Limited). Bill also carried out a full review of the book and made many invaluable comments.

Pearson and JTL would like to thank Richard Swann of Loughborough College and Andy Jeffery of Oaklands College for their comprehensive and painstaking review work, which has again made an invaluable contribution towards ensuring accuracy in this book.

Pearson and JTL would also wish to thank Paul, Rita and Jeff Hurt at P & R Hurt Education and Training for their patience, assistance, advice and support during the photo shoot.

Candidate handbook answers

The answers to all the questions in this book can be found on the Training Resource Disk. They can also be downloaded from Pearson's website at the following URL: **www.pearsonfe.co.uk/iestudentbookanswers**

If there are any problems downloading this PDF, please contact Pearson on 0845 6301111.

UNIT ELTK 01

Understanding health and safety legislation, practices and procedures

All employers, including the self-employed, have duties under the Health and Safety at Work Act 1974 to ensure the health and safety of themselves and others affected by what they do. This includes people working for employers and the self-employed (for example, part-time workers, trainees and subcontractors), those who use the workplace and equipment they provide, those who visit their premises, and people affected by their work (for example, neighbours or the general public).

This unit will cover the following learning outcomes:

■ understand how relevant health and safety legislation applies in the workplace

■ understand the procedures for dealing with health and safety in the work environment

■ understand the procedures for establishing a safe working environment

■ understand the requirements for identifying and dealing with hazards in the work environment.

K1. Understand how relevant health and safety legislation applies in the workplace

This outcome looks at the regulatory requirements and responsibilities laid down by official acts and regulations. All of the following legislation is law and is enforceable.

As a worker, these acts and regulations provide a legal framework that protects you. This includes the building in which you are working and the electrical supply that you are working with. They also ensure that you do not cause harm to yourself or others through bad practices.

Roles and responsibilities in regard to current legislation

Health and Safety at Work Act 1974 (HASAWA)

The Health and Safety at Work Act 1974 (HASAWA) is an 'Enabling Act' and sets out the basic principles by which health and safety at work are regulated.

Attached to the Act is a series of regulations covering the practical detail of how employers and employees create, maintain and operate a safe working environment. These regulations are also used by government authorities to control the standard of working conditions throughout industry.

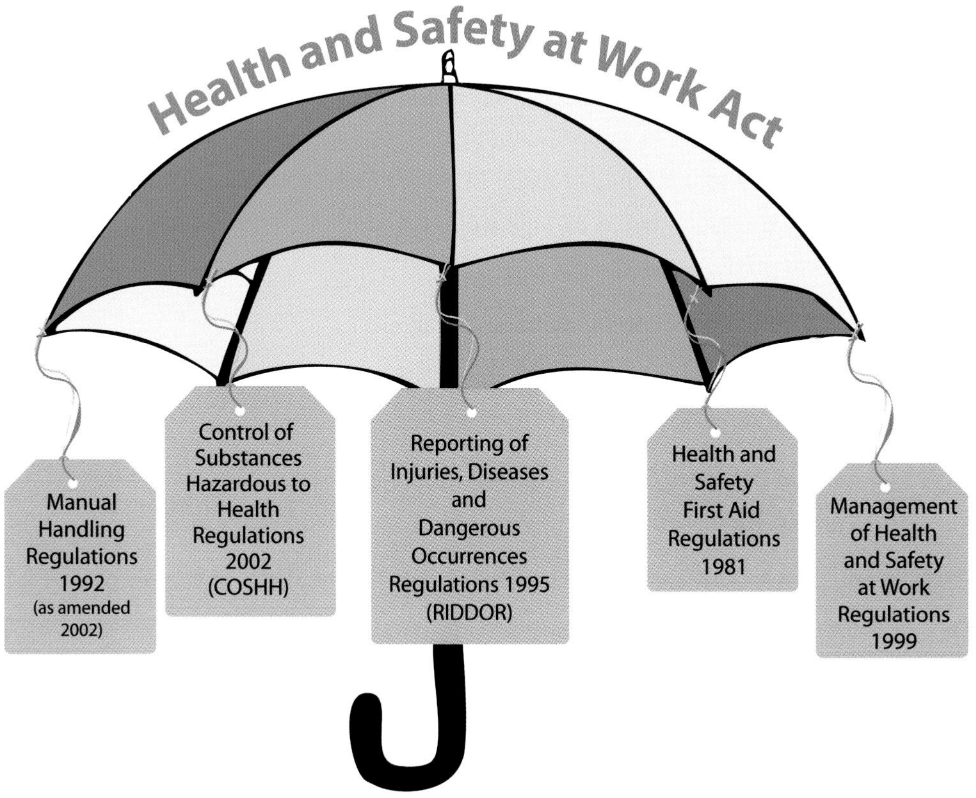

Remember

There are many sites and projects and consequently many acts and regulations. It would be impossible to predict every site that you will ever work on, but here you will look at the most important and the most commonly applied legislation.

Manual Handling Regulations 1992 (as amended 2002)

Control of Substances Hazardous to Health Regulations 2002 (COSHH)

Reporting of Injuries, Diseases and Dangerous Occurrences Regulations 1995 (RIDDOR)

Health and Safety First Aid Regulations 1981

Management of Health and Safety at Work Regulations 1999

Figure 1.01 The Health and Safety at Work Act is an 'umbrella' act covering a number of other regulations

The Act is statutory: it is binding in law, and criminal penalties can be imposed on people found guilty of malpractice and misconduct. All employers are covered by HASAWA. The Act places certain specific duties on both employers and employees, which must be complied with by law.

Each employee – and this includes you – is also required by law to assist and co-operate with their employer and others in making sure that safe working environments are maintained, that all safety equipment is fully and correctly used and that all safety procedures are followed. You should work safely and use common sense at all times.

There are many sections to the Act, but the main ones affecting you – Section 2 (duties of employers to their employees) and Section 7 (duties of employees) – are shown in Table 1.01.

Find out

Look up the Health and Safety at Work Act using a relevant website – you can download it for free. Prepare a brief report on it describing:

- what you would expect to find provided for you in a job role of your choice
- what your duties would be to your employer.

Section 2: Employers	Section 7: Employees
All employers have a duty to:	As an employee you have a duty to:
• care for the welfare, health and safety of their employees where it is practicable for them to do so	• look after your own safety and that of your colleagues and people around them
• provide and maintain safe equipment, tools and plant within the workplace	• not intentionally or recklessly interfere with or misuse anything provided for your health and safety
• ensure working conditions are safe and hygienic	• take reasonable care at work for the health and safety of yourself and others who may be affected by what you do or do not do • bring to your employer's attention any situation you consider dangerous
• provide proper **personal protective equipment** (**PPE**) and make sure it is used correctly	• use any PPE provided correctly
• make sure articles and substances are used, handled, stored and transported safely	• help the employer to meet their statutory obligations
• provide any necessary information, instruction, training and supervision to ensure the health and safety of employees • make sure everyone can get in and out of the workplace safely • provide adequate facilities and arrangements for welfare at work	• co-operate with your employer on health and safety matters • bring to your employer's attention any weakness in their welfare, health and safety arrangements

Table 1.01 Summary of the duties of employers and employees

The employer must also:

- draw up a health and safety policy statement if there are five or more employees

- carry out risk assessments for all the company's work activities and review these (and record the assessments if there are five or more employees)
- identify and implement control measures, and tell employees about them
- consult with any official trade union safety representatives
- establish a safety committee if requested in writing by any two or more safety representatives within three months of the request.

Management of Health and Safety at Work Regulations 1999

The Management of Health and Safety at Work Regulations 1999 explain what employers are required to do to manage health and safety under HASAWA. They apply to every work activity.

Employers must carry out a risk assessment (and, for companies with five or more employees, record any significant findings), implement any control measures identified (appointing competent people to do this) and arrange for appropriate information and training for employees.

Risk assessment will be covered later in more detail on pages 40–46.

Electricity at Work Regulations 1989 (EAWR)

These Regulations impose general health and safety requirements to do with electricity at work on employers, self-employed people and employees.

Every employer and self-employed person has a duty to comply with the provisions of the Regulations, as far as they relate to matters within their control. Every employee has the duty to co-operate with their employer, as far as necessary to enable the Regulations to be complied with. Because these are statutory regulations, penalties can be imposed on people found guilty of malpractice or misconduct.

The Regulations refer to a person as a '**duty holder**' in respect of systems, equipment, and conductors. The Regulations clearly define the various duty holders. Table 1.02 shows the definitions of the duty holders.

EAWR are designed to take account of the responsibilities that many employees in the electrical trades and professions have to take on as part of their job. The level of responsibility you hold to make sure the Regulations are met depends on the amount of control you have over electrical safety in any particular situation.

Find out

Consider your own workplace and working practice. What risks do you think might need a formal assessment? For example, is any work done that involves working at height or manual handling? Make a list of six other areas that may need formal assessment.

Remember

As far as EAWR is concerned, a trainee like you is an employee – so you have the same duties and responsibilities as any other employee.

Remember

It is the duty of every duty holder to comply with the provisions of these Regulations in so far as they relate to matters within their control.

Employer	Any person or body who employs one or more individuals under a contract of employment or Apprenticeship or provides training under the schemes to which HASAWA applies
Self-employed	An individual who works for gain or reward other than under contract of employment, whether or not he or she employs others
Employee	Regulation 3(2)(b) repeats the duties placed on employees by HASAWA, which are equivalent to those placed on employers and self-employed people where these matters are within their control. This includes trainees like you, who are considered employees under the Regulations

Table 1.02 Duties under Electricity at Work Regulations 1989

Working life

Tom is working on an electrical system while his colleague Ahmed is preparing to do some work nearby. Tom energises the circuit to do his work but, as Ahmed is doing a different task, he doesn't think it is necessary to tell him exactly what he is doing.

- Why is Tom a duty holder in this situation? You will need to think about whether the circumstances are in Tom's control and what his responsibilities are.
- What are the dangers here?
- What should Tom do immediately?

Absolute/reasonably practicable

Duties in some regulations are regarded as either 'absolute' or 'reasonably practicable'. Table 1.03 offers a guide to how to interpret these terms.

Absolute	Where the requirement must be met regardless of cost or any other consideration
Reasonably practicable	Where the person doing a particular work process must take into account: • the amount of risk to health and safety involved • the amount of time, trouble, expense and effort it would take to reduce these risks

Table 1.03 Definitions of absolute and reasonably practicable

For example, in a home you would expect to find a fireguard in front of a fire to stop young children being injured. This is a cheap and effective way of preventing accidents, so it would be a reasonably practicable situation.

However, if the costs or technical difficulties of taking certain steps to prevent those risks are very high, it might not be reasonably practicable to take those steps. The greater the degree of risk, the less you should be concerned about the cost of measures needed to prevent that risk.

For these Regulations, the risk is often that of death from electrocution, yet the precautions are often simple and cheap to take. Therefore, the level of duty is likely to be an absolute duty in order to reduce the risk.

Did you know?

Using insulation around cables is one example of a simple, cheap precaution to reduce the danger of electrocution.

Table 1.04 summarises the particular Electricity at Work Regulations you are most likely to have to comply with, showing whether they are regarded as 'absolute' or 'reasonably practicable'.

Regulation number	Absolute or reasonably practicable?	What the Regulation says
Regulation 4	reasonably practicable	All electrical systems shall be constructed and maintained to prevent danger. All work activities are to be carried out so as not to give rise to danger
Regulation 5	absolute	No electrical equipment is to be used where its strength and capability may be exceeded so as to give rise to danger
Regulation 6	reasonably practicable	Electrical equipment sited in adverse or hazardous environments must be suitable for those conditions
Regulation 7	reasonably practicable	Permanent safeguarding or suitable positioning of live conductors is required
Regulation 8	absolute	Equipment must be earthed or other suitable precautions must be taken: for example, the use of residual current devices, double-insulated equipment, reduced voltage equipment, etc.
Regulation 9	absolute	Nothing is to be placed in an earthed circuit conductor that might, without suitable precautions, give rise to danger by breaking the electrical continuity or introducing high impedance
Regulation 10	absolute	All joints and connections in systems must be mechanically and electrically suitable for use
Regulation 11	absolute	Suitable protective devices should be installed in each system to ensure all parts of the system and users of the system are safeguarded from the effects of fault conditions
Regulation 12	absolute	Where necessary to prevent danger, suitable means shall be available for cutting off the electrical supply to any electrical equipment. Note: drawings of the distribution equipment and methods of identifying circuits should be readily available. Ideally, mains-signed isolation switches should be provided in practical work areas
Regulation 13	absolute	Adequate precautions must be taken to prevent electrical equipment that has been made dead in order to prevent danger from becoming live while any work is carried out
Regulation 14	absolute	No work can be carried out on live electrical equipment unless this can be properly justified. This means that risk assessments are required. If such work is to be carried out, suitable precautions must be taken to prevent injury
Regulation 15	absolute	Adequate working space, adequate means of access and adequate lighting shall be provided at all electrical equipment on which or near which work is being done in circumstances that may give rise to danger
Regulation 16	absolute	No person shall engage in work that requires technical knowledge or experience to prevent danger or injury, unless he or she has that knowledge or experience, or is under appropriate supervision

Table 1.04 Examples of absolute and reasonably practicable duties in the Electricity at Work Regulations 1989

Workplace (Health, Safety and Welfare) Regulations 1992

The Workplace (Health, Safety and Welfare) Regulations 1992 aim to ensure that workplaces meet the health, safety and welfare needs of all members of a workforce, including people with disabilities.

Several of the Regulations require things to be 'suitable'. Regulation 2 (3) makes it clear that things should be suitable for anyone, and this includes people with disabilities. Where necessary, parts of the workplace – especially doors, passageways, stairs, showers, washbasins, lavatories and workstations – should be made accessible for people with disabilities.

Here are some of the definitions used in this legislation.

- **Workplace** These Regulations apply to a wide range of workplaces: not only factories, shops and offices but also schools, hospitals, hotels and places of entertainment. 'Workplace' includes the common parts of shared buildings, private roads and paths on industrial estates and business parks, and temporary worksites (except workplaces involving construction work on construction sites).
- **Work** Here 'work' can be as an employee or self-employed person.
- **Premises** 'Premises' includes any place, including an outdoor place.
- **Domestic premises** Under these Regulations, this means a private dwelling. These Regulations do not apply to domestic premises, and exclude home-workers. However, they do apply to hotels, to nursing homes and to parts of workplaces where 'domestic' staff are employed, such as the kitchens of hostels.
- **Disabled person** This means a disabled person as defined by section 1 of the Disability Discrimination Act 1995.

Health in the workplace

This section of the Regulations looks at the general working environment of people in the workplace. Table 1.05 shows the considerations that the Regulations say need to be made in the workplace for some of the key health issues.

Find out

How does the Disability Discrimination Act 1995 define a disabled person? See if you can find out, on the Internet or in your local library.

Did you know?

Windows or other openings may provide sufficient ventilation but, where necessary, mechanical ventilation systems should be provided and regularly maintained.

Health issue	What to expect/provide
Ventilation	If you are working inside, your workplace should be adequately ventilated with fresh, clean air, but not draughty.
Indoor temperatures	Individuals have different **thermal comfort**, making it hard to create an environment that will satisfy everyone. For workplaces where the activity is mainly seated, such as offices, the temperature should be at least 16°C; for more physical work, it should be at least 13°C (unless other laws require lower temperatures). If working in extreme temperatures, your employer should provide the correct clothing and equipment. You should also be medically fit to carry out the work and take precautions to remain so: for example, drinking water regularly in a very hot environment. Work can also be organised to minimise exposure to more extreme environments.
Lighting	Every workplace should have sufficient, suitable lighting and be supplemented by natural light. Automatic emergency lighting, powered by an independent source, should be provided where sudden loss of light would create a risk.
Cleanliness and waste materials	Every workplace and the furniture, furnishings and fittings should be kept clean, and it should be possible to keep the surfaces of floors, walls and ceilings clean. Cleaning and waste removal should be regular and proper facilities should be provided for storing waste.
Space	Workplaces shouldn't be crowded – there should be enough room for you to move around freely and safely, including the height of ceilings. The Regulations suggest that the amount of space for each person should not be less than 11 cubic metres.
Workstations and seating	Workstations should be suitable for the people using them and the work they do. You should be able to leave your workstation swiftly in an emergency, as well as work comfortably at it. Seating should give adequate support for the lower back, and footrests should be provided for workers who cannot place their feet flat on the floor.

Table 1.05 Health issues in the workplace

Safety in the workplace

Table 1.06 shows the considerations that will need to be made in the workplace for some of the key safety issues.

Safety issue	What to expect/provide
Maintenance	The workplace, certain equipment, devices and systems should be maintained in efficient working order. Maintenance is required for any equipment and devices that could cause a risk to health, safety or welfare if a fault occurred. The condition of buildings needs to be monitored to ensure they are stable and solid enough to use. There could be risks from the normal running for the work process (for example, vibration or floor loadings) as well as foreseeable risks (such as fire in a cylinder store).
Floors and traffic routes	Workplaces should be organised so that pedestrians and vehicles are separated, so that each can move safely. Appropriate speed limits should be set and route markings clearly set down.
Escalators and staircases	Open sides of staircases should be fenced with an upper rail at 900 mm or higher, and a lower rail. A handrail should be provided on at least one side of every staircase. Escalators and moving walkways should be maintained and equipped with necessary safety devices and emergency stop controls that are easy to identify and get to.
Windows, doors and gates	These should be cleaned and maintained and capable of being opened, closed and adjusted safely. Doors and gates should open both ways. Power-operated doors and gates should have safety features to prevent people getting stuck or trapped, along with a stop button that is easy to identify and get to. If transparent or translucent, these should be made from safety material and protected against breakage. They should be marked to make them visible.

Table 1.06 Safety issues in the workplace

Safety tip

It is often difficult for drivers to see behind their vehicle when they are reversing. As far as possible, traffic routes should be planned so that drivers do not need to reverse, by using one-way systems and drive-through loading areas.

Welfare

Table 1.07 shows the considerations that will need to be made in the workplace for some of the key welfare issues.

Welfare issue	What to expect/provide
Sanitary conveniences and washing facilities	Suitable and sufficient sanitary conveniences and washing facilities should be provided at readily accessible places. They should be clean, with running hot and cold water and drying equipment. If required, showers should also be provided. Men and women should have separate facilities, unless each facility is in a separate room with a lockable door and is for use by only one person at a time.
Drinking water	There should be an adequate supply of high-quality drinking water.
Accommodation for clothing and changing	Adequate, suitable and secure space should be provided to store workers' own clothing and special clothing and provide an opportunity to dry clothes if necessary. There should also be a place to change privately if you need to and store clothing securely.
Rest and eating meals	Suitable and sufficient, readily accessible rest facilities should be provided with seating. Similar facilities should be provided where workers regularly eat meals at work. Facilities should also be provided where food would otherwise be likely to be contaminated. Suitable rest facilities should be provided for pregnant women and nursing mothers. They should be near to sanitary facilities and, where necessary, include the facility to lie down.

Table 1.07 Welfare issues in the workplace

Control of Substances Hazardous to Health Regulations 2002 (COSHH)

COSHH is the law that requires employers to control substances that are hazardous to health.

A hazardous substance is anything that can harm your health when you work with it if it is not properly controlled. Hazardous substances are found in nearly all workplaces, including factories, shops, mines, construction sites and offices.

Hazardous substances include:

- substances used directly in work activities, such as glues, paints, cleaning agents
- substances generated during work activities, such as fumes from soldering and welding
- naturally occurring substances, such as grain dust, blood and bacteria.

The COSHH regulations cover the following:

- chemicals, products containing chemicals, fumes, dusts, vapours, mists and gases and biological agents (germs)
- asphyxiating gases
- germs that cause diseases such as leptospirosis or legionnaires' disease, and germs used in laboratories.

Figure 1.02 shows the steps employers can take to prevent or reduce workers' exposure to hazardous substances.

For step 3, if it is reasonably practicable, exposure must be prevented by changing the process so that the hazardous substance is not required or generated. If prevention is not reasonably practicable, exposure should be adequately controlled by one or more of measures outlined in the Regulations. Engineering controls and respiratory protective equipment have to be examined and tested at regular intervals.

Work at Height Regulations 2005 (amended 2007)

The Work at Height Regulations 2005 apply to all work at height where there is a risk of a fall liable to cause personal injury. The Regulations place a duty on employers, the self-employed, and any person that controls the work of others (for example, facilities managers or building owners who may contract others to work at height). The later Work at Height (Amendment) Regulations 2007 added a duty for those who work at height providing instruction or leadership to one or more people engaged in caving or climbing or any other similar activity in Great Britain.

Step 1: Find out what hazardous substances are used in the workplace and the risks these substances pose to people's health

Step 2: Decide what precautions are needed before any work starts with hazardous substances

Step 3: Prevent people being exposed to hazardous substances or, where this is not reasonably practicable, control the exposure

Step 4: Make sure control measures are used and maintained properly, and that safety procedures are followed

Step 5: If required, monitor exposure of employees to hazardous substances

Step 6: Carry out health surveillance where assessment has shown that this is necessary, or where COSHH makes specific requirements

Step 7: If required, prepare plans and procedures to deal with accidents, incidents and emergencies

Step 8: Make sure employees are properly informed, trained and supervised

Figure 1.02 COSHH safety procedures

Did you know?

COSHH does not cover lead, asbestos or radioactive substances, as these have their own specific regulations.

Here are some of the definitions used in these Regulations.

- **At height** This means at a height where a person could be injured falling from it, even if it is at or below ground level.
- **Work** Work here includes moving around at a place of work (except by a staircase in a permanent workplace), but not travel to or from a place of work.

Working life

While working on site, another operative from a ceiling company wants to borrow your mobile tower platform as he says your supervisor has said that it was okay to use it. A loud scream is heard later and you hear that the operative has been badly hurt due to falling through the open floor access point.

- What should have been done on receiving that request?
- Who's responsibility is it to check the condition of the platform?
- What training is required before using the platform?

If you are an employee or working under someone else's control, Regulation 14 says you must:

- report any safety hazard to them
- use the equipment supplied (including safety devices) properly, following any training and instructions (unless you think that it would be unsafe to do so, in which case you should seek further instructions before continuing).

Regulation 6(3) states that 'the employer must do all that is reasonably practicable to prevent anyone falling'.

The Regulations then set out a simple hierarchy for managing and selecting equipment for work at height (see Figure 1.03).

Duty holders' responsibilities

The Regulations require duty holders to ensure that:

- all work at height is properly planned and organised
- all work at height takes account of weather conditions that could endanger health and safety
- those involved in work at height are trained and competent
- the place where work at height is done is safe
- equipment for work at height is appropriately inspected
- the risks from fragile surfaces are properly controlled
- the risks from falling objects are properly controlled.

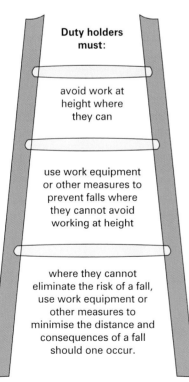

Duty holders must:

avoid work at height where they can

use work equipment or other measures to prevent falls where they cannot avoid working at height

where they cannot eliminate the risk of a fall, use work equipment or other measures to minimise the distance and consequences of a fall should one occur.

Figure 1.03 The role of duty holders

Planning

Regulations 4 and 6 require employers to:

- make sure that no work is done at height if it is safe and reasonably practicable to do it other than at height
- make sure that the work is properly planned, appropriately supervised, and carried out in as safe a way as is reasonably practicable
- plan for emergencies and rescue
- take account of the risk assessment carried out under Regulation 3 of the Management of Health and Safety at Work Regulations.

Weather

Regulation 4 states the employer 'must ensure that the work is postponed while weather conditions endanger health or safety'. However, this does not apply to emergency services acting in an emergency.

Staff training

Regulations 5 and 6 require the employer to ensure that:

- everyone involved in the work is competent (or, if being trained, is supervised by a competent person). This includes involvement in organisation, planning, supervision, and the supply and maintenance of equipment
- where other precautions do not entirely eliminate the risk of a fall, those who will be working at height are trained on how to avoid falling (as far as it is reasonably practicable), and how to avoid or minimise injury to themselves if they do fall.

The place where work is done

Regulation 6 states that employers must ensure that the place where work is done at height (including the means of access) is safe and has features to prevent a fall, unless this would mean that the worker was then unable to carry out the work safely. Here you are meant to take into account the demands of the task, the equipment and the working environment.

Equipment, temporary structures, and safety features

Regulations 6, 7, 8 and 12 require an employer who is selecting equipment for work at height to:

- use the most suitable equipment giving collective protection measures (such as guard rails) priority over personal protection measures (such as safety harnesses)
- take account of the working conditions and risks to the safety of all those at the place where the work equipment is to be used
- ensure that all equipment, temporary structures (such as scaffolding) and safety features comply with the detailed requirements of Schedules 2 to 6.

Inspections

'Inspection' is defined by Regulation 12(10) as 'such visual or more rigorous inspection by a competent person as is appropriate for safety purposes … (including) any testing appropriate for those purposes'.

Employers must ensure that:

- as far as it is reasonably practicable, each individual place at which work is to be done at height is checked on every occasion before that place is used. This involves checking the surface and every parapet, permanent rail, etc.
- any item of collective fall prevention (for example, guard rails and toe boards), working platforms, collective fall arrest (for example, nets, airbags, etc.), personal fall protection (for example, work restraints, work positioning, fall arrest and rope access), ladders and stepladders, is inspected:
  - after it is assembled or installed (or after it has been assembled and installed if both are required), if its safety depends on how it is assembled or installed
  - as often as is necessary to ensure safety, and in particular to make sure that any deterioration can be detected and remedied in good time
- before using any equipment that has come from another business, and before any equipment leaves your business, it is accompanied by an indication (clear to everyone involved) that the last inspection required by these Regulations has been carried out

- any platform used for (or for access to) construction work and from which a person could fall more than 2 metres is inspected in place before use (and not more than seven days before use). Where it is a mobile platform, inspection at the site is sufficient without re-inspection every time it is moved
- the person inspecting a platform:
  - prepares a report before going off duty
  - gives the report (or a copy) within 24 hours of completing the inspection to the person for whom the inspection was done (for example, the employer or site manager)
  - keeps the report of a platform at the construction site until the work is completed and then at an office for another three months.

Fragile surfaces

Regulation 9 requires employers to:

- ensure that no one working under their control goes onto or near a fragile surface unless that is the only reasonably practicable way for the worker to carry out the work safely, having regard to the demands of the task, equipment, or working environment
- ensure (as far as reasonably practicable) that, if anyone does work on or near a fragile surface, suitable platforms, coverings, guard rails and the like are provided (and used) to minimise the risk
- do all that is reasonably practicable, if any risk of a fall remains, to minimise the distance and effect of a fall
- ensure that, if anyone working under their control may go onto or near a fragile surface, they do all that is reasonably practicable to make them aware of the danger, preferably by prominent warning notices fixed at the approaches to the danger zone.

Falling objects

Regulations 10 and 11 require that, where it is necessary to prevent injury, employers must do all that is reasonably practicable to prevent anything falling. If it is not reasonably practicable, they must ensure that no one is injured by anything falling.

Employers must ensure that:

- nothing is thrown or tipped from height if it is likely to injure anyone, or stored in such a way that its movement is likely to injure anyone
- if the workplace contains an area in which there is a risk of someone being struck by a falling object or person, that the area is clearly indicated and that (as far as reasonably practicable) unauthorised people are unable to reach it.

Personal Protective Equipment at Work Regulations 1992

PPE is defined in these Regulations as 'all equipment (including clothing affording protection against the weather) which is intended to be worn or held by a person at work and which protects them against one or more risks to their health and safety'. Table 1.08 shows responsibilities under these Regulations.

Hearing protection and respiratory protective equipment provided for most work situations are not covered by these Regulations because other regulations apply to them. However, these items need to be compatible with other PPE provided.

The main point of these Regulations is that PPE is to be supplied by the employer free of charge and used at work wherever there are risks to health and safety that cannot be adequately controlled in other ways. The Regulations also require that PPE is:

- properly assessed before use to ensure it is suitable
- maintained and stored properly
- provided with instructions on how to use it safely
- used correctly by employees.

> **Remember**
>
> Common examples of PPE include safety helmets, gloves, eye protection, high-visibility clothing, safety footwear and safety harnesses.

> **Did you know?**
>
> An employer cannot ask for money from an employee for PPE, whether it is returnable or not.

> **Remember**
>
> There are other ways in which a risk can be adequately controlled, such as engineering controls. PPE should be used as a last resort in these situations.

Employers	Employees
Must train employees and give information on maintaining, cleaning and replacing damaged PPE	Must use PPE provided by their employer, in accordance with any training in the use of the PPE concerned
Must provide storage for PPE	Must inform employer of any defects in PPE
Must ensure that PPE is maintained in an efficient state and in good repair	Must comply with safety rules
Must ensure that PPE is properly used	Must use safety equipment as directed

Table 1.08 PPE responsibilities of employers and employees

Manual Handling Operations Regulations 1992

These Regulations aim to reduce the very large incidence of injury and ill health caused by the manual handling of loads at work.

The Manual Handling Operations Regulations 1992 apply to the transporting or supporting of loads by hand or by bodily force, involving a wide range of manual handling activities, including lifting, lowering, pushing, pulling or carrying either animate (a person) or inanimate (a box etc.) objects.

Under these Regulations employers must:

- avoid the need for hazardous manual handling, so far as is reasonably practicable
- assess the risk of injury from any hazardous manual handling that can't be avoided
- reduce the risk of injury from hazardous manual handling, so far as is reasonably practicable.

However, employees also have duties too, and must:

- follow appropriate systems of work laid down for their safety
- make proper use of equipment provided for their safety
- co-operate with their employer on health and safety matters
- inform the employer if they identify hazardous handling activities
- take care to ensure that their activities do not put others at risk.

You will find more detail on safe lifting and moving on pages 85–90.

Did you know?

More than one in four of all reportable injuries are caused by manual handling. These accidents do not include injury done over a longer period, particularly to the back, which can lead to physical impediment or even permanent disablement.

Working life

You arrive at a site in a company vehicle with heavy equipment. There is no one around to help and your supervisor says you must help him. You tell him that you have forgotten your safety boots. He says not to worry, it will be alright. While unloading the equipment it traps your foot causing severe swelling, which necessitates you going to hospital where the diagnosis reveals a broken foot.

- What should the supervisor have done on arrival at the site?
- What are the PPE requirements relating to this task?
- Who should be notified following this accident?

Remember

Manual handling implies that an attempt is being made to move a load. If a girder being moved manually is dropped and fractures an employee's foot, it is a manual handling accident; if the girder is inadvertently knocked over and causes a similar injury, this would not be due to manual handling.

Provision and Use of Work Equipment Regulations 1998 (PUWER)

Under PUWER, equipment provided for use at work must be:

- suitable for the intended use
- safe for use, maintained in a safe condition and, in certain circumstances, inspected to ensure this remains the case
- used only by people who have received adequate information, instruction and training
- accompanied by suitable safety measures, such as protective devices, markings and warnings.

Generally, any equipment used by an employee at work is covered, including hammers, knives, ladders, drilling machines, power presses, circular saws, lifting equipment (including lifts), dumper trucks and motor vehicles, and even photocopiers. If you are allowed to provide your own equipment, this will also be covered by PUWER, and the employer will need to make sure it complies.

These Regulations cover places where HASAWA applies – including factories, offshore installations, offices, shops, hospitals, hotels and places of entertainment. PUWER also applies in common parts of shared buildings and temporary places of work, such as construction sites. While the Regulations cover equipment used by people working from home, they do not apply to domestic work in a private household.

The Regulations do not apply to equipment used by the public, such as compressed-air equipment used in a garage forecourt. However, such circumstances are covered by HASAWA.

Health and Safety (Display Screen Equipment) Regulations 1992

These Regulations require employers to:

- minimise the risks in VDU work by ensuring that workplaces and jobs are well designed
- analyse workstations to assess and reduce risks
- ensure workstations meet specified minimum requirements
- plan work activities so that they include breaks or changes of activity
- provide eye and eyesight tests on request, and special spectacles if needed
- provide information and training.

Did you know?

As an employee, you do not have duties under PUWER. However, you do have general duties under HASAWA and the Management of Health and Safety at Work Regulations 1999 to take reasonable care of yourself and others who may be affected by your actions, and to co-operate with others.

The Regulations apply where staff normally use VDUs as a significant part of their normal work. People who use VDUs only occasionally are only covered by the workstation requirements of these Regulations. However, their employers still have general duties to protect them under other health and safety at work legislation.

Control of Asbestos Regulations 2006

These Regulations prohibit the importation, supply and use of all forms of asbestos. They continue the ban introduced for blue and brown asbestos in 1985 and for white asbestos in 1999. They also continue to ban the second-hand use of asbestos products, such as asbestos cement sheets and asbestos boards and tiles; this includes panels that have been covered with paint or textured plaster-containing asbestos.

The Regulations require mandatory training for anyone liable to be exposed to asbestos fibres at work. This includes maintenance workers and others who may come into contact with or who may disturb asbestos (such as cable installers) as well as those involved in asbestos removal work.

For more on asbestos, see pages 114–16.

Chemicals (Hazard Information and Packaging for Supply) Regulations 2009

These Regulations, also known as CHIP 4, apply to suppliers of dangerous chemicals. They aim to protect people and the environment from the effects of those chemicals by requiring suppliers to provide information about the dangers and to package them safely.

CHIP 4 requires the supplier of a dangerous chemical to:

- identify the hazards or dangers of the chemical (a process known as 'classification')
- give information about the hazards to their customers, usually provided on a label on the package itself
- package the chemical safely.

See page 91 for some of the most common warning signs you will find on chemicals you work with.

Remember

The ban applies to new use of asbestos. If existing asbestos-containing materials are in good condition, they may be left in place, with their condition monitored and managed to ensure they are not disturbed.

Remember

By June 2015, the CHIP 4 regulations will be replaced by the European Regulation on Classification, Labelling and Packaging of Substances and Mixtures – known as the CLP Regulation. Until then, suppliers must classify substances according to both CHIP 4 and CLP, but must label and package them according to CLP only.

Construction (Design and Management) Regulations (CDM) 2007

The key aim of CDM is to integrate health and safety into the management of a project and to encourage everyone involved to work together to:

- improve the planning and management of projects from the start
- identify risks early on
- target effort where it can do the most good in terms of health and safety
- discourage unnecessary bureaucracy.

CDM aims to focus attention on planning and management throughout construction projects, from design concept onwards. The idea is that health and safety considerations be treated as an essential but normal part of a project's development – not as an afterthought.

The Regulations are divided into five Parts, as shown in Table 1.09.

Remember

Under Part 4, this does not mean everyone involved with design, planning or management of the project must legally ensure all the requirements of this section are complied with. They only have duties if they exercise significant control over the actual working methods, safeguards and site conditions.

Part	What it covers
Part 1	Interpretation and application; apply to both employers, employees and self-employed on all construction work in Great Britain and through HASAWA, its territorial seas
Part 2	General management duties on all construction projects including non-notifiable projects
Part 3	Additional management details for projects above the notification threshold that require particular appointments/documents to be completed to assist health and safety management from concept to completion
Part 4	Physical safeguards provided to prevent danger and apply to all construction work. Duties are held by contractors who carry out the work and those who control the way the work is done. The extent of the duty is in proportion to the degree of control the individual or organisation has over the work. Contractors cannot let work begin until safeguards are in place
Part 5	Civil liability; transitional provisions which apply during the period where the Regulations come into force and amendments and revocations of other legislation

Table 1.09 CDM Regulations Parts

Notification

Except where the project is for a domestic client, the Health and Safety Executive (HSE) must be notified of projects where construction work is expected to either:

- last more than 30 working days, or
- involve more than 500 person days (for example, 50 people working for over 10 days).

When a small project requires a short extension, or a short-term increase in the number of people, there is no need to notify the HSE. However, if the work or the scope changes significantly, so that it becomes notifiable, the HSE should be informed.

Did you know?

The information to be sent to the HSE is set out in Schedule 1 to CDM 2007. The HSE form (F.10) can be completed and submitted online or downloaded. You do not have to use the F.10 form, as long as you provide all of the specified information.

Summary of the duties under the Regulations

Table 1.10 gives a summary of the duties and how they are applied.

	All Construction projects (Part 2 of the Regulations)	Additional duties for notifiable projects (Part 3 of the Regulations)
Clients (excluding domestic clients)	Check competence and resources of all appointees Ensure there are suitable management arrangements for the project's welfare facilities Allow sufficient time and resources for all stages Provide pre-construction information to designers and contractors	Appoint CDM co-ordinator* Appoint principal contractor* Make sure that the construction phase does not start unless there are suitable: • welfare facilities, and • construction phase plans in place Provide information relating to the health and safety file to the CDM co-ordinator * = There must be a CDM co-ordinator and principal contractor until the end of the construction phase
CDM co-ordinators		Advise and assist the client with their duties Notify the HSE Co-ordinate health and safety aspects of the design work and co-operate with others involved with the project Facilitate good communication between client, designers and contractors Liaise with principal contractor regarding ongoing design Identify, collect and pass on pre-construction information Prepare/update health and safety file
Designers	• Check client is aware of their duties • Eliminate hazards and reduce risks during design • Provide information about remaining risks	• Check CDM co-ordinator has been appointed • Provide any information needed for the health and safety file

Table 1.10 Duties under the CDM Regulations

	All Construction projects (Part 2 of the Regulations)	Additional duties for notifiable projects (Part 3 of the Regulations)
Principal contractors		• Plan, manage and monitor construction phase in liaison with the contractor • Prepare, develop and implement a written plan and site rules (initial plan completed before the construction phase starts) • Give contractors relevant parts of the plan • Make sure suitable welfare facilities are provided from the start and maintained throughout the construction phase • Check competence of all appointees • Ensure all workers have site inductions and any further information and training needed for the work • Consult with the workers • Liaise with CDM co-ordinator regarding ongoing design • Secure the site
Contractors	• Check client is aware of their duties • Plan, manage and monitor own work and that of workers • Check competence of all their appointees and workers • Train own employees • Provide information to their workers • Comply with the specific requirements in Part 4 of the Regulations • Ensure there are adequate welfare facilities for their workers	• Check a CDM co-ordinator and a principal contractor have been appointed and HSE notified before starting work • Co-operate with principal contractor in planning and managing work, including reasonable directions and site rules • Provide details to the principal contractor of any contractor whom they engage in connection with carrying out the work • Provide any information needed for the health and safety file • Inform principal contractor of problems with the plan • Inform principal contractor of reportable accidents, diseases and dangerous occurrences
Everyone	• Check own competence • Co-operate with others and co-ordinate work so as to ensure the health and safety of construction workers and others who may be affected by the work • Report obvious risks • Comply with requirements in Schedule 3 and Part 4 of the Regulations for any work under their control • Take account of and apply the general principles of prevention when carrying out duties	

Table 1.10 Duties under the CDM Regulations (cont.)

Responsibilities of the client

The client has one of the biggest influences over the way a project is run. However, the Regulations recognise clients might not know much about health and safety, so clients aren't expected to plan or manage the project themselves. In these cases, for notifiable projects, the client must appoint a competent CDM co-ordinator to advise them on meeting their duties. The CDM co-ordinator will need the client's support to work effectively. The client remains responsible for ensuring that client duties are met.

Domestic clients have no client duties under CDM 2007, which means that there is no legal requirement for the appointment of a CDM co-ordinator or principal contractor when such projects reach the notification threshold.

Control of Major Accidents and Hazards (Amendment) Regulations 2005 (COMAH)

The Control of Major Accident Hazards (Amendment) Regulations 2005 are largely concerned with hazards involving dangerous substances.

The COMAH Regulations apply mainly to the chemical industry, but also to some storage activities, explosives and nuclear sites and other industries where threshold quantities of dangerous substances identified in the Regulations are kept or used.

The main aim of these Regulations is to prevent and mitigate the effects of major accidents involving dangerous substances such as chlorine, liquefied petroleum gas, explosives and arsenic pentoxide, which can cause serious damage to the environment and harm to people. The general duty on all operators is to take all measures necessary to prevent major accidents and limit their consequences to people and the environment for all establishments within their scope.

Requiring measures both for prevention and mitigation acknowledges that all risks cannot be completely eliminated. The phrase 'all measures necessary' indicates that a judgment must be made about the measures in place. Where hazards are high, high standards will be required to make sure that risks are acceptably low.

Changes were made in 2005 to reflect lessons learned from major accidents in Europe and take account of the results of EC working groups on carcinogens and substances dangerous for the environment.

In particular the 2005 changes:

- added some new named substances
- changed some existing named substances and generic categories of substances, including revised qualifying quantities
- changed the aggregation rule.

The application of COMAH at mines, quarries, boreholes and landfill sites was also broadened slightly.

The COMAH Regulations are enforced by a competent authority (CA) consisting of the Health and Safety Executive and the Environment Agency, operating jointly.

The Regulations place duties on the CA to inspect activities subject to COMAH and prohibit the operation of an establishment if there is evidence that measures taken for prevention and mitigation of major accidents are seriously deficient. The CA also has to examine safety reports and inform operators about the conclusions of its examinations within a reasonable time period.

Progress check

1 List some of the key responsibilities that both employers and employees have placed upon them by the Health and Safety at Work Act 1974.

2 What are the main risks to operatives covered by the Electricity at Work Regulations 1989 and what is normally done on site to ensure those risks are minimised?

3 Manual Handling Operations Regulations 1992 guidance should be followed when deciding whether a load should be handled or not. What are the three main steps the Regulations set out?

4 Which regulations are most appropriate in setting out the requirements for working at heights? What should duty holders do where work at heights is unavoidable?

K2. Understand the procedures for dealing with health and safety in the work environment

You have seen that there is a lot of legislation to read and interpret correctly if the workplace is to be as safe as possible. Keeping up to date with and understanding the implications of this legislation can be daunting, so how do you and your employers manage it?

There is no set method for distributing new or amended health and safety requirements to employers. It is up to every employer to ensure that they are meeting legal requirements, with lack of knowledge not regarded as an excuse in a court of law.

Companies keep up to date in different ways, often depending on their size and structure.

In smaller companies the owner or manager is likely to have responsibility, but pressure on their time might be very high. Many smaller companies choose to hire the services of an external health and safety consultant.

The consultant will normally visit the employer and carry out an inspection of the main premises and types of site and activities undertaken. The consultant will then produce the necessary paperwork – for example, risk assessments, safety policy, Control of Substances Hazardous to Health Regulations (COSHH) assessments – and provide a manual for the employer to refer to and implement.

Some employers use a consultant on a one-off basis and then maintain the information and systems provided themselves. However, they must ensure that the manual remains current with all developments. Other employers engage a consultant on an ongoing basis, where the consultant monitors all legislation, codes of practice and relevant information, and continually provides the employer with updated information, documentation and services.

> **Did you know?**
>
> 152 people were killed in workplace accidents in Britain in 2009/10. There were also 26,061 reported major injuries and 95,369 reported injuries that caused three or more days absence from work. (Source: The Health and Safety Executive)

Figure 1.04 Displaying a health and safety notice in the workplace is compulsory

> **Find out**
>
> Check your own health and safety manuals. Are they up to date? How can you tell if they are? If they are not up to date, what could you do about this?

Large companies often have their own specialist staff with responsibility for health and safety.

Health and Safety Executive (HSE)

The Health and Safety Executive is responsible for the regulation of almost all the risks to health and safety arising from work activity in Britain. Its mission is to protect people's health and safety by making sure that risks in the workplace are controlled.

Among other things, the HSE looks after health and safety in nuclear installations and mines, factories, farms, hospitals and schools, offshore gas and oil installations, the gas grid and railways, as well as with the movement of dangerous goods and substances.

HSE Inspectors

To help enforce the law, HSE Inspectors have a range of statutory powers. They can, and sometimes do, visit and enter premises without warning. Here are the courses of action an Inspector can take if there is a problem.

- **Issue an improvement notice** This will say what needs to be done, why and by when. The time period within which to take the remedial action will be at least 21 days to allow the duty holder time to appeal to an Employment Tribunal. The notice also contains a statement of the Inspector's opinion that an offence has been committed.

- **Issue a prohibition notice** Where an activity involves, or will involve, a risk of serious personal injury, the Inspector can serve a prohibition notice to stop the activity immediately or after a specified time period, and not allow it to be resumed until remedial action has been taken. The notice will explain why the action is necessary.

- **Take legal action** The Inspector may also wish to start legal proceedings. Improvement and prohibition notices and written advice may be used in court proceedings.

HSE Inspectors apply 'proportionality' before deciding on a course of action. This means: they make sure that the degree of enforcement to be taken will be in proportion to the degree of risks they have discovered.

Penalties for health and safety offences

The Health and Safety Offences Act 2008:

- raised the maximum fine which may be imposed in the lower courts to £20,000 for most health and safety offences
- made imprisonment an option for more health and safety offences in both the lower and higher courts
- made certain offences, once triable only in the lower courts, triable in either the lower or higher courts.

Failing to comply with an improvement or prohibition notice, or a court remedy order will incur a range of penalties, as high as £20,000 (figure for 2011). In a higher court the fine could be unlimited.

Approved Codes of Practice (ACOPs)

ACOPs give practical advice on how to comply with the law and are approved by the HSE, with the consent of the Secretary of State. Failing to follow an ACOP is not an offence in itself. However, in any criminal or civil proceedings, the court will use an employer's non-compliance or contravention of any relevant ACOP to show an offence. Here the employer's only defence would be to prove that he or she was using another appropriate method or system.

Procedures for accidents and emergencies

Everyone hopes that, with careful planning, there will never be an accident or emergency. However, you need to be prepared for dealing with such an event.

Reporting of Injuries, Diseases and Dangerous Occurrences Regulations 1995 (RIDDOR)

The Reporting of Injuries, Diseases and Dangerous Occurrences Regulations 1995 (RIDDOR) place a legal duty on employers, the self-employed and those in control of premises to report work-related accidents, diseases and dangerous occurrences to the relevant enforcing authority (the HSE or local authority). This can help enforcing authorities to identify where and how risks arise, as well as to investigate serious accidents.

The law requires the following work-related incidents to be reported:

- deaths
- injuries where an employee or self-employed person has an accident that is not major, but results in the injured person being away from work or unable to do the full range of their normal duties for more than three days (including any days they wouldn't normally be expected to work such as weekends, rest days or holidays), not counting the day of the injury itself
- injuries to members of the public where they are taken to hospital
- work-related diseases
- dangerous occurrences where something happens that does not result in a reportable injury, but which could have done.

Incident Contact Centre

The Incident Contact Centre is a central point for employers to report any incident, irrespective of whether it is the HSE or the local authority that enforces the business involved. The Centre completes the report for the employer and sends it to the correct enforcing authority. A final copy of the report will also be sent to the employer. The number is 0845 300 99 23.

Accident reporting

Employers are legally obliged to keep a record of any reportable injury, disease or dangerous occurrence. Each record must include the date and method of reporting, the date, time and place of the event, personal details of those involved and a brief description of the nature of the event, along with any supporting evidence (such as photographs). Employers can keep the record in any form they wish.

Tables 1.11 and 1.12 list the major injuries and dangerous incidents that must be reported.

Did you know?

Gas Safe registered gas fitters must also report dangerous gas fittings they find, and gas conveyors and suppliers must report some flammable gas incidents.

Find out

If you or your employer needed to contact the Incident Contact Centre how would you do this? Discuss this with your tutor or employer.

Category	Injury
Injuries to bones, arms and legs	fracture other than to fingers, thumbs or toes
	amputation
	dislocation of the shoulder, hip, knee or spine
Eye injuries	loss of sight (temporary or permanent)
	chemical or hot metal burn to the eye or any penetrating injury to the eye
Injuries resulting in unconsciousness, resuscitation or hospitalisation	injury resulting from an electric shock or electrical burn leading to unconsciousness or requiring resuscitation or admittance to hospital for more than 24 hours
	any other injury leading to hypothermia, heat-induced illness or unconsciousness; or requiring resuscitation; or requiring admittance to hospital for more than 24 hours
	unconsciousness caused by asphyxia or exposure to a harmful substance or biological agent
Acute illness	acute illness requiring medical treatment, or loss of consciousness arising from absorption of any substance by inhalation, ingestion or through the skin
	acute illness requiring medical treatment where there is reason to believe that this resulted from exposure to a biological agent or its toxins or infected material
Diseases and infections	reportable diseases and some skin diseases, such as occupational dermatitis, skin cancer, chrome ulcer, oil folliculitis and acne
	lung diseases including occupational asthma, farmer's lung, pneumoconiosis, asbestosis, mesothelioma
	infections such as leptospirosis, hepatitis, tuberculosis, anthrax, legionellosis (legionnaires' disease) and tetanus
Other	occupational cancer, certain musculoskeletal disorders, decompression illness and hand–arm vibration syndrome
	certain poisonings

Table 1.11 Reportable major injuries

Dangerous occurrences that need to be reported wherever they occur	Dangerous occurrences that need to be reported <u>unless in an offshore workplace</u>
collapse, overturning or failure of load-bearing parts of lifts and lifting equipment	the unintended collapse of any building or structure under construction, alteration/demolition where over five tonnes of material falls, a wall or floor in a place of work and any false-work
explosion, collapse or bursting of any closed vessel or associated pipework	any explosion or fire causing suspension of normal work for over 24 hours
failure of any freight container in any of its load-bearing parts	the sudden, uncontrolled release in a building of 100 kg or more of a flammable liquid; 10 kg or more of a flammable liquid above its boiling point; or 10 kg or more of a flammable gas; or 500 kg of these substances if the release is in the open air
plant or equipment coming into contact with overhead power lines	the accidental release of any substance that may damage health
electrical short circuit or overload causing fire or explosion	
any unintentional explosion, misfire, failure of demolition to cause the intended collapse, projection of material beyond a site boundary, injury caused by an explosion	
accidental release of a biological agent likely to cause severe human illness	
failure of industrial radiography or irradiation equipment to de-energise or return to its safe position after the intended exposure period	
malfunction of breathing apparatus while in use or during testing immediately before use	
failure or endangering of diving equipment, the trapping of a diver, an explosion near a diver, or an uncontrolled ascent	
collapse or partial collapse of a scaffold over five metres high, or erected near water where there could be a risk of drowning after a fall	
unintended collision of a train with any vehicle	
dangerous occurrence at a well (other than a water well)	
dangerous occurrence at a pipeline	

Table 1.12 Reportable dangerous occurrences

Dangerous occurrences that need to be reported wherever they occur	Dangerous occurrences that need to be reported <u>unless in an offshore workplace</u>
failure of any load-bearing fairground equipment, or derailment or unintended collision of cars or trains	
a road tanker carrying a dangerous substance overturns, suffers serious damage, catches fire or the substance is released	
a dangerous substance being conveyed by road is involved in a fire or released	

Table 1.12 Reportable dangerous occurrences (cont.)

For small, non-reportable incidents and injuries, most employers have an accident book on site. The HSE has produced a new accident book to help businesses comply with the Data Protection Act 1998. This records accidents, but stores individuals' details in a secure location. It contains information on first aid and how to manage health and safety information to help prevent accidents from happening.

Figure 1.05 Accident report book

Summoning the emergency services

The official emergency telephone number in the UK is 999. This number can be used to summon assistance from the three main emergency services:

- police
- fire and rescue
- ambulance

as well as more specialist services, such as HM Coastguard.

An emergency can be:

- a person in immediate danger of injury or their life is at risk
- suspicion that a crime is in progress
- another serious incident that needs immediate emergency service attendance.

When you make a 999 call, ask for the service you need then provide the following information:

- your telephone number
- the location of the incident and your location
- the type of incident
- the number, gender and ages of any casualties
- details of any injuries that may have occurred
- any information you may have observed about hazards, such as power cables, fog, ice or gas leaks.

Assessment of work

Employers and the self-employed need to assess the first-aid requirements of their work. Make sure there are enough suitably trained first-aiders and facilities so any casualties can get immediate help. Procedures should be in place for contacting emergency services, service providers and environmental agencies.

An assessment of the working area should be carried out before work begins, taking into account:

- the nature of the work
- the past history and consequences of accidents
- the nature and distribution of the workforce
- the remoteness of the site from the emergency services, including location, terrain and weather conditions
- working on shared or multi-occupied sites
- holidays and other absences of first-aiders

- the presence of trainees and the public
- the possibility of medical conditions or allergies – consider MedicAlert® a charity providing an identification system of bracelets or necklets for those with hidden medical conditions or allergies.

Carry a personal first-aid kit on you while at work – Figure 1.06 shows what your kit should contain. A worksite first-aid kit should be kept in a central location.

You should avoid working alone. If you must, make arrangements for someone to check on you at regular intervals: the greater the risk, the more frequent the checks should be. As a minimum requirement, always inform your contact when work starts and finishes.

Figure 1.06 Contents of a first-aid kit

For any emergency procedures to work well, everyone must be aware of them and have had the opportunity to test them. Emergency procedures should be tested, evaluated and modified, to ensure they are working. When setting up a project, anticipate problems that could exist in getting to a casualty. Identify the personnel and equipment that need to be on site and establish

Safety tip

If you are part of a team scattered across an area, everyone in the team should arrange to meet at agreed times throughout the day.

Ensure everyone on site can be alerted in an emergency.
Ensure everyone working on site knows the signal for an emergency and knows what to do.
Ensure someone has been trained in and will take responsibility for co-ordinating procedures.
Ensure emergency routes are available, clear, signed and adequately lit. If site is not adequately lit for all periods when people are at work, provide lighting that will come on automatically in an emergency.
Ensure there are arrangements for calling the emergency services. It is good practice to let the Fire Brigade know about any work where specialist rescue equipment may be needed.
Ensure there is adequate access to the site and that access does not become blocked.
Ensure arrangements for treating and recovering injured people are in place.
If an emergency does arise, post someone at the site entrance (or another prominent position) to direct emergency services.

Table 1.13 Tips for forward planning

how to quickly reach anything additional that may become necessary. Everyone on site should know, and have practised, any evacuation procedures.

Insurance

Employers are responsible for the health and safety of their employees while they are at work. An employee may be injured at work, or a former employee may become ill as a result of their work while in employment. The employee might try to claim compensation from the employer if they believe the employer is responsible. The Employers' Liability (Compulsory Insurance) Act 1969 ensures that employers have at least a minimum level of insurance to cover against any such claims.

Public liability insurance is different. It covers an employer for claims made against it by members of the public or other businesses, but not for claims by the employer's employees.

When an employer takes out employers' liability insurance, it will have an agreement with the insurer about the circumstances in which the latter will pay compensation. For example, the policy will cover the specific activities that relate to the business.

There are certain conditions that could restrict the amount of money an insurer might have to pay. Employers must make sure that their contract with the insurer does not contain any of these conditions.

The insurer cannot refuse to pay compensation purely because the employer:

- has not provided reasonable protection for employees against injury or disease
- cannot provide certain information to the insurer
- did something the insurer told it not to do (for example, said it was at fault)
- has not done something the insurer told it to do (for example, report the incident)
- has not met any legal requirement connected with the protection of the employees.

However, this does not mean employers can forget about their legal responsibilities to protect the health and safety of their employees. If an insurer believes that an employer has failed to meet its legal responsibilities for the health and safety of its employees and that this failure has led to the claim, the policy may enable the insurer to sue the employer to reclaim the cost of the compensation.

Appropriate responsible persons to report health and safety issues

Responsibility begins with senior management. Strong leadership is vital in delivering effective health and safety risk control. Everyone should know and believe that management is committed to continuous improvement in health and safety performance. Management should explain its expectations, and how the organisation and procedures will deliver them. Although health and safety functions can (and should) be delegated, legal responsibility for health and safety rests with the employer.

These general duties are expanded and explained in the Management of Health and Safety at Work Regulations 1999. Employers must:

- assess work-related risks for both employees and those not in their employ
- have effective arrangements in place for planning, organising, controlling, monitoring and reviewing preventive and protective measures
- appoint one or more competent people to help undertake the measures needed to comply with health and safety law
- provide employees with good, clear information on the risks they face and the preventive and protective measures that control those risks.

The key body for dealing with health and safety issues is the Health and Safety Executive (HSE). Take a look back at pages 26–27, which cover the HSE.

Now you will look at the responsibilities of employees and of other organisations you may need to report incidents to.

Employee rights, responsibilities and limitations

As an employee, you have rights, but you also have responsibilities for your own well-being and that of your colleagues. Many of these rights and responsibilities fall under HASAWA and were covered on page 3, or under CDM, covered on pages 21–22.

Find out

What is your employer's approach to health and safety? Can you see any ways in which it could be improved?

Here are some additional rights and responsibilities.

- If you have reasonable concerns about your safety, you can stop work and leave your work area, without being disciplined.

- You can get in touch with the HSE or your local authority if your employer won't listen to your concerns, without being disciplined.

- You have the right to take rest breaks during the working day, to have time off from work during the working week, and to have annual paid holiday.

- You must report any injuries, strains or illnesses you suffer as a result of doing your job (your employer may need to change the way you work).

- You must tell your employer if something happens that might affect your ability to work (for example, becoming pregnant or suffering an injury). Your employer has a legal responsibility for your health and safety, so they may need to suspend you while they find a solution to the problem, but you will normally be paid if this happens.

- If you drive or operate machinery, you have a responsibility to tell your employer if you take medication that makes you drowsy. They should temporarily move you to another job if they have one for you to do.

Limitations on your responsibilities

There may also be limitations on your responsibilities – things you should **not** do or try to do, because really they are somebody else's responsibility. You should be aware of the appropriate responsible persons to whom you should report any health and safety or welfare matters.

As an apprentice electrician, the most obvious starting point will be your site supervisor, but there are other people who could be involved, depending on the situation. You will look at some of these people below.

Health and safety policy statement

As you saw in the section on HASAWA (page 3), by law anyone employing five or more people must have a written health and safety policy, including arrangements for putting that policy into practice. This should be the key to achieving acceptable standards, reducing accidents and reducing the incidence of work-related ill health. A good health and safety policy also shows employees that their employer cares about them.

Remember

A health and safety policy should be more than just a legal requirement – it should show the commitment of an employer to planning and managing health and safety well.

The safety officer

The safety officer is a suitably qualified member of staff with delegated responsibility for all things related to health and safety, who is answerable to the company managers. A safety officer's role and responsibilities are likely to include:

- arranging internal and external training for employees on safety issues
- monitoring and implementing codes of practice and regulations
- updating and displaying information
- liaising with external agencies, such as the HSE
- carrying out and recording regular health and safety inspections and risk assessments
- advising on selection, training, use and maintenance of PPE
- maintenance of accident reports and records.

If there were an accident, the safety officer would lead the investigations, identify the causes and advise on any improvements in safety standards that need to be made.

> **Remember**
>
> What is written in the policy has to be put into practice. The true test of a health and safety policy is the actual conditions in the workplace, not how well the statement has been written.

The safety representative

The safety representative is often a trade union member. The representative's role is similar to that of the safety officer, and includes:

- making representations to the employer on behalf of members on any health, safety and welfare matter
- representing members in consultation with the HSE Inspectors or other enforcing authorities
- inspecting designated workplace areas at least every three months
- investigating any potential hazards, complaints by members and causes of accidents, dangerous occurrences and diseases
- requesting facilities and support from the employer to carry out inspections and receive legal and technical information
- paid time off to carry out the role and do union-approved training.

The Health and Safety Executive has issued guidance for employers who do not recognise independent trade unions. For all but the smallest companies, it recommends setting up a safety committee of members drawn from both management and

Did you know?

Under the Consultation with Employees Regulations, any employees not in groups covered by trade union safety representatives must be consulted by their employers. The employer can choose to consult with them directly or through the elected representatives.

employees. This will help employers to meet the conditions of Section 2 (Part 4) of HASAWA and the Health and Safety (Consultation with Employees) Regulations 1996.

Environmental health officers

Environmental health officers are employed by local authorities to inspect commercial businesses such as warehouses, offices, shops, pubs and restaurants within a borough area.

They have the right to enter any workplace without giving notice, though in practice they often do give notice. Normally, the officer looks at the workplace, work activities and management of health and safety, and checks that the business is complying with health and safety law.

Environmental health officers may offer guidance or advice to help businesses. They may also talk to employees and their representatives, take photographs and samples, serve improvement notices and take action if there is a risk to health and safety that needs to be dealt with immediately.

Progress check

1   Following a fatal accident on site, which agency would be involved with the investigation and what powers do they have? What range of penalties could be applied?

2   To maintain a safe working environment, companies often employ the services of a consultant to produce certain paperwork. Identify the items the consultant would be likely to supply.

3   While working on site one of your members of staff causes injuries to a member of the public who is taken to hospital. What are the legal requirements for reporting the accident and who is responsible to ensure the report is made?

4   During work on site you notice water leaking from the toilet area, onto the floor where you are working. What should you do in this situation?

K3. Understand the procedures for establishing a safe working environment

How do you prevent accidents?

Many accidents have 'environmental causes' as they relate to the environment that you are working in. Such causes can be unguarded machinery, defective tools and equipment, poor ventilation, excessive noise or workplaces that are poorly lit, overcrowded, untidy or dirty.

The other main cause of accidents is people. It would be nice to think that we all possess common sense, but the 'human' causes of accidents include:

- carelessness
- bad and foolish behaviour
- improper dress
- lack of training
- lack of experience
- poor supervision
- fatigue
- use of alcohol or drugs.

Health and safety in the workplace is something you need to take personal responsibility for. By thinking about what you are doing, or are about to do, you can avoid many potentially dangerous situations. Always report potentially dangerous situations, hazards or activity to the right people. Even if the hazard is something you can easily fix yourself – for example, moving a brick to prevent a trip hazard – still report it to your supervisor. The fact that the brick was there at all might indicate that someone is not doing their job properly and it could happen again.

The secret is to be aware of all possible danger in the workplace and have a positive attitude towards health and safety. Follow safe and approved procedures where they exist, and always act in a responsible way to protect yourself and others.

Hazard or risk?

A **hazard** is anything that can cause harm, such as chemicals, asbestos, electricity, working from ladders or scaffolding. A **risk** is the chance, high or low, that somebody will be harmed by the hazard.

Remember

Prevention is better than cure.

Safety tip

If the hazard is something that is easy to fix, and as long as it is not dangerous to yourself, then you should fix it – but make sure you report it to your supervisor.

Producing risk assessments

Accidents and ill health can ruin lives. They can also affect business if output is lost, machinery is damaged, insurance costs increase or employers have to go to court. This is why employers are legally required to assess the risks in their workplace.

A risk assessment is nothing more than a careful examination of what, during working activities, could cause harm to people, so that employers can weigh up whether they have taken enough precautions or should do more to prevent harm. The aim is to make sure that no one gets hurt or becomes ill.

The important questions to answer through a risk assessment are:

- what are the hazards?
- are they significant?
- are hazards covered by satisfactory precautions so that the risk is small?

For example, we know that electricity is a hazard that can kill, but the risk of it doing so in a tidy, well-run office environment is remote, provided that the installation is sound, 'live' components are insulated and metal casings are properly earthed.

How to assess risks in the workplace

Some years ago the HSE produced guidance for employers to help with the process, called 'Five Steps to Risk Assessment'. This has become an invaluable tool for grasping the essentials of risk assessment.

Here are the five steps.

- Step 1 Identify the hazard.
- Step 2 Decide who might be harmed and how.
- Step 3 Evaluate the risks and decide on precautions.
- Step 4 Record your findings and implement them.
- Step 5 Review your assessment and update it if necessary.

You can see two sample forms from Five Steps to Risk Assessment on pages 44–45.

To help you understand the concepts, as you read on, try to put yourself in the position of an employer carrying out a risk assessment.

Step 1: Identify the hazards

Unless you actively look for hazards, by the time you discover them, you could already be in danger. Here are some ways in which you can find hazards before they find you.

- First you need to work out how people could be harmed. Remember that when you work in a place every day it is easy to overlook some small hazards.

- Walk around the workplace and look at what could reasonably be expected to cause harm. Employers should ask employees or their representatives what they think, as they may have noticed things that are not immediately obvious.

- Visit the HSE website or call the HSE infoline (0845 345 0055), as they can guide you towards HSE guidance on where hazards occur and how to control them.

- If you are a member of a trade association, contact them. Many produce very helpful guidance.

- Manufacturers' instructions or data sheets can also help you spot hazards and put risks in their true perspective.

- Have a look back at your accident and ill-health records – these often help to identify the less obvious hazards. Remember to think about long-term hazards, such as noise, as well as safety hazards.

Step 2: Decide who might be harmed and how

For each hazard you need to be clear about who might be harmed; it will help you identify the best way of managing the risk. That doesn't mean listing everyone by name, but rather identifying groups of people, such as 'people working in the storeroom' or 'passers-by'.

In each case, identify how they might be harmed – what type of injury or ill health might occur. For example, shelf stackers may suffer back injury from repeated lifting of boxes.

Some workers have particular requirements. For example:

- new cleaners, visitors, contractors or maintenance workers who may not be in the workplace all the time and young workers

- new or expectant mothers and people with disabilities who may be at particular risk

- members of the public, or people you share your workplace with, if there is a chance they could be hurt by your activities.

Ask other staff if they can think of anyone that you may have missed.

Step 3: Evaluate the risks and decide on precautions

Evaluate the risks and decide whether existing precautions are adequate or more should be done. Consider how likely it is that each hazard could cause harm; this will determine whether more needs to be done to reduce the risk.

Even after all precautions, some risk usually remains. Decide for each hazard whether this remaining risk is high, medium or low.

- First, ask yourself whether you have done all the things that the law says you have to do. For example, there are legal requirements on prevention of access to dangerous parts of machinery.
- Then ask yourself whether generally accepted industry standards are in place.

Your real aim is to make all risks low by adding precautions as necessary. If something needs to be done, draw up an action plan and give priority to any remaining risks that are high or which could affect most people.

In taking action ask yourself:

- can I get rid of the hazard altogether?
- if not, how can I control the risks so that harm is unlikely?

In controlling risks apply the following principles, ideally in this order:

- try a less risky option
- prevent access to the hazard, for example by guarding and using barriers and notices
- organise work to reduce exposure to the hazard
- issue personal protective equipment, such as clothing or goggles
- provide welfare facilities such as washing facilities for the removal of contamination.

Improving health and safety need not cost a lot. For instance, it costs little to place a mirror on a dangerous blind corner to help prevent vehicle accidents, or to put non-slip material on slippery steps. Failure to take simple precautions can cost you a lot more if an accident does happen.

If the work you do is very varied, or you or your employees move from one site to another, identify the hazards you can reasonably expect and assess the risks from them.

Step 4: Record your findings and implement them

If you have fewer than five employees, you do not need to write anything down, though it is useful to keep a written record of what you have done.

If you employ five or more people, you must record the significant findings of your assessment – the significant hazards and conclusions – and you must tell your employees about your findings. When writing down your results, keep it simple. For example 'tripping over rubbish: bins provided, staff instructed, weekly housekeeping checks'.

The HSE does not expect risk assessments to be perfect, but they must be suitable and sufficient. You need to be able to show that:

- a proper check was made
- you asked who might be affected
- you dealt with all the obvious significant hazards, taking into account the number of people who could be involved
- the precautions are reasonable, and the remaining risk is low
- you involved your staff or their representative in the process.

Keep the written record for future reference or use; it can help you if an inspector asks what precautions you have taken, or if you become involved in any action for civil liability.

Step 5: Review your assessment and update it if necessary

Few workplaces ever remain the same. Sooner or later you will need to bring in new equipment, substances and procedures that could lead to new hazards. It makes sense to review what you are doing on an ongoing basis.

Every year or so, formally review where you are, to make sure that you are still improving, or at least not sliding back. If there is any significant change, add to the risk assessment to take account of the new hazard.

You do not necessarily need to amend your assessment for every trivial change or new job. However, you will want to consider any significant new hazards in their own right and do whatever you need to keep the risks down. In any case, it is good practice to review your assessment from time to time (no later than annually) to make sure that the precautions are still working effectively.

Safety tip

A written record can also remind you to keep an eye on particular hazards and precautions, and it helps to show that you have done what the law requires.

Safety tip

During the year, if there is a significant change, don't wait. Check your risk assessment and, where necessary, amend it. If possible it is best to think about the risk assessment when you are planning your change – that way you leave yourself more flexibility.

Remember

When you are running a business, it's all too easy to forget about reviewing your risk assessment – until something has gone wrong and it is too late. Why not set a review date for this risk assessment now? Write it down and make a note of it as an annual event.

Company name:

Step 1
What are the hazards?

Spot hazards by:

- walking around your workplace;
- asking your employees what they think;
- visiting the *Your industry* areas of the HSE website or calling HSE Infoline;
- calling the Workplace Health Connect Adviceline or visiting their website;
- checking manufacturers' instructions;
- contacting your trade association.

Don't forget long-term health hazards (such as noise).

Step 2
Who might be harmed and how?

Identify groups of people. Remember:

- some workers have particular needs;
- people who may not be in the workplace all the time;
- members of the public;
- if you share your workplace think about how your work affects others present.

Say how the hazard could cause harm.

Step 5 Review date:

Figure 1.07 Five Steps to Risk Assessment

Date of risk assessment:

Step 3
What are you already doing?

List what is already in place to reduce the likelihood of harm or make any harm less serious.

What further action is necessary?

You need to make sure that you have reduced risks 'so far as is reasonably practicable'. An easy way of doing this is to compare what you are already doing with good practice. If there is a difference, list what needs to be done.

Step 4
How will you put the assessment into action?

Remember to prioritise. Deal with those hazards that are high risk and have serious consequences first.

Action by whom   Action by when   Done

- Review your assessment to make sure you are still improving, or at least not sliding back.
- If there is a significant change in your workplace, remember to check your risk assessment and, where necessary, amend it.

Key terms

Young people – according to the Management of Health and Safety at Work Regulations, people who have not reached the age of 18

Child – according to the Education Act 1996, a person who is below Minimum School Leaving Age (MSLA). This will be 15 or 16 years old depending on when their birthday falls

Remember

There is no need for a new risk assessment each time a young person is employed, as long as the current assessment takes into account the characteristics of young people and activities which present significant risks to their health and safety. Remember to take into account young people's experience and capacity.

Risk assessments for young people

All risk assessments need to take into account certain features that apply to **young people**. It is important to remember that, when you are young, you will have little or no experience of hazardous substances and will work at a slower pace than others.

Before employing a young person, a health and safety risk assessment must take into account:

- the fitting-out and layout of the workplace
- the nature of any physical, biological and chemical agents, the length of time and the extent of the exposure
- the types of equipment to be used and how these will be handled
- the organisation of the work and processes involved
- the need to provide health and safety training and to assess this
- risks from particular agents, processes and work.

As a young person, generic risk assessments can be useful if you are likely to be doing temporary or transient work. Many electricians carry out small works and visit more than one job a day. These risk assessments could be modified to deal with particular work situations and any unacceptable risks. In all cases, the risk assessment will need to be reviewed if the nature of the work changes or if there is reason to believe that it is no longer valid.

There are restrictions on when and where you can be employed and what work you do too. These do not apply in 'special circumstances'. This could be if you are over the minimum school leaving age and are doing work necessary for your training, under proper supervision by a competent person in conditions reduced, as far as is reasonably practicable, to their lowest level of risk. Under no circumstances can children of compulsory school age do work involving these risks, whether they are employed or under training. This includes children on work experience.

Training includes government-funded training schemes for school leavers, modern apprenticeships, in-house training arrangements and work that qualifies for assessment for N/SVQs, such as craft skills.

Training and supervision

As a young person, you should:

- have training as soon as you start work
- be trained in the risks and hazards of the workplace, with the key messages checked understood
- know basic health and safety procedures including first aid, which should be part of your training.

You are also likely to need more supervision, This also allows your employers to monitor how effective your training has been as well as your occupational competence.

Procedures for working in accordance with risk assessments, method statements and safe systems of working

After risk assessment, an employer should look to determine what can be done, as far as is reasonably practicable, to remove any hazards. If any hazards remain, then safe systems of work should be developed to work around these.

The process to follow in developing a modern, safe system of work is as follows.

- Make a risk assessment.
- Determine what can be done so far as is reasonably practicable to remove the identified hazards.
- Should hazards remain, develop a safe system of work.
- Where necessary formalise these systems of work into procedures (method statements).
- Include in the procedures where necessary the use of permits to work coupled with physical lock-off systems for electrical supplies where necessary.
- Monitor the observance of all parts of the procedure.
- Feed back any information on weaknesses or failures in the system.
- Rectify these by modifying the system.
- Keep monitoring and modifying the systems as necessary.

A method statement takes information about significant risks from your risk assessment, and combines this with the job specification, to produce a practical, safe working method for the workers to follow on site. Method statements should be easy to read and specific to that job or site.

These should include the use of work permits and all parts of the procedure should be observed with feedback on any weaknesses so that the procedure can be improved if needed.

Permit to Work

A Permit to Work is a document that specifies the work to be done, the person(s) involved, when it is going to be done, the hazards involved and the precautions to be taken.

To understand how important a Permit to Work can be, read the Working life feature on page 48. This example is based on an area that continues to be of concern: the dangers of working in isolation.

Did you know?

A generic method statement may be acceptable in some circumstance, such as for minor repetitive jobs in identical situations, but this is generally not the case.

Remember

Generic information about frequently used company policies should form part of any general health and safety policy and be incorporated into induction training.

Roger, a painter, has gone off to paint some rooms one Saturday morning. Arriving on site, he says hello to a few people and goes off to start his work. Once there he realises that he also has a fairly large storage tank to paint. Many hours later, his body is found inside the storage tank. He has been killed by the fumes from the specialist paint he was using.

- What precautions should Roger have taken before starting this work?
- Who would have been involved in the investigation of this accident?
- What recommendations do you think they would make?

The Working life feature is just one example of when it can be dangerous to work in isolation, but there are many more, such as:

- in confined spaces
- in trenches
- near, or on, live sources or equipment
- at height
- near to unguarded machinery
- where there is a risk of fire or in hazardous atmospheres
- with toxic or corrosive substances.

Safety tip

There will always be circumstances where you will be involved in potentially dangerous tasks. For many situations, the answer is the use of the Permit to Work system.

One contributing factor to the painter's death in the Working life feature was that no one knew exactly where he was or what he was doing. The Permit to Work specifies the work to be done, the persons involved, the hazards involved, when it is to be done and precautions taken. If he had had a Permit it would only have been 'active' for a set period; if the painter had not returned by then, someone would have gone to investigate – and his life could have been saved.

As the Permit to Work has to be authorised, it is essential that the person doing the authorising is competent and fully able to understand the work to be done, the hazards and the proposed system of work and precautions needed. A sound knowledge of Regulations such as the Confined Spaces Regulations 1997 and COSHH is also essential.

Safety tip

Remember there are situations in construction and maintenance work where the nature of the task involves risks that are so high that you should not undertake those tasks alone.

The Permit to Work form aids communication between everyone involved, and employers must train staff in its use. Ideally the company issuing the permit should design it, taking into account individual site conditions and requirements. On certain sites, separate permit forms may be required for different tasks, such as hot work and entry into confined spaces, so that sufficient emphasis can be given to the particular hazards present and precautions required.

Provided documentation

You should bear in mind that the documentation and systems prepared and provided for any situation are not foolproof.

Imagine you are driving to a location using a Sat-Nav system. The voice and the map tell you to turn left after 200 m, but you can clearly see that doing so would take you the wrong way down a one-way street and into a river. Would you follow the instructions?

The point is that in life we all make mistakes, and situations change, so the information that is issued and received may not always be correct. Equally HASAWA makes it clear that both employers and employees have responsibilities for safety.

Be sensible. If any document you are issued with doesn't match the circumstances you are working in or sets alarm bells ringing in your mind, check with an appropriate person before you do anything.

PPE and its purpose

The use of PPE is governed by the Personal Protective Equipment at Work Regulations 1992, which are covered on page 16.

Eye protection

Every year thousands of workers suffer eye injuries, which result in pain, discomfort, lost income and even blindness. Following safety procedures correctly and wearing eye protection can prevent these injuries. There are many types of eye protection equipment available: for example, safety spectacles, box goggles, cup goggles, face shields and welding goggles. To be safe, you have to have the right type of equipment for the specific hazard you face.

Figure 1.08 Safety goggles

Figure 1.09 Helmet with visor face screen

Figure 1.10 Half-face mask

Foot and leg protection

Did you know?

Toe and foot accidents account for a large proportion of reported accidents in the workplace each year, and many more accidents go unrecorded.

Guarding your toes, ankles, feet and legs from injury also involves protecting your whole body from injury caused by improper footwear, for example an injury caused by electric shock.

Protective footwear can help prevent injury and reduce the severity of injuries that do occur.

PPE	Usage
Safety shoe, boot or trainer	Basic universal form of foot protection Safety clogs can also be worn, which give particularly good protection against hot asphalt
Spats, often made of leather and worn over the shoe	Protect the feet from stray sparks during welding
Gaiters, worn over the lower leg and top of the shoe	Used to give protection against foul weather/ splashing water
Leggings	Protect general clothing

Table 1.14 Types of leg and foot PPE

Hand protection

This involves the protection of two irreplaceable tools – your hands, which you use for almost everything, including working, playing, driving and eating. Unfortunately hands are often injured. One of the most common problems other than cutting, crushing or puncture wounds is **dermatitis**.

Skin irritation may be indicated by sores, blisters, redness or dry, cracked skin that is easily infected.

To protect your hands from irritating substances you need to:

- keep them clean by washing them regularly with approved cleaners
- wear appropriate personal protection when required
- make good use of barrier creams where provided.

Head protection

Head protection is important because it guards your most vital organ – your brain! Head injuries pose a serious threat to your brain and your life.

Here is a list of good safety practice to help you protect your head.

- Know the potential hazards of your job and what protective headgear to use.
- Follow safe working procedures.
- Take care of your protective headgear.
- Notify your supervisor of unsafe conditions and equipment.
- Get medical help promptly in the case of head injury.

There are several types of protective headwear for use in different situations; use them correctly and wear them whenever they are required.

Safety helmets

Here are a few important rules.

- Adjust the fit of your safety helmet so it is comfortable: all straps should be snug but not too tight.
- Don't wear your helmet tilted or back to front.
- Never carry anything inside the clearance space of a hard hat, like cigarettes, cards or letters.
- Never wear an ordinary hat under a safety helmet.

Key term

Dermatitis – inflammation of the skin normally caused by contact with irritating substances

Figure 1.11 Rigger gloves

Figure 1.12 Gauntlet gloves

Remember

A single injury can handicap a person for life, or even be fatal.

Figure 1.13 Safety helmet

- Do not paint your safety helmet as this could interfere with electrical protection or soften the shell.
- Only use approved types of identification stickers on your safety helmet, such as 'First-aider'.
- Do not use sticky tape or labelling tape as the adhesive could damage the helmet.
- Handle the helmet with care: do not throw it or drop it.
- Regularly inspect and check the helmet for cracks, dents or signs of wear, and if you find any, get your helmet replaced.
- Check the strap for looseness or worn stitching and also check your safety helmet is within its 'use-by' date.

Bump caps

For less dangerous situations, where there is a risk of bumping your head rather than things falling, or where space is restricted, bump caps, which are lighter than safety helmets, may be acceptable.

If you have to work outside in poor conditions, and a safety helmet is not a requirement, consider using a sou'wester and cape.

Working life

Fazal is a first-year apprentice who is starting his first day on a construction site. The electrician in charge gives Fazal a hard hat to wear. Fazal notices that the hat has a crack down one side and reports this to his supervisor. The supervisor tells Fazal that he must wear this hat because there are no more in the store and he needs to get on with his work.

- What should Fazal do in this situation?
- If an object hit Fazal on the head and he was injured because of the crack in the hat, who would be responsible for this injury?

Hearing protection

You may sometimes have to work in noisy environments. If you do, your employer must ensure compliance with Control of Noise at Work Regulations 2006. This would involve, if necessary, carrying out noise risk assessments and providing you with suitable hearing protection. Like any other sort of PPE, hearing protection must be

Figure 1.14 Ear plugs

Figure 1.15 Ear defenders

worn properly and you must check it regularly to make sure it is not damaged.

Common types of hearing protection are:

- ear plugs that fit inside the ear
- ear defenders that sit externally, such as headphones.

Lung protection

We all need clean air to live, and we need correctly functioning lungs to allow us to inhale that air. Fumes, dusts, airborne particles such as asbestos or just foul smells, such as in sewage treatment plants, can all be features of construction environments.

You can get a range of respiratory protection, from simple dust-protection masks to half-face respirators, full-face respirators and powered breathing apparatus. To be effective these must be carefully matched to the hazard involved and correctly fitted. You may also require training in how to use them properly.

Figure 1.16 Disposable dust respirator

Whole body protection

To complete your 'suit of armour' you need to protect the rest of your body. Usually this will involve overalls or similar, to protect against dirt and minor abrasions. However, you may also need:

- specialist waterproof or thermal clothing in adverse weather conditions
- high-visibility clothing on sites or near traffic
- chemical-resistant clothing, such as neoprene aprons, if you are working near or with chemical substances.

When working outdoors, sunscreen is now also a consideration.

Figure 1.17 High-visibility jacket

Figure 1.18 High-visibility waistcoat

Figure 1.19 Overalls

Remember

All items of PPE that are provided for your protection must be worn and kept in good condition.

When should I wear PPE?

Risk assessment is the key here (see pages 40-46) However, all construction sites that you work on will require you to wear the same basic level of PPE.

PPE	When worn
Hard hats	• where there is a risk of you either striking your head or being hit by falling objects
Eye protection	• drilling or chiselling masonry surfaces • grinding or using grinding equipment • driving nails into masonry • using cartridge-operated fixing tools • drilling or chiselling metal • drilling any material that is above your head
Ear protection	• close to noisy machinery or work operations
Gloves	• if there is a risk to the hands from sharp objects or surfaces • handling bulky objects to prevent splinters, cuts or abrasion • working with corrosive or other chemical substances
Breathing protection	• in dusty environments • working with asbestos • where noxious odours are present • where certain gases are present

Table 1.15 When to wear PPE

Other items

When working involves long periods of kneeling or having to take your weight on your elbows, you may be issued with specialist protectors for these areas. Other items you may use could include face masks, safety harnesses or breathing apparatus.

Personal hygiene

If necessary, use a barrier cream before starting work. This fills the pores of the skin with a water-soluble antiseptic cream so that, when you wash your hands, the dirt and germs are removed with the cream.

Always wash at the end of the work period, before and after using the toilet, and before handling food. Reapply barrier cream after washing.

Safety tip

Do not use solvents to clean your hands. They remove protective oils from the skin and can cause serious problems, such as dermatitis.

Change your overalls regularly before they get too dirty and become a health hazard themselves.

First-aid facilities that must be available in the work area

First aid can save lives, reduce pain and help an injured person make a quicker recovery. The Health and Safety (First Aid) Regulations 1981 require you to provide adequate and appropriate equipment, facilities and personnel to enable first aid to be given to your employees if they are injured or become ill at work. The minimum provision for all sites is:

- a first-aid box, with enough equipment to cope with the number of workers on site
- an appointed person to take charge of first-aid arrangements
- information telling workers the name of the appointed person or first-aider and where to find them.

An appointed person will take charge in the event of an injury and may not always be a trained first-aider. A first-aider is someone who has undergone a training course in administering first aid at work and holds a current first aid at work certificate. The number of qualified first-aiders needed depends on the risk of injury and ill health on site – Table 1.16 gives you a guide.

Numbers employed at any location	Number of first-aid personnel
Fewer than five	At least one appointed person
5 to 50	At least one first-aider
More than 50	One additional first-aider for every 50 employed

Table 1.16 Number of first-aiders needed

The first-aid arrangements should cover any shift working, night and weekend working. This may mean appointing or training several people.

First aid for electric shock

When someone receives an electric shock, the passage of electricity through the body may stun them, causing their heartbeat and breathing to stop. The severity of the shock will depend on the level of current and the length of time it is in contact with the body.

Did you know?

A notice in the site hut is a good way of giving information to workers about the appointed person or first-aider.

Remember

The contents of a first aid kit were shown on page 33 and include sterile plasters, sterile eye pads, triangular wrapped bandages, safety pins and sterile wound dressings (large and medium). If you use anything from a first aid kit, always replace it immediately.

Key term

Fibrillate – make rapid twitching movements

As a rough guide, low levels of current – about 1 to 10 milliamps – may be felt as only an unpleasant tingle, but could cause you to lose balance or fall over. A current of about 10 to 30 milliamps can give you muscular spasms and make you lose control. A current of 30 milliamps can cause your heart to **fibrillate**. At 50 milliamps or above, a period of just one second can be lethal.

✔ **Check for your own safety** – make sure that you do not put yourself at risk by helping the casualty.

✔ **Break the electrical contact** to the casualty by switching off the supply, removing the plug or wrenching the cable free (only if these are not damaged). If this is not possible break the contact with a piece of non-conductive material, for example, a piece of wood.

✔ **If the casualty is conscious, guide them down to the ground** making sure that they do not get any further injuries, such as banging their head.

✔ **If the casualty is unconscious, get help straight away**, then check the casualty's response. Talk to and gently shake the casualty to gauge their level of response. If the casualty appears unharmed, advise them to rest and then see a GP.

✔ **If there is no movement or sign of breathing, summon help immediately**. If there is someone with you, tell them to ring 999; if you are on your own, leave the casualty for a moment while you get help yourself. As soon as you return, begin CPR (cardio-pulmonary resuscitation).

Figure 1.20 Action checklist for electric shock

Basic Life Support (BLS)

This section will give brief guidance to BLS, but there is no substitute for having a first-aid qualification.

When giving basic life support, stay calm: you'll be more effective if you are. Then follow these steps.

Step 1
Make sure the casualty, any bystanders and you are safe.

Step 2
Check the victim for a response. Gently shake the shoulders and ask loudly, 'Are you all right?'

Step 3A – *If they respond*
Leave them in the position in which you found them provided there is no further danger.

Try to find out what is wrong with the victim and get help if needed.

Reassess the victim regularly.

Step 3B – *If they do not respond*
Shout for help.

Turn the victim onto their back and then open the airway using head tilt and chin lift:

- place your hand on his forehead and gently tilt their head back
- with your fingertips under the point of the victim's chin, lift the chin to open the airway as shown in Figure 1.21.

Figure 1.21 Opening the airway

Step 4
Keeping the airway open, look, listen, and feel for normal breathing.

Look for chest movement.

Listen at the victim's mouth for breath sounds.

Feel for air on your cheek.

In the first few minutes after cardiac arrest, a victim may be barely breathing, or taking infrequent, noisy, gasps. Look, listen, and feel for no more than 10 seconds to determine if the victim is breathing normally. If in doubt, act as if breathing is not normal.

Step 5A – *If the victim is breathing normally*

Turn them onto the recovery position (see the box and Figure 1.22).

Send or go for help, or call for an ambulance.

Check for continued breathing.

Placing someone in the recovery position

- Kneel beside the victim and make sure that both their legs are straight.

- Place the arm nearest to you out at right angles to their body, elbow bent with their hand palm uppermost.

- Bring the far arm across the chest, and hold the back of the hand against the victim's cheek nearest to you.

- With your other hand, grasp the far leg just above the knee and pull it up, keeping the foot on the ground.

- Keeping the victim's hand pressed against their cheek, pull on the far leg to roll the victim towards you onto their side.

- Adjust the upper leg so that both the hip and knee are bent at right angles.

- Tilt the head back to make sure the airway remains open.

- Adjust the hand under the cheek, if necessary, to keep the head tilted.

- Check breathing regularly.

- If the victim has to be kept in the recovery position for more than 30 minutes, turn them to the opposite side to relieve the pressure on the lower arm.

Figure 1.22 The recovery position

Step 5B – *If they are not breathing normally*

Ask someone to call for an ambulance or, if you are on your own, do this yourself. You may need to leave the victim.

Start chest compressions (see the box and Figure 1.23).

Giving chest compressions: CPR

- Kneel by the side of the victim.

- Place the heel of one hand in the centre of the victim's chest.

- Place the heel of your other hand on top of your first hand.

- Interlock the fingers of your hands and ensure that pressure is not applied over the victim's ribs. Don't apply any pressure over the upper abdomen or the bottom end of the bony sternum, the breastbone (see Figure 1.23).

- Position yourself vertically above the victim's chest and, with your arms straight, press down on the sternum to a depth of 4–5 cm.

- After each compression, release all the pressure on the chest without losing contact between your hands and the sternum.

- Repeat at a rate of about 100 times a minute (a little less than two compressions a second).

- Compression and release should take an equal amount of time.

Figure 1.23 Chest compression

Step 6A – *Combine chest compression with rescue breaths*
After 30 compressions, open the airway again using head tilt and chin lift.

Pinch the soft part of the victim's nose closed, using the index finger and thumb of your hand on their forehead. Allow the mouth to open, but maintain chin lift.

Take a normal breath and place your lips around the mouth, making sure that you have a good seal.

Blow steadily into the mouth while watching for the chest to rise; take about one second to make the chest rise as in normal breathing – this is an effective rescue breath.

Maintaining head tilt and chin lift, take your mouth away from the victim and watch for the chest to fall as air comes out.

Take another normal breath and blow into the victim's mouth once more to give a total of two effective rescue breaths. Then return your hands without delay to the correct position on the sternum and give a further 30 chest compressions.

Continue with chest compressions and rescue breaths in a ratio of 30:2.

Stop to recheck the victim only if they start breathing normally; otherwise do not interrupt resuscitation.

If your rescue breaths do not make the chest rise as in normal breathing, then before your next attempt:

- check the victim's mouth and remove any visible obstruction
- recheck that there is adequate head tilt and chin lift
- do not attempt more than two breaths each time before returning to chest compressions.

If there is more than one rescuer present, another should take over CPR every two minutes or so to stop you getting tired out. Ensure the minimum of delay during the changeover.

Step 6B – *Chest-compression-only CPR*
If you are unable or unwilling to give rescue breaths, give chest compressions only, at a continuous rate of 100 a minute.

Stop to recheck the victim only if they start breathing normally; otherwise do not interrupt resuscitation.

Step 7

Continue resuscitation until:

- qualified help arrives and takes over
- the victim starts breathing normally, or
- you become exhausted.

Choking

How to recognise choking

Choking occurs when someone's airway is blocked by a foreign body. Recognising when someone is actually choking is the key here: it is important not to confuse this emergency with any other that may cause sudden respiratory distress, cyanosis or loss of consciousness. Table 1.17 shows you how to distinguish choking from other problems.

General signs of choking	
Attack occurs while eating Victim may clutch their neck	
Signs of mild airway obstruction	**Signs of severe airway obstruction**
Response to question 'Are you choking?' Victim speaks and answers yes	*Response to question 'Are you choking?'* Victim unable to speak Victim may respond by nodding
Other signs Victim is able to speak, cough and breathe	*Other signs* Victim unable to breathe Breathing sounds wheezy Attempts at coughing are silent Victim may be unconscious

Table 1.17 Signs to see if someone is choking

Choking first-aid sequence

If the victim shows signs of mild airway obstruction

- Encourage them to continue coughing, but do nothing else.

Did you know?

This sequence is suitable for use in children over the age of 1 year, as well as for adults.

If the victim shows signs of severe airway obstruction and is conscious

- Give up to five back blows.
  - Stand to the side and slightly behind the victim.
  - Support the chest with one hand and lean the victim well forwards so that when the obstructing object is dislodged it comes out of the mouth rather than goes further down the airway.
  - Give up to five sharp blows between the shoulder blades with the heel of your other hand.
- Check to see if each back blow has relieved the airway obstruction. The aim is to relieve the obstruction with each blow rather than necessarily to give all five.
- If five back blows fail to relieve the airway obstruction, give up to five abdominal thrusts.
  - Stand behind the victim and put both arms round the upper part of his abdomen.
  - Lean the victim forwards.
  - Clench your fist and place it between the navel (tummy button) and the bottom end of the breastbone
  - Grasp this hand with your other hand and pull sharply inwards and upwards.
  - Repeat up to five times.
- If the obstruction is still not relieved, continue alternating five back blows with five abdominal thrusts.

Remember

Abdominal thrusts can save lives, but can cause serious internal injuries at the same time. All victims receiving abdominal thrusts should be examined for injury by a doctor.

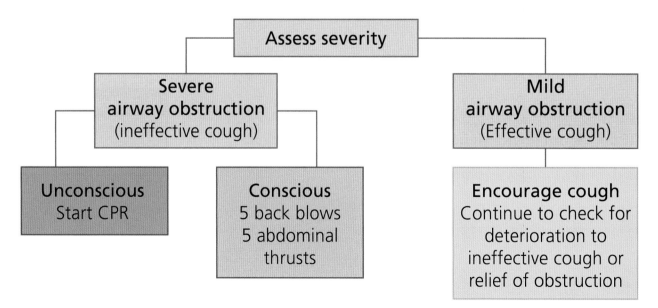

Figure 1.24 Treatment for adult choking

If the victim becomes unconscious

- Support the victim carefully to the ground.
- Immediately call an ambulance.
- Begin CPR (see instructions on page 59). Healthcare providers, trained and experienced in feeling for a carotid pulse, should start chest compressions even if a pulse is present in the unconscious choking victim.

Treatment for burns, shock and breaks

An electric shock may result in other injuries as well as unconsciousness. There may be burns at both the entry point and exit point of the current. The treatment for these burns is to flood the site of the injury with cold water for at least 10 minutes. This will halt the burning process, relieve the pain and minimise the risk of infection.

Shock is the medical condition where the circulatory system fails and insufficient oxygen reaches the tissues. If this shock is not treated quickly the vital organs can fail, leading ultimately to death. To treat shock you need to:

- stop external bleeding if there is any
- lay the casualty down, keeping the head low
- raise and support the legs but be careful if you suspect a fracture
- loosen tight clothing, braces, straps or belts to reduce constriction at the neck, chest and waist
- keep the casualty warm by wrapping them in a blanket or coat
- continue to check and record breathing, pulse and level of response
- be prepared to resuscitate if necessary.

Once you are satisfied that the casualty is in a stable condition, cover any burned skin with a loose, lint-free dressing, or even with loose sheets of cling film (do not wind any dressing tightly around the injured area).

If the casualty had sustained a broken bone, your first aim is to prevent movement. Do not move the casualty until the injury is secured and supported, unless they are in danger from further electric shock. You must arrange for the casualty's immediate removal to hospital, maintaining comfortable support during transport.

> **Remember**
>
> Be prepared to treat for shock at any time.

Treatment for smoke and fume inhalation

If you find someone suffering from the effects of fume inhalation or asphyxiation, then provided that it is safe to do so, get them outside into the fresh air as soon as possible. Loosen any clothes around their neck or chest that may impair their breathing. Call the emergency services and refer to the earlier guidance in this section regarding Basic Life Support (see pages 56–61).

Safe practices and procedures

Many of the safe practices and procedures you will need to use will be covered in the next section of this unit. These include:

- access equipment, including PASMA requirements (pages 107–109)
- portable power tools (pages 67–68)
- tools and materials storage facilities (pages 82–84)
- dangerous substances (page 91).

Safety signs and guarding

The Health and Safety (Safety Signs and Signals) Regulations 1996 require employers to use a safety sign where there is a significant risk to health and safety that has not been avoided or controlled by the methods required under the relevant law, provided the use of a sign can help reduce the risk.

Safety signs are not a suitable substitute for those other methods of controlling risks, such as engineering controls and safe systems of work. However, understanding their meaning and taking notice of them is vital to site safety.

There are four types of safety sign as shown in Figure 1.25. Figure 1.26 shows examples of each.

Remember

Working on construction sites, in factories or elsewhere, you will see a variety of signs and notices. It's up to you to learn and understand what they mean and to take notice of them. Signs are there for a reason and must not be ignored; ignoring safety signs could be very dangerous!

Safety tip

Each type of safety sign can be recognised by its shape and colour. Some signs are just symbols, while others have words or other information such as heights and distances. Make sure you read each sign carefully, and take account of it as you work.

Figure 1.25 Types of safety sign

Prohibition signs

Shape: Circular

Colour: Red borders and cross bar. Black symbol on white background

Meaning: Shows what must NOT be done

Example: No smoking

Mandatory signs

Circular

White symbol on blue background

Shows what must be done

Wear eye protection

Warning signs

Shape: Triangular

Colour: Yellow background with black border and symbol

Meaning: Warns of hazard or danger

Example: Danger: electric shock risk

Information or safe condition signs

Square or rectangular

White symbols on green background

Indicates or gives information on safety provision

First aid post

Figure 1.26 Safety signs

Fire safety signs are covered as a separate item in the Regulations. Figure 1.27 shows a typical example.

Figure 1.27 Fire sign

Progress check

1 To maintain a safe working environment and avoid accidents, hazards and the risk of being harmed by that hazard need to be identified. Before work commences briefly outline what should be done.

2 You are requested to fix a series of hanging brackets onto a concrete wall. Identify what PPE you would expect to use to keep you safe. If one of the items of PPE were damaged and you were injured as a result who would be responsible?

3 As a site operative you are required to know all the safety signs. What do the following safety signs look like: Prohibition, Mandatory, Warning and Information?

4 You hear a scream on site and upon investigation see your colleague clearly suffering from electric shock. What does the accident indicate? What would you do?

K4. Understand the requirements for identifying and dealing with hazards in the work environment

The construction industry presents many types of hazard, some general and some particular to the electrical trade. You will look at a range of these in this section.

Electricity (including temporary supplies and trailing leads and cables)

Electricity can kill. Even non-fatal shocks can cause severe and permanent injury. Shocks from faulty equipment may lead to falls from ladders, scaffolds or other work platforms. Those using electricity may not be the only ones at risk, as poor electrical installations and faulty electrical appliances can lead to fires.

The good news is that most of these accidents can be avoided through careful planning and straightforward precautions.

The main hazards are:

- contact with live parts causing shock and burns (mains voltage at 230 volts a.c. can kill)
- faults, which could cause fires
- fire or explosion where electricity could be the source of ignition in a potentially flammable or explosive atmosphere, for example, in a spray-paint booth.

Electrical equipment used on building sites (particularly power tools and other portable equipment and their leads) faces severe conditions and rough use and can become damaged and dangerous. Modern double-insulated tools are well protected, but their leads are still vulnerable to damage and should be regularly checked.

Where cables are needed for temporary lighting or mains-powered tools, run these at high level, particularly along corridors. Alternatively, use special abrasion-resistant or armoured flexible leads.

As with all aspects of health and safety, where possible, eliminate the risks.

Remember

The first person to alert about a health and safety issue should be your site supervisor or safety officer.

Safety tip

Many hazards can be reduced by precautions: safety guards and fences can be put on or around machines, safe systems of work introduced, safety goggles, helmets and shoes issued. Your employer should provide any additional PPE required, such as ear defenders, respirators, eye protection and overalls.

Remember

Electricity is dangerous. Always take precautions.

Did you know?

Each year about 1000 accidents at work involving electric shock or burns are reported to the Health and Safety Executive (HSE). Around 30 of these are fatal. Most of these fatalities arise from contact with overhead or underground power cables.

A joiner working on site asks if you can power up the socket outlets as he wishes to use his 230 V power tools.

- How should you respond to this request?
- What system should be adopted by the joiner?
- Whose responsibility would it be supply power for the joiner?

Using electrical power tools and lighting

Cordless tools and some other tools operate from a 110 V supply system, which is centre-tapped to earth so that the maximum voltage to earth should not exceed 55 V. This will help eliminate the risk of death and greatly reduce injury in the event of an electrical accident. For other purposes such as lighting, particularly in confined and wet locations, still lower voltages can be used and are even safer.

If mains voltage has to be used, the risk of injury is high if equipment, tools, leads, and so on are damaged, or there is a fault. Residual current devices (RCDs) with a rated tripping current no greater than 30 mA with no time delay will be needed to ensure the current is promptly cut off if contact is made with any live part.

RCDs must be kept free of moisture and dirt and protected against vibration and mechanical damage. They need to be properly installed and enclosed, including sealing of all cable entries. They should be checked daily by operating the test button. However, RCDs cannot give the assurance of safety that cordless equipment or a reduced low-voltage (such as 110 V) system provides.

Electrical systems should be regularly checked and maintained. Everyone using electrical equipment should know what to look out for. A visual inspection can detect about 95 per cent of faults or damage.

Before any 230 V hand tool, lead or RCD is used, check it against the safety checklist shown in Figure 1.28.

Safety tip

Check any equipment to be used in a flammable atmosphere – it must not cause ignition!

Remember

With lighting systems, provide protection for cabling in the same way as for tool supplies. Protect lamps against breakage. If breakage does occur, the exposed filaments may present a hazard.

- ❑ no bare wires visible
- ❑ cable covering not damaged and free from cuts and abrasions (apart from light scuffing)
- ❑ plug in good condition: casing not cracked, pins not bent, key way not blocked
- ❑ no taped or other non-standard joints (e.g. connector strips) in cable
- ❑ outer covering (sheath) of cable gripped at entry point
- ❑ coloured insulation of internal wires not visible
- ❑ equipment outer casing not damaged, all screws in place
- ❑ cables and equipment appropriate to the environment
- ❑ no overheating/burn marks on plug, cable or equipment
- ❑ RCDs working effectively – press 'test' button every day

Figure 1.28 Safety checklist for 230 V hand tools, leads and RCDs

Workers should be instructed to report any of these faults immediately and stop using the tool or cable as soon as any damage is seen. Managers should also arrange for a formal visual inspection of 230 V portable equipment on a weekly basis, and damaged equipment should be taken out of service as soon as the damage is noticed. Do not carry out makeshift repairs.

Checking new installations

With a new installation, ensure that it is installed to BS 7671 and that load characteristics and socket provision have been accurately calculated.

Once an installation has been put into service, make sure that it is adequately maintained. Simple preventative maintenance – a visual inspection supported by testing, if needed – can prevent most electrical risks.

The frequency of such inspection and testing depends upon the installation, the type of equipment, its frequency of use and the environment in which it is installed (BS 7671). Recording the results of these activities also helps to assess the effectiveness of the system.

Safety tip

Remember that overloading sockets by fitting adaptors can cause fires.

Safety tip

When you are working on an electrical installation, hazards can arise concerning the environment you are working in and the installation itself, whatever stage the installation is at.

Table 1.18 uses the example of the construction of a new leisure centre to illustrate the typical stages of an installation and some typical hazards that you may come across.

Project activity	Hazards
Initial preparation and planning on site	Set-up may involve craning in large cabins – access arrangements Delivery arrangements – site access, safe handling and storage
Install	Working at height – use of scaffold, ladders, etc. Presence of other trades – noise, fumes, traffic Electricity – temporary supplies Handling and storage of materials and equipment
Terminate and connect	As above, but live sources may now be present
Inspection and testing	As above, but live sources will now be present Other contractors may restrict access – inform them of your activities
Fault finding and maintenance	Work may be at height or in isolation Live sources are likely to be involved Presence of others may be dangerous – use barriers, communicate

Table 1.18 Hazards encountered in typical stages of an installation

Guidance to protect site workers

The following guidance is based on information published by The Electricity Safety Council, taking into account the requirements of the Electricity at Work Regulations. For more information on these Regulations, look back at pages 4–6. Here is a brief summary of the most important Regulations, and their implications when protecting workers.

- Regulation 12 – where necessary to prevent danger, suitable means be available to cut off the supply of electrical energy to equipment and the equipment must be 'isolated', defined as the secure disconnection of the electrical equipment from every source of electrical energy.
- Regulation 13 – the means of disconnection must be secured in the OFF position, with a warning notice or label at the point of disconnection, and proving 'dead' at the point of work with an approved voltage indicator.
- Regulation 14 – 'dead' working should be seen as the normal method of carrying out work on electrical equipment or

Remember

It is essential that site workers are not exposed to danger when working on or near live electrical systems or equipment.

Remember

Don't forget Regulation 16, which requires that no one shall engage in work with electricity unless they are competent to do so!

circuits. Live work should only be carried out in particular circumstances where it is unreasonable to work 'dead'.

Most people are aware that certain activities require the circuit to be live (such as when fault finding). There can also be commercial pressure to carry out work on, or near, live conductors. However, you must take precautions to ensure safety, and EAWR still applies.

Safe isolation

To ensure compliance with Regulations 12 and 13 of the EAWR, these working principles must be followed:

- the correct point of isolation has been identified and the appropriate means used
- ideally the point of isolation is under the control of the person carrying out the work on the isolated conductors
- warning notices should be applied at the point(s) of isolation
- conductors must be proved dead at the point of work before they can be touched (see GS 38)
- the supply cannot be re-energised while the work is in progress.

Means of isolating a complete installation

Sometimes circumstances mean that you have to isolate either a whole installation or large parts of an installation. When this is necessary, the normal method is to use the main switch or the DB switch disconnector (see Figure 1.29) mounted within the DB.

In either case, the locking device should be locked with a unique key or combination that remains in the possession of the person carrying out the work.

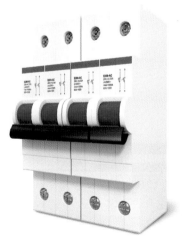

Figure 1.29 DB switch disconnector

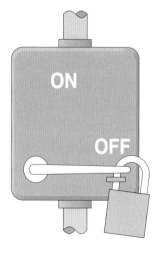

Figure 1.30 Device locked off with padlock

If locking off facilities don't exist on the relevant switch, you can also use a locked DB that prevents access to that switch, provided that it is locked with a device using a unique lock/combination as described above.

Means of isolating individual circuits or items of equipment

Obviously there will be circumstances where it is impractical to isolate a whole section of a building just to work on one item of equipment: for example, to repair a wall light in a hospital ward, you would not isolate the whole ward, just the relevant circuit. If you isolated the whole ward, you would endanger patients on various support systems.

The types of equipment used in circuits to provide switching and isolation of the circuits, and even complete installations, can be categorised as having one or more of the following functions:

- control
- isolation
- protection.

Figure 1.31 gives a simple example of the control, isolation and protection functions. This shows a one-way lighting circuit supplied from a distribution board with a mains switch and circuit breakers.

Did you know?

All live conductors must be isolated before work can be carried out, including the neutral conductor as this is a live conductor. This may mean removing the conductor from the neutral block in the distribution board. Not all distribution boards are fitted with double-pole isolators, so the connecting sequence for neutral conductors needs to be verified and maintained.

Did you know?

Table 53.2 of BS 7671 gives comprehensive guidance on the selection of protective, isolation and switching devices.

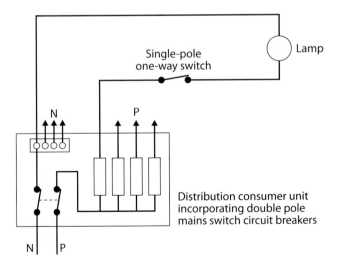

Figure 1.31 One-way lighting circuit with mains switch and breaker

Look at Figure 1.31 and you can see that:

- the distribution consumer unit combines all three functions of control, protection and isolation
- the main double-pole switch can provide the means for switching off the supply and, when locked off, complete isolation of the installation
- the circuit breakers provide protection against faults and over-currents in the final circuits
- when switched off and locked off, the circuit breakers can provide isolation of each individual circuit
- the one-way switch has only one function, which is to control the circuit enabling the luminaire to be switched 'on' or 'off'.

On- and off-load devices

Not all devices are designed to switch circuits 'on' or 'off'. It is important to know that, when a current is flowing in a circuit, the operation of a switch (or disconnector) to break the circuit will result in a discharge of energy across the switch terminals.

You may have seen this when you have entered a dark room, switched on the light and seen a blue flash from behind the switch plate. This is actually the **arcing** of the current as it dissipates and makes contact across the switch terminals. Similar arcing occurs when circuits are switched off or when protective devices operate, breaking fault current levels.

An isolator is designed as an off-load device and is usually only operated after the supply has been made dead and there is no load current to break. An on-load device can be operated when current is normally flowing and is designed to make or break load current.

A circuit breaker is an example of an on-load device. This is not only designed to make and break load current, but has to withstand high levels of fault current. Remembering the three previous functions, it is important to install a device that meets the needs of a particular part of a circuit or installation. Some devices can meet the needs of all three functions. However, some devices may only be designed to meet a single function.

All portable appliances should be fitted with the simplest form of isolator – a fused plug. When unplugged from the socket outlet, this provides complete isolation of the appliance from the supply. For equipment isolation, such a device should be mounted local to the equipment and be fitted with a means of being locked off.

Key term

Arcing – a plasma discharge as the result of current flowing between two terminals through a normally non-conductive media (such as air), producing high light and heat

---

Remember

A plug is not designed to make or break load current. The appliance should be switched off before removing the plug.

The means of isolation could also be an adjacent local isolation device, such as a plug and socket, fused connection unit, circuit breaker or fuse. However, for this to be allowed it must be under the direct control of the competent person carrying out the work and must be visible by them at all times to stop anyone interfering with it.

Sometimes more than one person can work on different circuits supplied from the same DB. Here you should ideally isolate and lock off each individual circuit using the appropriate devices. When this is not possible, you should use a multi-lock hasp on the main switch or DB switch disconnector (see Figure 1.32). This holds more than one padlock (one for each person working, each with a unique key or combination) and cannot be removed until all of the locks have been removed by those people.

If the facility doesn't exist to isolate the circuit like this, then it is permissible to disconnect the circuit from the DB, provided that the disconnected conductors are made safe against inadvertent re-energising of the circuit. Suitable labelling is essential. However, work carried out inside a live DB is classed as live working when there is access to exposed live conductors. In this case, you should take the appropriate precautions (HSG85 with respect to Regulation 14 of the EAWR).

Individual circuits protected by circuit breakers

Where circuit breakers are used as the means of isolation, they should be locked using an appropriate locking-off clip and padlock that can only be operated with a unique key or combination, both of which must be retained throughout by the individual working on the circuit. A warning notice should also be fitted at the point of isolation.

Individual circuits protected by fuses

Where a fuse is the means of isolation, it must be removed and retained by the person carrying out the work. Also, a lockable fuse insert should be fitted in the gap remaining and should be locked using a padlock that can only be operated with a unique key or combination. A warning notice must also be fitted. If lockable fuse inserts are not available then consider either:

- fitting a 'dummy' fuse (a holder with no fuse in)
- padlocking the DB door (retain unique key, fit notice)
- disconnection of the circuit (fit warning notice).

> **Did you know?**
>
> When there is no local means of isolation, the preferred method of isolating circuits or equipment is to use the main switch or DB switch disconnector as if we were isolating a whole installation. It helps if each circuit has the facility to have a suitable locking device and padlock fitted.

Figure 1.32 A switch fixed with a locking device. A multi-lock would use several padlocks

> **Remember**
>
> The practice of placing insulating tape over a circuit breaker to prevent inadvertent switching on is not a safe means of isolation!

Things to watch out for

Circuit identification

For a new installation, make sure all protective devices are correctly identified at the DB before the circuit is energised. Equally, on older installations, make sure all records of the installation are available and that circuits have been correctly identified.

Automatically controlled circuits

'I know it's dead,' he said, 'because I've just tested all the terminals on the equipment are dead.' And then five minutes later this person was just as dead.

Although the terminals were dead when he tested them, they were on a circuit that was controlled by a time switch that suddenly kicked in. You have been warned!

Neutral conductors

Despite the fact that BS 7671 forbids it, the practice of using the neutral of one circuit to supply another still remains, with lighting and control circuits being the favourite culprits.

Be aware that in BS 7671 a neutral is referred to as a 'live conductor' because, in the situation above, a neutral can be live if disconnected and the circuit borrowed from has its load energised.

Proving isolated equipment or circuits are dead

Just because you think you have isolated something doesn't mean that you have. You should never assume that, just because you locked what you thought was the right circuit in the off position, it is actually dead. Always assume and treat something as being live until you have proved otherwise.

Test instruments

All test equipment must be regularly checked to make sure it is in good and safe working order. If you have any doubt about an instrument or its accuracy, ask for assistance: test instruments are very expensive, so avoid causing any unnecessary damage.

Guidance Note GS 38

Published by the Health and Safety Executive, GS 38 is for electrical test equipment used by electricians. It gives guidance to

Remember

You must ensure that your test equipment has a current calibration certificate, which indicates that the instrument is working properly and providing accurate readings. If you do not do this, test results could be void.

electrically competent people involved in electrical testing, diagnosis and repair. Electrically competent people may include electricians, electrical contractors, test supervisors, technicians, managers or appliance repairers.

Voltage-indicating devices

Instruments used solely for detecting a voltage fall into two categories:

- detectors that rely on an illuminated lamp (test lamp) or a meter scale (test meter). Test lamps are fitted with a 15 watt lamp and should not give rise to danger if the lamp is broken. A guard should also protect them

- detectors that use two or more independent indicating systems (one of which may be audible) and limit energy input to the detector by the circuitry used. An example is a two-pole voltage detector: that is, a detector unit with an integral test probe, an interconnecting lead and a second test probe.

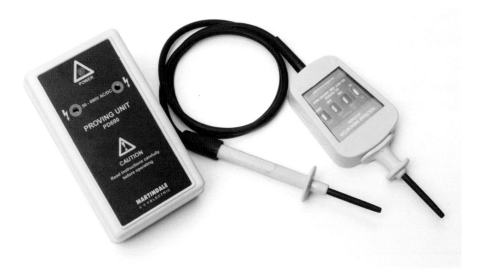

Figure 1.33 Voltage-indicating device

Both these types of detector are designed and constructed to limit the current and energy that can flow into the detector. This limitation is usually provided by a combination of the circuit design using the concept of protective impedance, and current-limiting resistors built into the test probes.

The detectors are also provided with in-built test features to check the functioning of the detector before and after use. The interconnecting lead and second test probes are not detachable components.

These types of detector do not need additional current-limiting resistors or fuses provided that they are made to an acceptable standard and the contact electrodes are shrouded.

It is recommended that test lamps and voltage indicators be clearly marked with the maximum voltage which may be tested by the device and any short-time rating if applicable. This rating is the recommended maximum current that should pass through the device for a few seconds, as these devices are generally not designed to be connected for more than a few seconds.

Restoration of the supply

After the fault has been rectified, which may have resulted in either parts being replaced or simple reconnection of conductors, it is important that the circuit is tested for functionality. These tests may be simple manual rotation of a machine or the sequence of tests as prescribed in BS 7671.

For example, a simple continuity test will check resistance values, open and closed switches and their operation.

You will now look at two safe isolation flow charts: one for isolating a complete installation, and the other when isolating an individual circuit or piece of equipment.

Remember

On 1 April 2004 the insulation colours changed. However for many years to come, you will continue to find existing installations with the old colours. These were:

- single-phase – red, black
- three-phase – red, yellow and blue
- three-phase and neutral – red, yellow, blue and black.

Working life

A plumber protests to you, that he has received an electric shock while connecting the boiler on site. On investigation it is clear that the supply is on.

- What requirements do the Electricity at Work Regulations 1989 place upon operatives working on circuits?
- What procedure should have been followed?
- What essential steps would be taken to ensure the circuit is made safe for work to be carried out?

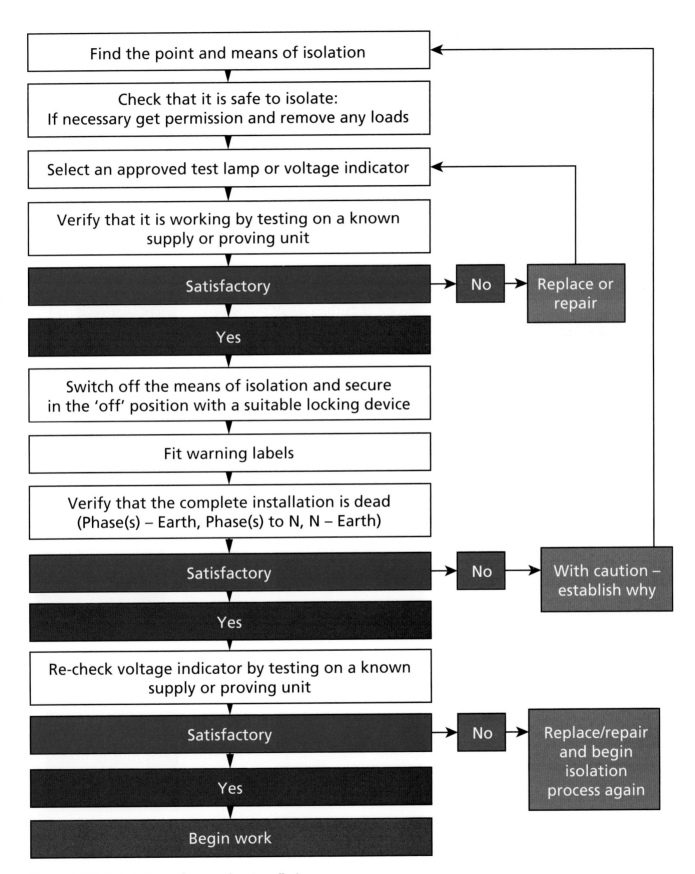

Figure 1.34 Safe isolation for complete installation

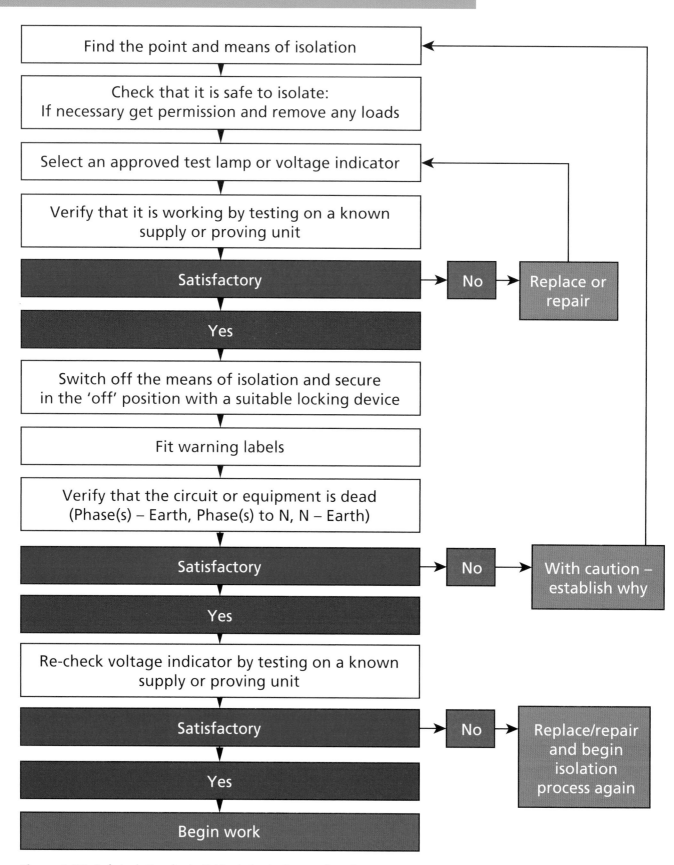

Figure 1.35 Safe isolation for individual circuits/items of equipment

Slippery or uneven surfaces

A simple slip or trip is the single largest cause of injury on construction sites, with more than 1000 major injuries being reported each year. The main causes of this type of accident are:

- having to walk over uneven ground, particularly when carrying unwieldy objects
- tripping over building materials or waste that has been left lying around
- tripping over trailing cables
- slipping caused by wet surfaces or poor ground conditions
- trips caused by small changes in level.

Each of these could easily be avoided, but in reality this can be difficult on a constantly changing construction site. Site managers must exercise good control, and everybody on site must take responsibility for ensuring they do not create a risk for others.

The checklist shown in Figure 1.36 (page 80) gives you some simple rules to help prevent accidents from slips and trips.

Presence of dust and fumes

On construction projects, dust and fumes can come from a wide range of sources, including:

- unmade roads
- drilling operations
- working in trenches
- welding
- painting
- woodworking
- cleaning activities
- plumbing work
- using chemicals as part of the activity.

Employers must decide whether breathing in fumes, vapours or dust is likely to harm anyone's health. Here are some questions that need to be asked.

- Does the manufacturer's information say there is a risk from inhaling the substance?
- Are large amounts of the substance being used?
- Is the work being done in a way that results in heavy air contamination, such as spray application?
- Is the work to be done in an area that is poorly ventilated, such as a basement or container?
- Does the work itself generate a hazard? For example, hot cutting metal covered with lead causes lead fumes.

Safety tip

Common reasons for slips, trips and falls include leaving trailing cables across corridors and leaving waste materials in stairwells. Make sure you don't!

Did you know?

You might want to consider providing wheelie bins or skips for people to put their rubbish in when working on site.

Remember

COSHH requires you to substitute harmful products with less harmful ones or use adequate control measures if this is not possible. All control measures must be in good working order, including mechanical, administrative and operator controls.

Remember

It is important to remember to assess both immediate risks, such as being overcome by fumes in a confined space, and longer-term health risks, such as asthma.

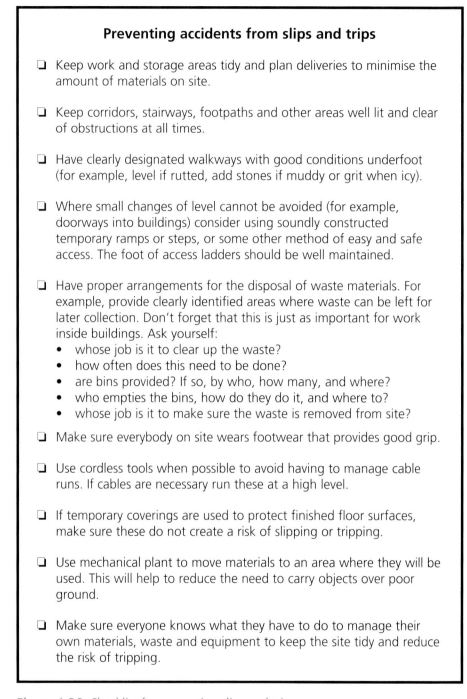

Preventing accidents from slips and trips

❏ Keep work and storage areas tidy and plan deliveries to minimise the amount of materials on site.

❏ Keep corridors, stairways, footpaths and other areas well lit and clear of obstructions at all times.

❏ Have clearly designated walkways with good conditions underfoot (for example, level if rutted, add stones if muddy or grit when icy).

❏ Where small changes of level cannot be avoided (for example, doorways into buildings) consider using soundly constructed temporary ramps or steps, or some other method of easy and safe access. The foot of access ladders should be well maintained.

❏ Have proper arrangements for the disposal of waste materials. For example, provide clearly identified areas where waste can be left for later collection. Don't forget that this is just as important for work inside buildings. Ask yourself:
  • whose job is it to clear up the waste?
  • how often does this need to be done?
  • are bins provided? If so, by who, how many, and where?
  • who empties the bins, how do they do it, and where to?
  • whose job is it to make sure the waste is removed from site?

❏ Make sure everybody on site wears footwear that provides good grip.

❏ Use cordless tools when possible to avoid having to manage cable runs. If cables are necessary run these at a high level.

❏ If temporary coverings are used to protect finished floor surfaces, make sure these do not create a risk of slipping or tripping.

❏ Use mechanical plant to move materials to an area where they will be used. This will help to reduce the need to carry objects over poor ground.

❏ Make sure everyone knows what they have to do to manage their own materials, waste and equipment to keep the site tidy and reduce the risk of tripping.

Figure 1.36 Checklist for preventing slips and trips

Breathing in certain dusts, gases, fumes and vapours within the workplace can cause asthma. Asthma is a serious health problem that can ruin lives. Shortness of breath, wheezing and painful coughing are just some of its symptoms. One of the main causes is rosin-based solder flux fume, caused when soldering: another cause is hardwood, softwood and wood composite dust that arise from sanding when wood machining.

Chronic Obstructive Pulmonary Disease (COPD) is a long-term illness that makes breathing difficult. The lungs and breathing tubes are damaged, making it difficult to get air in and out. COPD is slow to develop – the symptoms tend only to start becoming a problem from your late forties onwards – so many people do not realise they have the disease. Research findings suggest that for Great Britain:

- 15 per cent of COPD may be caused or made worse by work
- 4000 COPD deaths every year may be related to work exposures.

Once COPD develops, the damage to the lungs cannot be reversed. It can be prevented from getting worse by reducing exposure to the dust, fume and irritating gases at work that are causing the problem and avoiding smoking.

Safety tip

The risk of developing COPD is greatly increased if you breathe in dust or fumes in the workplace and you smoke. There is increasing research evidence that COPD can be caused or made worse by dusts, fumes and irritating gases at work.

Prevention

If harm from the substance is possible, first try to avoid it completely by not using it at all. This will mean either:

- doing the job in a different way (for example, instead of using acids or caustic soda to unblock a drain, use drain rods, damp down dusty areas)
- using a substitute substance (for example, instead of using spirit-based paints, use water-based ones, which are generally less hazardous).

However, always check that one hazard is not simply being replaced by another.

Control

If the substance has to be used because there is no alternative, or because use of the least hazardous alternative still leads to significant risk, try to control exposure. Some of the ways this could be done include:

- transferring liquids with a pump or siphon (not one primed by mouth) rather than by hand, keeping containers closed except when transferring
- rather than spraying solvent-based materials, use a roller with a splash guard or apply by brush
- using as little of the hazardous substances as possible – don't take more to the workplace than is needed
- using cutting and grinding tools and blasting equipment fitted with exhaust ventilation or water suppression to control dust

- ensuring good ventilation in the working area by opening doors, windows and skylights – mechanical ventilation equipment might be needed in some cases
- using a Permit to Work system.

If, and only if, exposure cannot be adequately controlled by any combination of the measures above, then you need to use personal protective equipment (PPE).

If you are using a respirator with replaceable cartridges, make sure the correct type is fitted, that they have not become exhausted or clogged and that they are still in date (many filters have a limited shelf life). Always have replacement filters available.

Handling, transport and storage of tools, equipment and materials

You will use a variety of materials, tools and pieces of equipment in your work as an electrician. All of them are potentially dangerous if misused or neglected. Instruction in the proper use of materials, tools and equipment will form part of your training and you should always follow safe working methods.

You have already looked at COSHH, which relates to substances, but hand tools and manually operated equipment are often misused too. You should always use the right tool for the job and never just make do with whatever tool you may have to hand. For example, never use a hammer on a tool with a wooden handle as you may damage the wooden handle and create flying splinters.

Here are some general guidelines.

- Keep cutting tools, saws, chisels and drills sharp and in good condition.
- Make sure handles are properly fitted and secure, and free from splinters.
- Check that the plugs and cables of hand-held electrically powered tools are in good condition. Replace frayed cables and broken plugs.
- Electrically powered tools of 110 volts or 230 volts must be portable appliance tested in accordance with your employer's procedures.

You may also come into contact with high-pressure airlines. Used carelessly, compressed air can be dangerous, causing explosions or blowing aside tools, equipment and debris. Never use an airline to blow dust away; never aim it at any part of your body and never point it at somebody else. If high-pressure air enters the body through a cut or abrasion or through one of the body's orifices it can cause an air **embolism**, which is very painful and can be fatal.

Equally, any piece of equipment – for example, a toolbox, portable floodlight or conduit bender – can deteriorate with use. Equipment should always be visually inspected before use, used correctly and stored appropriately when not in use. If it is damaged or broken, it should not be used, as it could cause an accident or injury.

When using equipment such as grinders or drills, make sure that:

- any required guards are in place
- the equipment is appropriate, undamaged and fit for purpose (for example, correct grinding wheel or drill fitted)
- you do not exceed vibration exposure limits
- you wear suitable PPE (for example, gloves, eye protection, respirators, ear defenders) where required
- you have been trained to operate the equipment
- suitable ventilation is available
- signs and barriers to prevent unauthorised access are provided where required.

Figure 1.37 A portable circular saw, one of the different types of portable tools you may encounter on site

Dangerous occurrences and hazardous malfunctions of equipment

Dangerous occurrences are conditions or actions that have the potential to cause hazardous malfunctions. Hazardous malfunctions are failures of objects or assemblies that could cause injury to operators or bystanders.

Both situations must be reported, even if no one is injured. If any incident is not reported, no one can take action to prevent it happening again, perhaps with more serious consequences.

Untidy working and storage

Trailing cables and air hoses, spilt oil and so on can cause people to slip, trip or fall. Clutter and debris should be cleared away to prevent fire hazards. Tools and equipment left lying around are also targets for thieves, making it difficult for an employer to maintain effective levels of insurance cover.

You should plan how the site will be kept tidy and how housekeeping will be managed.

- After any work, all tools should be cleared away and the workplace left in a safe condition.
- Keep walkways and stairways free of tripping hazards such as trailing cables, building materials and waste.
- Keep inside floor areas clean and dry.
- Outdoor footpaths should be level and firm and should not be used for storing materials.
- Keep all storage areas tidy, whether in an agreed storage area or on the site itself.
- Try to plan deliveries to keep the amount of materials on site to a minimum.

Designate storage areas for plant, materials, waste, flammable substances (for example, foam plastics, flammable liquids and gases such as propane) and hazardous substances (for example, adhesives or cutting compounds). Flammable materials will usually need to be stored away from other materials and protected from accidental ignition.

Do not store materials where they obstruct access routes or where they could interfere with emergency escape. For example, do not store flammable materials under staircases or near to doors or fire exits.

If materials or equipment are stored at height and people could fall when stacking or collecting them, make sure that necessary guard rails are in place and that suitable access equipment is used.

> **Remember**
>
> It is especially important to keep emergency routes clear. Make sure that all flammable waste materials (such as packaging and timber offcuts) are cleared away regularly to reduce fire risks.

Figure 1.38 An untidy work area can present many trip hazards

Lifting and moving

Many people suffer long-term injury from regularly lifting or carrying items that are heavy or awkward to handle, such as paving slabs, bagged products like cement or large distribution boards. Within the context of the electrical industry, manual handling can involve items such as scaffolding, tools, equipment, switchgear and motors.

To avoid problems, it is essential that you plan all material handling properly. Where possible, avoid having to lift materials at all. Where lifting is unavoidable, provide mechanical handling aids wherever possible, such as a conveyor, a pallet truck, an electric or hand-powered hoist or fork-lift truck. Make sure all equipment used for lifting is in good condition and used by trained and competent workers.

The Manual Handling Operations Regulations 1992, as amended in 2002, apply to a wide range of manual handling activities, including lifting, lowering, pushing, pulling or carrying.

Employer requirements	Employee responsibilities
Reduce the need for hazardous manual handling, so far as is reasonably practicable	Follow appropriate systems of work laid down for employee safety
Assess the risk of injury from any hazardous manual handling that cannot be avoided	Make proper use of equipment provided for employee safety
Reduce the risk of injury from hazardous manual handling so far as is reasonably practicable	Co-operate with the employer on health and safety matters
	Inform the employer if you identify hazardous handling activities
	Ensure that your activities do not put others at risk

Table 1.19 Requirements and responsibilities for manual handling

Did you know?

Many construction workers are killed or seriously injured during lifting operations because of accidents such as cranes overturning, material falling from hoists or slinging failures.

Find out

Where possible, avoid manual handling. Think about a manual handling task you may need to carry out. Does a large work piece really need to be moved, or can the activity be done safely where the item already is?

Safety tip

Automated plant still needs cleaning and maintaining, and fork-lift trucks must be suited to the work and have properly trained and certified operators. The movement of loads by machine requires careful planning to identify potential hazards.

This planning involves carrying out a risk assessment. Table 1.19 shows the areas such a risk assessment should look at.

Problems to look for when making an assessment	Ways of reducing the risk of injury
Do the tasks involve: • holding loads away from the body? • twisting, stooping or reaching upwards? • large vertical movement? • strenuous pushing or pulling? • repetitive handling? • insufficient rest or recovery time? • a work rate imposed by a process?	**Can you:** • use a lifting aid? • improve workplace layout to improve efficiency? • reduce the amount of twisting and stooping? • avoid lifting from floor level or above shoulder height, especially heavy loads? • reduce carrying distances? • avoid repetitive handling? • vary the work, allowing one set of muscles to rest while another is used? • push rather than pull?
Are the loads: • heavy, bulky or unwieldy? • difficult to grasp? • unstable or likely to move unpredictably? • harmful (for example, sharp or hot)? • awkwardly stacked? • too large for the handler to see over?	**Can the load be made:** • lighter or less bulky? • easier to grasp? • more stable? • less damaging to hold? If the loads come in from elsewhere, have you asked the supplier to help, for example, to provide handles or smaller packages?
Does the working environment have: • constraints on posture? • bumpy, obstructed or slippery floors? • variations in levels? • hot/cold/humid conditions? • gusts of wind or other strong air movements? • poor lighting conditions? • restrictions on movements or posture from clothes or personal protective equipment (PPE)?	**Can you:** • remove obstructions to free movement? • provide better flooring? • avoid steps and steep ramps? • prevent extremes of hot and cold? • improve lighting? • provide protective clothing or PPE that is less restrictive? • ensure your employees have the right clothing and footwear for their work?

Table 1.20 Making a manual handling risk assessment

Problems to look for when making an assessment	Ways of reducing the risk of injury
Does the job: • require unusual capability (for example, above average strength or agility)? • endanger those with a health problem? or learning/physical disability? • endanger pregnant women? • call for special information or training?	**Can you:** • pay particular attention to those who have a physical weakness? • take extra care of pregnant workers? • give your employees more information, (for example, about the range of tasks they are likely to face)? • provide more training? • get advice from an occupational health adviser if you need to.
With handling aids: • is the device the correct type for the job? • is it well maintained? • are the wheels on the device suited to the floor surface? • do the wheels run freely? • is the handle height between the waist and the shoulders? • are the handle grips in good order and comfortable? • are there any brakes and do they work?	**Can you:** • provide equipment that is more suitable for the task? • carry out planned preventative maintenance to prevent problems? • change the wheels, tyres and/or flooring so that equipment moves easily? • provide better handles and handle grips? • make the brakes easier to use, reliable and effective?
Some other questions • Is the work repetitive or boring? • Is the work machine or system-based? • Do workers feel the demands of the work are excessive? • Have workers little control of the work and working methods? • Is there poor communication between managers and employers?	**Can you:** • change tasks to reduce the monotony? • make more use of workers' skills? • make workloads and deadlines more achievable? • encourage good communication and teamwork? • involve workers in decisions? • provide better training and information?

Table 1.20 Making a manual handling risk assessment (cont.)

Some practical tips for safe manual handling

Step 1

Plan the lift

Think before lifting/handling, plan the lift

- Can handling aids be used?
- Where is the load going to be placed?
- Will help be needed with the load?
- Remove obstructions such as discarded wrapping materials.
- For a long lift, consider resting the load midway on a table or bench to change grip.

Step 2

Keep the load close to the waist

- Keep the load close to the body for as long as possible while lifting.
- Keep the heaviest side of the load next to the body.
- If a close approach to the load is not possible, try to slide it towards the body before attempting to lift it.

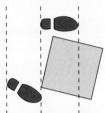

A stable position

Adopt a stable position

- Your feet should be apart, with one leg slightly forward to maintain balance (alongside the load, if it is on the ground).
- Test the weight of the load by pushing it with your foot.
- Be prepared to move feet during the lift to maintain stability.
- Avoid tight clothing or unsuitable footwear, which may make this difficult.

Step 3

A good hold

Get a good hold

Hug the load as close as possible to the body. This may be better than gripping it tightly with hands only.

Start in a good posture

At the start of the lift, slight bending of the back, hips and knees is preferable to fully flexing the back (stooping) or fully flexing the hips and knees (squatting).

Step 4

Lift without flexing

Don't flex the back any further while lifting

- This can happen if you start to straighten your legs before starting to raise the load.
- Avoid twisting your back or leaning sideways, especially while your back is bent.
- Keep your shoulders level and facing in the same direction as your hips.
- Turning by moving your feet is better than twisting and lifting at the same time.
- Keep your head up when handling.
- Look ahead, not down at the load, once you are holding it securely.

Step 5

Smooth movement

Move smoothly

Don't jerk or snatch the load as this can make it harder to keep control and can increase the risk of injury.

Don't lift or handle more you can manage easily.

There is a difference between what people can lift and what they can **safely** lift. If in doubt, seek advice or get help.

Step 6

Put down first

Put down, then adjust

If you need to position the load precisely, put it down first, then slide it into the desired position.

Pushing and pulling

Here are some practical points to remember when pushing and pulling loads.

Handling devices

Aids such as trolleys should have handle heights between the shoulder and waist and be of high quality and well maintained, with large-diameter, smooth-running wheels.

Force

The amount of force needed to move a load over a flat, level surface using a well-maintained handling aid is at least two per cent of the load weight. Try to push rather than pull when moving a load, provided you can see over it, steer it and stop it.

Slopes

Get help whenever necessary as pushing and pulling forces can be very high. For example, if you want to move a load of 400 kg up a slope of 1 in 12 (about 5 degrees), it will take over 30 kg of force, even in ideal conditions. This is just above the guideline weight for men, and well above the guideline weight for women.

Uneven surfaces

On uneven surfaces, the force needed to start the load moving could increase to 10 per cent of the load weight, and to even more on soft ground. Larger wheels may help here.

Stance and pace

To make it easier to push or pull, keep your feet well away from the load and go no faster than walking speed. This will stop you becoming too tired too quickly.

General risk assessment guidelines

There is no such thing as a completely 'safe' manual handling operation. However, working within the guidelines shown in Figure 1.39 will reduce the risk.

Use the diagram to make a quick and easy assessment. Each box contains a guideline weight for lifting and lowering in that zone. As you can see, the guideline weights are reduced if handling is done with arms extended, or at high or low levels, as that is when injuries are most likely to occur.

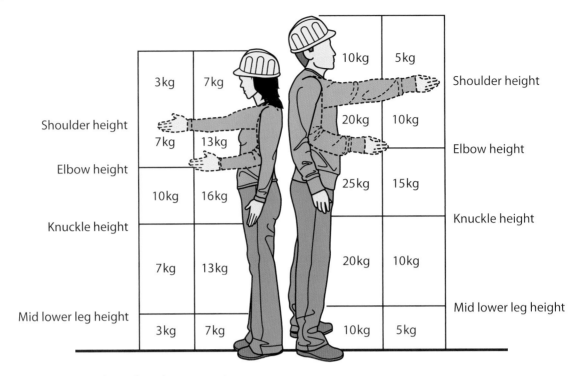

Figure 1.39 Lifting chart for men and women

Observe the work activity you are assessing and compare it to the diagram. First, decide which box or boxes the lifter's hands pass through when moving the load. Then, assess the maximum weight being handled.

- If it is less than the figure given in the box, the operation is within the guidelines.
- If the lifter's hands enter more than one box during the operation, use the smallest weight.
- Use an in-between weight if the hands are close to a boundary between boxes.

The guideline weights assume that the load is readily grasped with both hands and that the operation takes place in reasonable working conditions, with the lifter in a stable body position.

Working life

A number of items of electrical equipment are to be delivered on site. Before the electrical operatives are involved in moving these to their locations there are a number of duties that need to be performed.

- What should the employer consider before the work is undertaken?
- How should employees act once tasks have been delegated?
- If you are unsure as to your capability to lift a particular load how should you proceed?

Contaminants and irritants

Every year, thousands of workers are made ill by hazardous substances, contracting lung diseases such as asthma, cancer, and skin disease such as dermatitis. You should identify any hazardous substances, or processes that may produce hazardous materials, and assess the risk to workers or members of the public.

Ideally, project designers should eliminate hazardous materials from their designs. However, where this is not possible, they should specify the least hazardous product that will perform satisfactorily.

Generally there are two categories of hazard: contaminant and irritant.

- A contaminant is a biological, chemical, physical, or radiological substance which, in a sufficiently concentrated form, can adversely affect living organisms, entering them through air, water, soil and/or food.

- An irritant is a chemical substance which, although not itself corrosive, may cause inflammation of living tissue (such as eyes, skin, or respiratory organs) by a chemical reaction at the point of contact. The effects vary depending on the degree of exposure. Typical irritants include cement, solvents, petrol, diesel and detergents.

The most common sets of substances are shown in Figure 1.40.

Did you know?

Contractors often have detailed knowledge of alternative, less hazardous materials. Designers and contractors can help each other to identify hazardous materials and processes and suggest less hazardous alternatives.

Remember

Look back at the CHIP 4 Regulations on page 19. They are there to protect people and the environment from the effects of chemicals by requiring suppliers to provide information about the dangers and to package them safely.

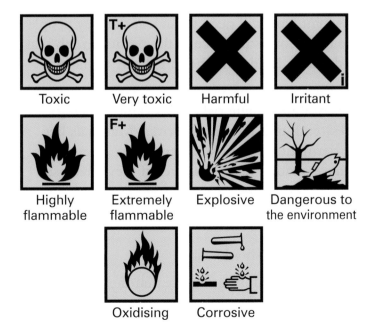

Figure 1.40 Most common contaminants and irritants

Each category has subsets. For example, under 'Toxic' (a chemical that damages health at low levels) there is the subset 'Very Toxic' (a chemical that damages health at very low levels). Note the inclusion of T+ in the graphic to define the subset.

Other subsets under 'Toxic' include:

- Carc Cat 1 Carcinogen – chemicals that cause cancer or increase its likelihood of occurring
- Muta Cat 1 Mutagen – chemicals that induce genetic defects or increase their likelihood of occurring.

If workers use or are exposed to hazardous substances as a result of their work, COSHH makes it a legal duty to assess the health risks involved and prevent exposure or adequately control it.

Identify the substance

People may be exposed to hazardous substances from handling, direct use (for example, cement and solvents) or because their work creates the substance (for example, scabbling concrete generates silica dust). You will need to identify and assess both kinds of hazard.

Manufacturers and suppliers have a legal duty to provide information on hazardous substances they produce. Read the label on the container and/or the safety data sheet. Approach the manufacturer or supplier directly for more information if you need it.

Assessment

To make an assessment, you will need to look at the way people are exposed to the hazardous substance in the particular job that they are about to do, and decide whether it is likely to harm anyone's health.

You should assess both immediate risks (for example, being overcome by fumes) and long-term health risks. Materials like cement can cause dermatitis. Sensitising agents like isocyanates can make people have sudden reactions, even if they have used the substance many times before.

Inhalation

Here are some questions you should ask.

- Does the manufacturer's information say that there is a risk from inhaling the substance?
- Is the work being done in a way that results in heavy air contamination, such as spraying?

Did you know?

There are separate regulations for asbestos and lead: the Control of Asbestos at Work Regulations and the Control of Lead at Work Regulations.

Safety tip

Some hazardous substances may be on site before any work starts: for example, sewer gases or ground contaminants. You should assess these risks in the same way as for other hazardous substances.

Remember

Information to help identify these risks may be available from the client, the design team or the principal contractor and should be contained in the pre-construction health and safety plan.

- Is the work to be done in a poorly ventilated space, such as a basement or enclosed space?

Direct contact with skin

Here are some questions you should ask.

- Does the manufacturer's information say there is a risk from direct contact?
- How severe is this risk – for example, are strong acids or alkalis being used?
- Does the method of work make skin contact likely, such as from splashes when pouring from one container to another, or from the method of application?

Ingestion

Some materials can contaminate the skin and hands. This contamination can then be passed to the mouth when the person eats or smokes. This is a particular problem when handling lead and sanding lead-based paints.

Prevention

If the substance is harmful, try to avoid it completely by not using it. This will mean either doing the job in a different way or using a substitute substance, for example, water-based paints instead of solvent-based. Always check one hazard is not being replaced by another.

> **Safety tip**
>
> When using chemicals of any sort you must understand the dangers involved, you must follow all safety procedures recommended and you must know what to do in an emergency.

Control

Ultimately, if the substance has to be used because there is no alternative, or because use of the least hazardous alternative still leads to a significant risk, the next step is to try to control the situation.

As an example, rather than spraying a substance, can it be applied with a brush? Can you use exhaust ventilation or water suppression to control dusts?

Legislation

The current European legislation CHIP 4 (see page 19) will be replaced in June 2015 by the globally harmonised Classification and Labelling of Chemicals (CLP). There is a transition period that allows suppliers to use the current CHIP 4 labelling. However, some suppliers are already starting to use the new CLP symbols.

Figure 1.41 shows the symbol used under the new system to illustrate a 'health hazard' that may cause breathing difficulties. This covers a range of things, including Category 1 Carcinogen and Category 1 Mutagen:

- Respiratory sensitisation, category 1
- Germ cell mutagenicity, categories 1A, 1B, 2
- Carcinogenicity, categories 1A, 1B, 2
- Reproductive toxicity, categories 1A, 1B, 2
- Specific target organ toxicity following single exposure, categories 1, 2
- Specific target organ toxicity following repeated exposure, categories 1, 2
- Aspiration hazard, categories 1, 2

Figure 1.41 Symbol used to show a health hazard

Liquid petroleum gas (LPG)

LPG is widely used in construction and building work as a fuel for burners, heaters and gas torches. Even small quantities of LPG mixed with air create an explosive mixture (LPG is a gas above -42°C).

LPG comes in cylinders and containers and is highly flammable. It needs careful handling and storage. Here are some guidelines.

- Everyone using LPG should understand the procedures to be adopted in case of an emergency.
- Appropriate firefighting extinguishers (dry powder) should always be available.
- Cylinders must be kept upright whether in use or in storage.
- When not in use the valve should be closed and the protective dust cap should be in place.
- When handling cylinders do not drop them or allow them to come into violent contact with other cylinders.

Did you know?

One litre of LPG when boiled or evaporated becomes 250 litres of gas. This is enough to make an explosive mixture in a large shed, room, store or office.

- When using a cylinder with an appliance, make sure it is connected properly, in accordance with the instructions you have been given, and that it is at a safe distance from the appliance or equipment that it is feeding.

LPG is heavier than air: if it leaks, it will not disperse in the air but sink to the lowest point and form an explosive concentration that could be ignited by a spark. LPG should not be used in excavations as the gas cannot flow out of these areas.

Fire and fire extinguishers

As an electrician, you may work in what are known as 'explosive atmospheres', and fire is always a risk when dealing with electrical installation.

So what exactly is a fire?

Essentially, fire is very rapid **oxidation**.

Rusting iron and rotting wood are common examples of slow oxidation. Fire, or combustion, is rapid oxidation as the burning substance combines with oxygen at a very high rate. Energy is given off in the form of heat and light. Because this energy production is so rapid, we can feel the heat and see the light as flames.

How fire happens

All matter exists in one of three states: solid, liquid or gas (vapour). The atoms or molecules of a solid are packed closely together, and those of a liquid are packed loosely. The molecules of a gas are not really packed together at all and are free to move about.

In order for a substance to oxidise, its molecules must be well surrounded by oxygen molecules. The molecules of solids and liquids are packed too tightly for this to happen, and therefore only gases can burn.

When a solid or liquid is heated, its molecules move about rapidly. If enough heat is applied, some molecules break away from the surface to form a gas just above the surface. This gas can now mix with oxygen. If there is enough heat to raise the gas to its ignition temperature, and if there is enough oxygen present, the gas will oxidise rapidly and it will start to burn.

What we call burning is the rapid oxidation of millions of gas molecules. The molecules oxidise by breaking apart into

> **Safety tip**
>
> On no account should you smoke in areas where LPG is in use.

> **Key term**
>
> **Oxidation** – a chemical process in which a substance combines with oxygen. During this process, energy is given off, usually in the form of heat

individual atoms and recombining with oxygen into new molecules. It is during the breaking-recombining process that energy is released as heat and light. The heat that is released is radiant heat, which is pure energy. It is the same sort of energy that the sun radiates and that we feel as heat. It radiates (travels) in all directions. Therefore, part of it moves back to the seat of the fire, to the 'burning' solid or liquid (the fuel). The heat that radiates back to the fuel is called radiation feedback.

Part of this heat releases more gas, and part of it raises the gas to the ignition temperature. At the same time, air is drawn into the area where the flames and gas meet. The result is that there is an increase in flames as the newly formed gas begins to burn.

Figure 1.42 Newly formed gas begins to burn

The fire triangle

The three things needed for combustion to take place are:

- fuel (to vaporise and burn)
- oxygen (to combine with fuel vapour)
- heat (to raise the temperature of the fuel vapour to its ignition temperature).

The fire triangle shows us that fire cannot exist without all three together:

- if any side of the fire triangle is missing, a fire cannot start
- if any side of the fire triangle is removed, the fire will go out.

Classes of fire

Combustible and flammable fuels have been broken down into five categories:

Figure 1.43 The fire triangle

Remember

A fire can be extinguished by breaking the fire triangle. If fuel, oxygen or heat is removed, the fire will die out. If the chain reaction is broken, the resulting reduction in vapour and heat production will put out the fire.

- Class A – those involving organic solids, such as paper or wood
- Class B – those involving flammable liquids
- Class C – those involving flammable gases
- Class D – those involving metals
- Class F – those involving cooking oils.

Electrical fires have no classification, as electricity is a source of ignition that will feed a fire until switched off or isolated. However these are sometimes known as Class E fires.

Fire legislation

Since 2006 there has been one simple piece of legislation: the Regulatory Reform (Fire Safety) Order 2005. Fire certificates have been abolished, with a risk-based approach to fire safety on community, industrial and business premises put in its place. This requires the responsible person (usually the employer, owner or occupier) to:

- carry out a fire safety risk assessment
- implement appropriate fire precautionary and protection measures
- maintain a fire management plan.

Fire and Rescue Authorities are the principal enforcers and have a statutory duty to enforce the requirements of the legislation; employers are solely responsible for fire safety within their workplaces.

The main rules for the responsible person under the Order are to:

- carry out a fire-risk assessment, identifying any possible dangers and risks
- consider who may be especially at risk
- get rid of, or reduce, the risk from fire as far as is reasonably possible
- provide general fire precautions to deal with any possible risk remaining
- make sure that everyone on the premises, or nearby, can escape safely if there is a fire
- take other measures to make sure there is protection if flammable or explosive materials are used or stored
- create a plan to deal with any emergency and, in most cases, keep a record of the findings
- review findings when necessary.

The Order applies to virtually all premises and covers nearly every type of building, structure and open space, including factories, offices, shops, care homes, hospitals, community halls, places of worship, pubs, clubs, restaurants, schools and sports centres. However, it does not apply to private homes, including individual flats within a block or house.

Fire prevention

Fires can spread rapidly. Once established, even a small fire can generate sufficient heat energy to spread and accelerate the fire to surrounding combustible materials. Fire prevention is largely a matter of common sense and good housekeeping. Keep the workplace clean and tidy. If you smoke, don't throw lit cigarettes on to the ground, and don't leave flammable materials lying around or near sources of heat or sparks.

From an electrical perspective, make sure all leads are in good condition, fuses have the correct rating and circuits are not overloaded. Any alterations or repairs to electrical installations must be carried out only by qualified personnel and to the standards laid down in the IET Regulations (BS 7671).

Firefighting

A fire safety officer once said: 'People should only use a portable fire extinguisher to break the window so that they can escape from the building.' The point he was making is that it is dangerous to try to fight a fire, and the use of fire extinguishers should only be considered as a first-response measure: for example, where the fire is very small or where it is blocking your only means of exit. Firefighting is a job for the professional emergency service.

Firefighting equipment

Normally available firefighting equipment includes portable appliances such as extinguishers, buckets of sand or water and fire-resistant blankets. In larger premises you will find automatic sprinklers, hose reels and hydrant systems.

Fire extinguishers

There are many types of fire extinguisher, each with a specific set of situations in which it may or may not be used. All fire extinguishers have different colour stripes that indicate their type and their uses – always be sure you have the right type before use.

Table 1.21 shows four common types of fire extinguisher.

Remember

The new fire regulations state that fire-fighting equipment in the workplace is there only to enable a small fire to be attacked, to prevent it from spreading.

Safety tip

In the past, fire extinguishers were coloured according to their type. Today this difference is less obvious. However, many of the coloured extinguishers remain in place.

Standard/multi-purpose dry powder

Stripe colour	Blue
Application	Powder 'knocks down' the flames. Safe to use on most kinds of fire. Multi-purpose powders are more effective, especially on burning solids; standard powders work well only on burning liquids
Dangers	Powder does not cool the fire well. Fires that seem to be out can re-ignite. Doesn't penetrate small spaces, such as those inside burning equipment. Jet could spread burning fat or oil around
How to use	Aim jet at the base of the flames and briskly sweep it from side to side

Water

Stripe colour	Red
Application	Water cools the burning material. Only use water on solids, like wood or paper. Never use water on electrical fires or burning fat or oil
Dangers	Water can conduct electricity back to you. Water actually makes fat or oil fires worse – they can explode as the water hits them
How to use	Aim the jet at the base of the flames and move it over the area of the fire

CO_2

Stripe colour	Black
Application	Displaces oxygen with CO_2 (a non-flammable gas). Good for electrical fires as it doesn't leave a residue
Dangers	Pressurised CO_2 is extremely cold. DO NOT TOUCH. Do not use in confined spaces
How to use	Aim the jet at the base of the flames and sweep it from side to side

Foam/AFFF (Aqueous Film Forming Foam)

Stripe colour	White or cream
Application	Foam forms a blanket or film on the surface of a burning liquid. Conventional foam works well only on some liquids, so not good for use at home. Very effective on most fires except electrical and chip-pan fires
Dangers	'Jet' foam can conduct electricity back to you though 'spray' foam is much less likely to do so. Foam could spread burning fat or oil around
How to use	For solids, aim jet at the base of the flames and move it over the area of the fire. For liquids, don't aim foam straight at the fire – aim it at a vertical surface or, if the fire is in a container, at the inside edge of the container

Table 1.21 Types of fire extinguisher

Figure 1.44 Sprinkler

When considering using a fire extinguisher:

- never use a fire extinguisher unless you have been trained to do so
- do not use water extinguishers on electrical fires due to the risk of electric shock and explosion
- do not use water extinguishers on oils and fats as this too can cause an explosion
- do not touch the horn on a CO_2 extinguisher as this can freeze burn the hands
- do not use the CO_2 extinguisher in a small room as this could cause suffocation
- read the operating instructions on the extinguisher before use.

Sprinkler systems

To reduce the risk of dying in a fire, you should get fire sprinklers installed. These can be individually heat-activated, so the whole system doesn't go off at once, and they rarely get set off accidentally as they need high temperatures to trigger them. They operate automatically, whether you're in the building or not. Sprinklers should sound the alarm when they go off – so they alert you and also tackle the fire.

Smoke alarms

Smoke alarms will alert you to slow-burning, smoke-generating fires that may not create enough heat to trigger a sprinkler.

General fire safety

Regular fire drills must be held and all personnel must be familiar with normal and alternative escape routes. Fire routes should be clearly marked and emergency lighting signs fitted above each exit, where applicable. Make sure you know where your assembly point is.

If you discover a fire:

- raise the alarm immediately
- leave by the nearest exit
- call the fire service out
- if in doubt, stay out
- close, but don't lock, the windows to help starve the fire of oxygen
- go to your assembly point and report to your supervisor
- do not return to the building until you are authorised to do so.

Working at height

Ladders and stepladders

Ladders, stepladders and trestles are perhaps the most commonly used access equipment on sites – and also perhaps the most misused. Where work at height is necessary, you need to use risk assessments and a hierarchy of controls to decide the best method. Here are some factors you need to consider.

Is it a suitable activity?

'Suitable' here is to do with the type of work and how long it goes on for. As a guide, the HSE advises only using a ladder or stepladder:

- in one position for a maximum of 30 minutes
- for 'light work'.

They are not suitable for strenuous or heavy work. If a task involves a worker carrying more than 10 kg while working at height it will need to be justified by a detailed manual handling assessment:

- where a handhold is available on the ladder or stepladder
- where you can maintain three points of contact (between hands and feet) at the working position.

On a ladder where you cannot maintain a handhold for long, other measures will be needed to prevent, or reduce the consequences of, a fall. On stepladders where a handhold is not practicable a risk assessment will have to justify whether it is safe or not.

On a ladder or stepladder do not:

- overload – do not exceed the highest load stated on the ladder
- overreach – keep your belt buckle (navel) inside the stiles (the long side parts of the ladder that hold the rungs) and both feet on the same rung throughout the task (see Figure 1.45).

> **Did you know?**
>
> According to the HSE, falls from height are the most common cause of fatal injury and the second most common cause of major injury to employees, accounting for 15 per cent of all such injuries.

Figure 1.45 User maintaining three points of contact with ladder

When working on stepladders avoid work that imposes a side loading, such as side-on drilling through solid materials (such as bricks or concrete). Instead, have the steps facing the work activity.

Where side-on loadings cannot be avoided, take steps to prevent the steps from tipping over – such as tying the steps to a suitable point – or use a more suitable type of access equipment

You should also avoid holding items when climbing. If using a ladder and you must carry something, leave one hand free to grip the ladder, or use a tool belt.

On a stepladder where you cannot maintain a handhold you will need to take into account:

- the height of the task
- a safe handhold still being available on the stepladder
- whether it is light work
- whether it avoids side-loading
- whether it avoids overreaching
- whether your feet are fully supported
- whether you can tie the stepladder.

Is it a safe place to use a ladder or stepladder?

As a guide, only use a ladder or stepladder:

- **on firm ground** or where you can spread the load (e.g. use a board)

- **on level ground** For stepladders refer to the manufacturer's instructions; for ladders the maximum safe ground slopes on a suitable surface (unless the manufacturer states otherwise) are:
    - side slope 16°, but the rungs still need to be levelled (see Figure 1.46)
    - back slope 6° (see Figure 1.46)
- **on clean, solid surfaces**, such as paving slabs. These need to be free of loose material so the feet can grip. Shiny floor surfaces can be slippery even without contamination
- **where it has been secured**.

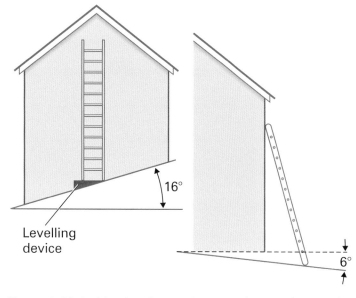

Figure 1.46 Ladder showing maximum angles at 16° on a side slope and 6° on a back slope

Here are some options for securing a ladder, which you should try in the order shown.

- Tie the ladder to a suitable point, making sure both stiles are tied as shown in Figures 1.47–1.50.
- Use a safe, unsecured ladder or a ladder supplemented with an effective ladder stability device.
- Securely wedge the ladder, for example against a wall.
- As a last resort, get another worker to foot the ladder. Footing should be avoided, where reasonably practicable, by the use of other access equipment.

Figure 1.47 Ladder tied at top stiles (correct for working on, not for access)

Figure 1.48 Tying part way down

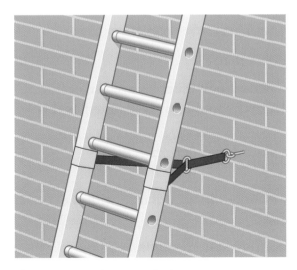

Figure 1.49 Tying near the base

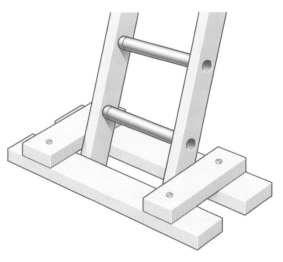

Figure 1.50 Securing at the base

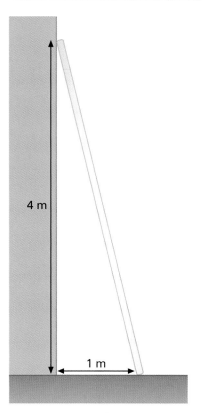

Figure 1.51 Ladder showing correct 1 in 4 angle

Ladders used for access to another level should be securely tied and extend at least 1 metre above the landing point to provide a secure handhold. Stepladders should not be used like this unless they have been designed for the purpose.

Only use ladders or stepladders where:

- they will not be struck by vehicles – protect them with barriers or cones
- pedestrians can't walk under or near them – stop them with cones or barriers or a person standing guard
- they will not be pushed over by doors or windows – secure doors and windows if possible, have a person standing guard, or tell people not to open them
- ladders can be put up at the correct angle of 75° – use the angle indicator marked on the stiles or the 1 in 4 rule as shown in Figure 1.51 (4 units up for each unit out)
- the restraint devices on stepladders can be fully opened – engage any locking devices.

On a ladder or stepladder:

- don't work within 6 m horizontally of any overhead power lines – unless they are dead or protected with temporary insulation
- always use a non-conductive ladder or steps for live electrical work
- don't rest it against weak upper surfaces, such as glazing or guttering – use spreader bars or stand-offs.

Is the ladder or stepladder safe to be used?

Only use ladders or stepladders that:

- have no visible defects – **pre-check** them before use each working day – wooden ladders should not be painted as this hides defects
- have a current **detailed visual inspection**
- are suitable for work use
- have been maintained and stored in accordance with the manufacturer's instructions.

Ladders that are part of a scaffold system still have to be inspected every seven days.

Does the user know how to use it safely?

You may think that anyone knows how to use a ladder – but you should only use a ladder, stepladder or stability device if you are competent – you should have been trained and instructed to use it safely. Here are some things you should know.

Don't:

- use the top three rungs of a ladder, or the top two of a stepladder without a suitable handrail (see Figures 1.52–53)
- use the top three steps of swing-back or double-sided stepladders, where a step forms the very top of the stepladder (see Figure 1.54)
- use them during inclement weather, e.g. strong or gusting winds – follow safe working practices
- move it while standing on the rungs/steps
- support it by the rungs or steps at the base
- slide down the stiles
- stand it on moveable objects, such as pallets, trucks or bricks
- extend a ladder while standing on the rungs.

Figure 1.52 Stand-off device and working maximum height on a ladder

Figure 1.53 Two clear rungs from the top

Figure 1.54 Three clear steps – don't work higher than this on a stepladder

Do:

- wear robust, sensible, clean footwear in good repair without dangling laces
- know how to prevent other workers and the public from using them
- make sure you are fit – certain medical conditions or medication, alcohol or drug abuse could mean you should not use a ladder, so check with a health professional
- know how to tie a ladder or stepladder properly.

Ladders used for access should project at least 1 m above the landing point and be tied, or have a safe and secure handhold available. The rungs or steps must be level – use a levelling device if you can't tell by eye.

Trestles

A 'trestle scaffold' consists of a pair of 'A' frames or adjustable steel trestles, spanned by scaffolding boards, providing a simple working platform. Trestles are used less these days due to the development of lightweight steel platforms and podiums.

Trestles must:

- be erected on a firm, level base with the trestles fully opened
- use a scaffold board at least 600 mm wide and no higher than two-thirds of the way up the A frame
- use scaffolding boards of equal length and thickness, not overhanging the trestle by more than four times their own thickness
- be spaced 1 m apart for 32 mm-thick boards, 1.5 m for 38 mm-thick boards and 2.5 m for 50 mm-thick boards
- use toe boards, guard rails and a separate access ladder for heights over 2 m.

Trestles must not be used where anyone can fall more than 4.5 m, and tie any trestles above 3.5 m tall to the building structure.

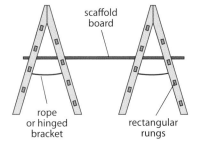

scaffold board

rope or hinged bracket

rectangular rungs

Figure 1.55 Trestles

Podium steps and mobile platforms

Podium steps and mobile platforms have become increasingly popular with tradespeople on site as they are portable and lightweight and can often fold away.

Mobile platforms are simple step-ups, generally 300 mm or 520 mm high and 300 mm, 450 mm or 700 mm wide. Accessories such as guard rails are available.

Podium steps offer a number of advantages over more traditional access equipment because they:

- offer easy compliance with the Work at Heights Regulations 2005
- are compact, lightweight and foldaway, so are easy to store and transport
- are easy to erect, with guard rails that deploy automatically
- come in a range of sizes and platform heights (2.5 m models are available)
- offer 360° working access without you having to move the podium
- have a wide range of accessories, such as tool trays
- can be easily combined with other variations, such as stair decks, to provide quick, safe and easy access to traditionally difficult working areas, such as staircases.

> **Did you know?**
>
> Many sites now advocate the use of podium steps and mobile platforms in place of stepladders as they are ideal for working at lower heights.

Mobile scaffold towers

Tower scaffolds – or scaffold towers – are widely used and are involved in many accidents each year. These usually happen because the tower has not been properly erected or used. Made from light aluminium tube, towers are light and can easily overturn. They are built by slotting sections together and rely on all the parts being in place to ensure adequate strength; they can easily collapse if sections are left out.

Scaffold towers can be mobile (with wheels) or static with plates. They are useful for installing long runs, such as factory lighting installations.

Towers should be erected following a safe method of work. There are two approved methods recommended by the Prefabricated Access Suppliers' and Manufacturers' Association (PASMA), which have been developed in conjunction with the HSE.

> **Safety tip**
>
> Beware of the first step up on a step-up, as it can be quite high.

Figure 1.56 Mobile scaffold tower

Figure 1.57 Mobile scaffold tower

- **Advance guard rail method** This uses specially designed temporary guard rail units, locked in place from the level below and moved up to the platform level. The temporary guard rail units provide fall prevention and are in place before the operator accesses the platform to fit the permanent guard rails.

- **'Through-the-trap' (3T)** This allows the person erecting the tower to position themselves at minimum risk during the installation of guard rails to the next level. The operator takes up a working position in the trap door of the platform, from where they can add or remove components that act as guard rails on the level above the platform. It is designed to prevent the operator standing on an unguarded platform.

Whichever method you choose, when you are working with a tower scaffold, you must:

- make sure the person erecting the tower is competent
- make sure the tower is resting on firm, level ground, with the wheels or feet properly supported – do not use bricks or blocks to take the weight of any part of it
- remember that the taller the tower, the more likely it is to become unstable. As a guide, if towers are used in exposed conditions or outside, the height of the working platform should be no more than three times the minimum base dimension; if the tower is to be used inside, on firm, level ground, the ratio may be extended to 3.5.

Here are some more important points on how to use a tower scaffold safely.

- Before use, check the scaffold is vertical and any wheel brakes are on.
- There must be a safe way to get to and from the work platform. Do not climb up the end frames of the tower except where the frame has an appropriately designed built-in ladder or a purpose-made ladder can be attached safely on the inside.
- Suitable edge protection should be provided where a person could fall more than two metres. Guard rails should be at least 910 mm high and toe boards at least 150 mm high. An intermediate guard rail or suitable alternative should be provided so that the unprotected gap does not exceed 470 mm.
- When moving a tower, check there are no overhead obstructions. Check the ground is firm and level. Push or pull only from the base; never move it while there are people or materials on the upper platforms or in windy conditions.

- Outriggers can increase stability by effectively increasing the area of the base, but must be fitted diagonally across all four corners of the tower and not on one side. When outriggers are used they should be clearly marked (for example, with hazard marking tape) to indicate a trip hazard is present.

- When towers are used in public places, extra precautions may be needed such as minimising the storage of materials and equipment on the working platform, erecting barriers at ground level to prevent people from walking into the tower or work area, and removing or boarding access ladders to prevent unauthorised access if the tower is to remain in position unattended.

- Before you use a tower on a pavement, check whether you need a licence from the local authority.

- Tower scaffolds must be inspected by a 'competent person' before first use and following substantial alteration or any event likely to have affected their stability. If a tower remains erected in the same place for more than seven days, it should also be inspected at regular intervals (not exceeding seven days) and a written report be made. Any faults found should be put right.

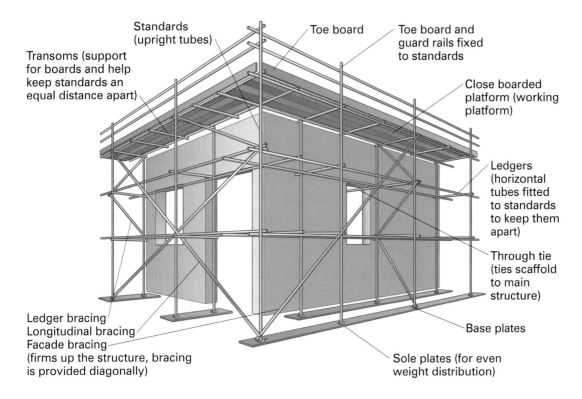

Figure 1.58 An independent scaffold

Scissor and boom lifts

This access equipment is often referred to as mobile elevating platforms (MEWPs) or 'powered platforms'. Those mounted on trucks are often referred to as 'cherry pickers'.

There is a wide range of machines available (including rough-terrain applications) so they can be used for various installations from large factories to external lighting. However, scissor lifts can only extend upwards. Sometimes it is necessary to 'reach' over objects to carry out work: for example, repairing a street-lighting column. Here a telescopic boom platform – a cherry picker – is likely to work better.

Figure 1.59 Scissor lift

In both types of lift, it is essential that workers wear a safety harness. This must be attached to the lift and never to the structure being worked on. All operatives must be suitably trained.

Roof work and fragile surfaces

Almost one in five workers killed in construction accidents is doing roof work. The main causes of accidents are falling off the edges of roofs and falling through holes, roof lights and other fragile surfaces. All roof work requires a risk assessment and, if the work is extensive, a method statement setting out a safe system of work.

Figure 1.60 Boom lift

If work is going to be done on any roof, make sure there is:

- safe access onto the roof, such as a general access scaffold, tower scaffold (preferably of the stairway design) or mobile access equipment
- a safe means of moving across the roof, such as using proprietary staging or purpose-made roof ladders
- a means of preventing falls when working on the roof, such as edge protection consisting of guard rails and toe boards, a proprietary access system or a MEWP

- measures to prevent falls through fragile materials (such as barriers or covers) and to mitigate the consequences should a fall occur (for example, nets).

Independent scaffolds that provide safe access onto the roof, a safe working platform and the capacity for material storage are the ideal solution. However, irrespective of the type of edge protection used, safe access onto the roof and a safe way of lifting materials to roof level must be provided and maintained.

On sloping roofs, workers should not work directly on the tiles or slates. Roof ladders and proprietary staging should be used to enable safe passage across a roof. They must be designed for the purpose, of good construction, properly supported and, if used on a sloping roof, securely fixed with a ridge hook over the ridge. Roof ladders should be used in addition to eaves-level edge protection; if the work requires access within 2 metres of the gable ends, edge protection will be needed there too.

Short-duration work means tasks are measured in minutes rather than hours. It includes jobs like inspection, replacing a few tiles or adjusting television aerials. However short-duration work is still dangerous and appropriate safety measures are essential.

For short-duration work it may not be reasonably practicable to provide full edge protection, but you will need to provide something in its place. The minimum requirements for short-duration work on a roof are:

- a safe means of access to roof level
- a safe means of working on the roof:
  - on a sloping roof, a properly constructed roof ladder
  - on a flat roof, a harness attached to a secure anchorage and fitted with as short a lanyard as possible.

Safety harnesses

If work at height cannot be avoided, measures should be put in place to prevent falls. Harnesses and lanyards can be used to secure workers to the platforms. For this, the lanyard is kept as short as possible while allowing operators to reach their place of work. This prevents them from getting into a fall position, as they are physically unable to get close enough to the open edge (see Figure 1.61). This is acceptable for light, short-duration work and inspection and maintenance.

> **Safety tip**
>
> Always check with the scaffold designer before stacking material at roof level.

> **Remember**
>
> The maximum load that should be carried up a ladder is 10 kg.

Figure 1.61 A harness and lanyard can prevent a worker from falling to the ground

Everyone who uses a harness must know how to check, wear and adjust it before use and how to connect themselves to the structure or anchor point.

Safety nets and soft landing systems

If falls cannot be prevented, use equipment that mitigates the consequences of a falls. Safety nets and soft landing systems (such as airbags) are now being used as cutting-edge protection. They should not be treated as a substitute for fall prevention measures, but can be used with them.

Nets are a complex energy-absorbing system. Before gaining access to height, you must decide whether nets can be installed at ground level.

Figure 1.62 Safety netting is used when working at the highest point

Working in excavations

Many of the same rules for working at height also apply to working in excavations. Every year people are killed or seriously injured when working in excavations. They are at risk from:

- excavations collapsing and burying or injuring people
- material falling from the sides into any excavation
- people or plant falling into excavations.

Excavation work has to be properly planned, managed, supervised and carried out to prevent accidents. Before digging any excavations, it is important to consider and plan for a whole range of potential dangers, as shown in Figure 1.63.

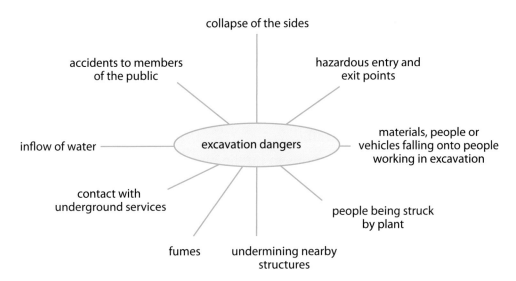

Figure 1.63 Dangers of excavation

Before work starts:

- make sure the necessary equipment is available
- use locators to trace any services, and mark the ground accordingly – also look for valve covers or patching of a road surface
- provide good ladder access or other safe ways of getting into and out of the excavation
- put on a hard hat!

Make sure that:

- a competent person supervises the installation at all times
- the supervisor has service plans and knows how to use them
- any plant operators are competent
- everyone knows about safe digging practices and emergency procedures
- the edges of the excavation are protected against falling materials, using toe boards if necessary
- the sides and ends are battered to a safe angle or supported with timber, sheeting or proprietary support systems
- excavations do not undermine the footings of scaffolds or the foundations of nearby structures.

Wherever possible:

- keep vehicles away from excavations
- take steps to prevent people falling into excavations, if excavation is 2 m or more then provide substantial barriers (e.g. guard rails and toe boards)
- keep workers separate from moving plant.

If necessary:

- put in extra support for the sides (if the excavation is 2 m or more deep)
- use stop blocks to prevent tipping vehicles from overrunning
- use safe systems of work to prevent people being struck by moving plant or traffic
- fence off all excavations in public places
- take extra precautions where children might get onto a site out of hours.

Never:

- go into an unsupported excavation
- work ahead of the supports
- store spoil close to the sides of excavations as it can fall in or destabilise the sides.

Procedures for dealing with presence of asbestos

Asbestos is a mineral found naturally in certain rock types. When separated from rock it becomes a fluffy, fibrous material that has had many uses in the construction industry. In the past it has commonly been used in cement wall and roof panels, ceiling tiles, textured coatings such as Artex™ and insulation lagging around boilers and flash guards inside electrical distribution boards.

Safety tip

The removal of asbestos must only be carried out by specialist contractors. Consider your health at all times. Don't work with asbestos under any circumstances. After all, it's your health, and nothing is more important.

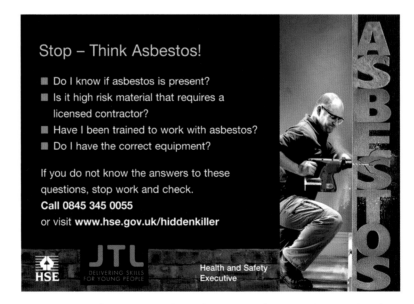

Figure 1.64 The HSE provides information about working with asbestos

Asbestos becomes a health hazard if you inhale the dust; some of the fine rod-like fibres may work their way into your lung tissue and remain embedded for life. This will become a constant source of irritation and can lead to lung diseases (mainly cancers), particularly if you are repeatedly exposed over a number of years. The inhaling of asbestos can lead to serious lifelong health problems, even though symptoms can take years to develop. The health problems caused by asbestos can be fatal.

The use of asbestos is now banned but there are still a number of buildings containing asbestos. Any building built or refurbished before the year 2000 is likely to have asbestos materials somewhere in the building.

Asbestos is perfectly safe in its solid form, but it begins to pose a risk when it breaks down or is disturbed. It is estimated that around 20 tradespeople die every week from asbestos-related disease. As an apprentice electrician working in the construction trade, it is likely that you will come across this hidden killer at some point in your career. Remember: in its solid form it poses no risk, and where possible it should be left undisturbed or removed by a **licensed contractor**.

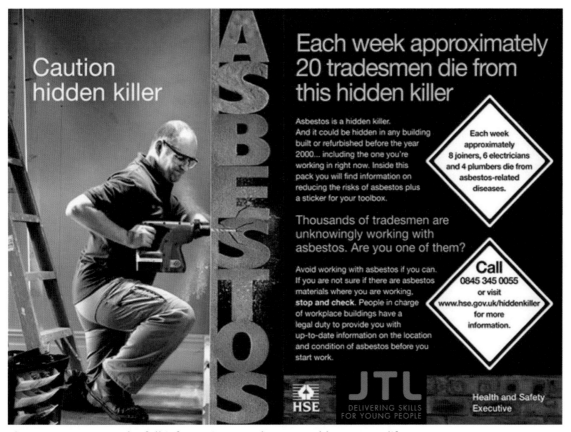

Figure 1.65 Having the full information on asbestos could save your life

To protect yourself you should:

- avoid working with asbestos if possible – if you are unsure whether asbestos is present don't start work until your supervisor has confirmed it is safe to do so

- don't work if the asbestos materials are in sprayed coatings, boards or lagging on pipes and boilers – only licensed contractors should work on these

- know the hazards, avoid exposure and always follow recommended controls

- wear and maintain any personal protective equipment provided

- practise good housekeeping – use special vacuums and dust-collecting equipment

- report any hazardous conditions, such as unusually high dust levels, to your supervisor.

As an apprentice, if you are required to work with asbestos you should be supervised at all times until you have gained enough experience. To minimise the risk from asbestos you should:

- minimise dust – use hand tools instead of power tools, keep materials damp but not wet, clean up as you go using a class H vacuum cleaner, double-bag asbestos waste and label the bag properly

- wear a properly fitted, suitable mask (for example, disposable FFP3 type), suitable disposable overalls (type 5) and boots without laces or disposable boot covers

- wipe down your overalls after work with a damp rag and remove them before removing your mask

- dispose of overalls and so on properly – as with asbestos waste, don't take overalls home to wash

- not smoke, eat or drink in the work area.

Progress check

1. Why do we use reduced voltage equipment on building sites?
2. Describe how the safe isolation process works, and explain why this is so important.
3. What are the recommended points that you need to look for when choosing test probes?
4. Describe the process used to lift a heavy object and explain what risk assessments you may need to carry out before you do it.
5. What are the five different categories of fire, and what types of fire extinguisher should you use to combat each of them?
6. What checks should you make before using or ladder and a stepladder?
7. Why might you choose to use trestles rather than a ladder?
8. What are the main risks associated with working in excavations?

Getting ready for assessment

Working in the Building Services sector is challenging and rewarding. However there are a number of dangers and hazards that you will encounter in your daily working life. Whatever role you work in, you will need to understand these and know the practices and procedures you should use to minimise or eliminate these risks, as far as is reasonably practicable.

For this unit you will need to be familiar with:

- relevant health and safety legislation and how it applies in the workplace
- procedures for dealing with health and safety in the work environment
- procedures for establishing a safe working environment
- requirements for identifying and dealing with hazards in the work environment

For each Learning Outcome, there are several skills you will need to acquire, so you must make sure you are familiar with the assessment criteria for each outcome. For example, for Learning Outcome 2 you will need to be able to state the procedures that should be followed in the case of accidents, specifying what procedures should be followed in a range of emergency situations. You should also be able to state the limitations of your responsibilities, state what actions should be taken in situations that exceed your authority and who the responsible persons are that you should report situations to.

It is important to read each question carefully and take your time. Try and complete both progress checks and multiple choice questions without assistance to see how much you have understood. Refer to the relevant pages in the book for subsequent checks. Always use correct terminology as used in BS7671. There are some simple tips to follow when writing answers to exam questions:

- **Explain briefly** – usually a sentence or two to cover the topic. The word to note is 'briefly' meaning do not ramble on. Keep to the point.
- **Identify** – refer to reference material, showing which the correct answers are.
- **List** – a simple bullet list is all that is required. An example could include, listing the installation tests required in the correct order.
- **Describe** – a reasonably detailed explanation to cover the subject in the question.

Your English skills will be particularly important, as you will need to make sure that you are following all the details of any instructions you receive. This will be the same for any instructions you receive as part of any practical assignment, as it will for any specifications you might use in your professional life.

Before you start work, always remember to have a plan of action. You will need to know the clear sequence for working in order to make sure you not making any mistakes as you work and that you are working safely at all times.

This unit has explained the dangers you may face when working. Understanding these dangers, and the precautions that can be taken to help prevent them, will not only aid you in your training but will help you remain safe and healthy throughout your working life.

Good luck!

CHECK YOUR KNOWLEDGE

1. Reduced voltage is normally utilised on site. Which of the following is the normal voltage used?
   a) 25V
   b) 110V
   c) 230V
   d) 400V

2. When checking to ensure a circuit is isolated what should you do?
   a) Test with a lampholder and flex
   b) Use an approved voltage indicator
   c) Ask a colleague
   d) Switch off only

3. Which of the following fire extinguishers **must not** be used if there was a fire involving electrical equipment?
   a) Dry powder
   b) Carbon dioxide
   c) Halon
   d) Water

4. When checking a person following an accident what is the ABC of resuscitation?
   a) Accident, Breathing, Circulation
   b) Accident, Bleeding, Casualty
   c) Airways, Breathing, Circulation
   d) Airways, Bleeding, Casualty

5. CDM regulations **do not** apply to which of the following people?
   a) Electricians
   b) Designers
   c) Apprentices
   d) Clients

6. Who is responsible for health and safety on construction sites?
   a) Designer
   b) Client
   c) Main contractor
   d) Employees and employers

7. What colours are mandatory signs on site?
   a) Red and white
   b) Blue and white
   c) Green and white
   d) Black and yellow

8. Who should check and inspect ladders before use on site?
   a) Manufacturer
   b) User
   c) Site manager
   d) Client

9. The flexible cord attached to an electric drill you wish to use has insulation damage with conductors showing. What should you do?
   a) Tape over the damage and report later
   b) Put the drill away and use another
   c) Do not use and Inform your site supervisor
   d) Strip the faulty cable

10. Before lifting or moving any load, what should you always do?
   a) Ensure you have safety footwear on
   b) Bend your knees
   c) Assess the weight of the load
   d) Maintain a straight back

UNIT ELTK 02

Understanding environmental legislation, working practices and the principles of environmental technology systems

This unit is about environmental awareness – understanding why it is important to conserve energy, dispose of waste properly and prevent wastage of materials. Environmental awareness is currently very topical, for government, local authorities and individuals alike – and it is something that anyone working in construction needs to be aware of.

Building Regulations have been improved and updated to ensure that buildings are more energy efficient and that building materials are more effectively used.

This unit will cover the following learning outcomes:

■ understand the environmental legislation, working practices and principles which are relevant to work activities

■ understand how work methods and procedures can reduce material wastage and impact on the environment

■ understand how and where environmental technology systems can be applied.

K1. Understand the environmental legislation, working practices and principles which are relevant to work activities

The word 'environment' relates to our surroundings. It includes:

- the built environment – man-made items such as houses, roads and electricity supplies
- the natural environment – natural items such as plants, birds, forests, rivers and rocks
- the social and cultural environment – the culture that someone has been educated in or lives in, and the people and institutions with whom that person interacts.

In this unit, the focus is the effect of the built environment on the world we live in.

Climate change

'Climate change' is the term used for the long-term alteration of the world's weather patterns. You no doubt come across words like 'climate change', '**ecological**' and 'environmental' in the newspapers or on television often, because these are very current issues.

There is still some debate about just how far human activity is responsible for climate change. In 1988, the United Nations set up the Intergovernmental Panel on Climate Change (IPCC), a body of scientists from all parts of the world who assess the best available scientific and technical information on climate change. Their 2007 report warned of an increase in average global temperatures of up to 6.4°C by the end of this century, depending on future levels of emissions. It also said that such changes to the climate were 90 per cent likely to be the result of human activity.

In all parts of the construction industry, climate change is a live issue, and one you need to take into account. Tackling climate change needn't damage the economy, but industry will have to adapt, and jobs and working practices will change. It may even be that more jobs are created overall as we look at the way that we interact with our environment.

Key term

Ecological or **eco-** – linked with ecology, the branch of biology that looks at the relationship between organisms and their environments

Current legislation

This country's early effort at controlling the polluting effects of the Industrial Revolution began way back in the 19th century. Nowadays, European environmental legislation plays its part in shaping UK policy and aims to control the impact of industry on the environment.

The Environmental Protection Act 1990

The Environmental Protection Act 1990 (EPA) aims to protect the environment from the results of 'the release into any environmental medium (e.g. air, land, water), from any process, of substances which are capable of causing harm to man or any other living organisms supported by the environment'.

This Act brought together a wide range of environmental legislation, and replaced most of the Control of Pollution Act 1974 (COPA). EPA defines the fundamental structure and authority for waste management and control of emissions into the environment.

There are effectively nine parts to the legislation, covering these topics:

- Part 1: Integrated pollution control and air pollution control by local authorities
- Part 2: Waste on land
- Part 2a: Contaminated land
- Part 3: Statutory nuisances and clean air (such as smoke, fumes, insects, noise and artificial light)
- Part 4: Litter, etc.
- Part 5: Amendment of the Radioactive Substances Act 1960
- Part 6: Genetically modified organisms
- Part 7: Nature conservation in Great Britain and Countryside Matters in Wales
- Part 8: Miscellaneous (including sea pollution, control of dogs and burning of hay)
- Part 9: General (including corporate offences).

Figure 2.01 Smog is just one of many ways that our environment is being damaged

The Pollution Prevention and Control Act 1999

The Environmental Protection Act 1990 introduced a regime of controlling industries that emit significant levels of pollution to the environment (air, land and water). This has now been superseded by a new regime under the Pollution Prevention and Control Act 1999.

Under this new regime, local authorities are required to regulate the smaller industries, termed Part A2 and Part B installations; the Environment Agency regulates the larger industries, which are known as Part A1 installations.

Emissions to the environment (air, land and water) must be controlled from Part A1 and A2 installations and such installations are also required to account for energy-efficiency and to control against **noise pollution**. Part B installations on the other hand are only regulated for emissions to air.

For A1 installations, the system of control is called Integrated Pollution Prevention and Control (IPPC), with the Environment Agency as the regulator; for A2 installations, it is called Local Authority Integrated Pollution Prevention and Control (LA-IPPC), with local authorities as the regulators.

The Control of Pollution Act 1989

This legislation requires carriers of controlled waste to register with the Environment Agency (EA) or Scottish Environment Protection Agency (SEPA) and outlines the penalties (including seizure and disposal) for vehicles shown to have been used for illegal waste disposal. The Controlled Waste (Registration of Carriers and Seizure of Vehicles) Regulations 1991 then introduced a registration system for carriers of such controlled waste.

The Environment Act 1995

The Environment Act 1995 created the EA in England and SEPA in Scotland. Both of these are designed to protect the environment and manage resources, as well as setting new standards for environmental management.

The Act gives the EA and SEPA responsibility for regulating pollution control, water, general environmental and recreational duties, environmental duties relating to sites of special scientific interest (SSSIs), regional and local fisheries, flood prevention and control. It also sets out a system for identifying and repairing

> **Key term**
>
> **Noise pollution** – excessive noise from any source, but particularly industrial sources, that spoils people's experience of the environment; examples could be noise from machinery, plant or power tools

contaminated land and requires local councils to prepare reviews of air quality.

The Hazardous Waste Regulations 2005 and The List of Waste Regulations 2005

The Hazardous Waste Regulations (HWR) and the List of Waste Regulations (LoWR) introduced the requirements of the European Hazardous Waste Directive (HWD) into England and Wales. The European HWD's main aim is to define hazardous waste and to make sure it is properly managed and regulated.

The HWR control waste that can harm human health or the environment, or is difficult to handle. They aim to make sure that hazardous waste is properly managed at all times.

The HWR:

- define hazardous waste in England and Wales
- require producers or consignors of hazardous waste to register their premises
- restrict mixing of wastes and require separation of wastes where appropriate
- make sure that companies document the movement of hazardous waste
- require consignees receiving hazardous waste to keep thorough records and provide the EA with information on the disposal and recovery of hazardous waste every three months.

The LoWR:

- introduced the 'List of Wastes', also known as the European Waste Catalogue
- explain the List, giving help on choosing the code for a specific waste
- show how waste is classified as either hazardous or non-hazardous
- show limits for certain hazardous properties.

Hazardous or non-hazardous?

In the List of Wastes, each type of waste is given a six-digit code. Wastes categorised as hazardous are marked with an asterisk. Some waste types are classed as hazardous waste outright and are known as absolute entries. In the EA consolidated guidance they are highlighted in red and marked with an 'A'.

> **Did you know?**
>
> If you leave materials on site when your work is complete, you may be discarding them. If they are discarded they will be 'waste' and, as the producer of the waste, you will be responsible for it.

Figure 2.02 Would this become hazardous waste?

Remember

You must no longer use the HSC 'Approved Supply List' to classify hazardous wastes, instead you must use Table 3.2 of Annex VI of the CLP Regulations (see pages 93–94).

Did you know?

Across the world, countries have different rules on classification and labelling. For example, a chemical could be classified as 'toxic' or 'explosive' in one country, but not in another. Different symbols are also used to indicate the same hazards.

Waste needs to be assessed to see if it contains dangerous substances. Examples of hazardous materials in waste include lead acid batteries, fluorescent tubes, solvents, used transformer oils, asbestos, chemicals, mercury switches and some IT equipment. There are specialist provisions for explosives, mineral wastes from mining and quarrying and radioactive wastes.

Countries in the United Nations, including those in the European Union, have developed a shared classification and labelling system that can be used worldwide, called the Globally Harmonised System of Classification and Labelling of Chemicals, or GHS. The GHS provides a single system to identify hazards and to communicate them in transporting and supplying chemicals across the world. This is not law, but an international agreement.

The European Union (EU) member states agreed to adopt the GHS through the CLP Regulations.

Dealing with waste

Current government policy regarding this could be summarised as 'protection of human health and the environment by producing less waste and by using it as a resource wherever possible'. The government want to encourage more sustainable waste management, such as reduction, re-use, recycling, composting and using waste as a source of energy, to break the link between economic growth and the production of waste.

This introduces a hierarchy of waste management shown in Figure 2.03, where waste prevention is the preferred option and disposal the least favourite.

Figure 2.03 Hierarchy of waste management

Waste prevention

Preventing waste in the first place is the best option of all. Examples would be using less product packaging, restricting junk mail, using re-usable shopping bags or sustainability in building design.

Re-use or recycle?

There are two aspects to the broader idea of recycling:

- re-using a product you no longer want, even if that product contained a hazardous material – called re-use
- taking something and turning it, or parts of it, into something new – usually what most people mean by recycling.

Re-use

One example could be a half-full tin of paint. It can't be turned into something new, but it could be taken to a recycling plant and carefully distributed for use by others, in community projects and the like, where perhaps the money does not exist to purchase new tins of paint.

Equally, a school or youth club may be grateful to receive that old PC monitor that you were going to throw on the scrap heap. You might give items you don't want to a charity shop or community project.

Recycle

Items made out of materials that can be turned into something new provide the basic concept for recycling.

Most paper-based products such as magazines, newspapers, fliers and Christmas and birthday cards can be processed via **pulping** and turned into new paper-based products. Juice cartons and the like can go through a similar process.

Industrial oil can be usefully recovered and used as a fuel. Larger quantities of used oil, such as hydraulic fluid or lubricants from lorries, buses or mechanical plant, should be stored securely before collection by a registered waste carrier. Specialist companies will collect used cutting oils, and then treat and recover the oil.

Cooking oils from commercial users, such as caterers and fish and chip shops, must not be disposed of with the general waste stream. The waste oils are not hazardous but are subject to a duty of care in terms of their disposal. Such disposal can see waste oils collected by specialist contractors and they can even be reused as fuel for diesel engine cars.

Find out

Find out what waste disposal facilities there are locally. Which of these might you use in your work as an electrician, and what for?

Key term

Pulping – where paper-based products are dropped into a huge tank and blitzed with water, which separates the paper from any impurities

Remember

Emulsified cutting oils are highly polluting in water, so take great care when disposing of them. Contact a specialist contractor for advice.

Food and drink cans are made up of up to 50 per cent steel. You may only be able to drink that cola or eat those beans once but, by recycling the can, you can use the same steel again and again. Steel in Europe contains 56 per cent recycled steel and is fully recyclable.

Energy recovery

Energy recovery (sometimes known as 'waste to energy') systems create energy in the form of electricity or heat from the incineration of waste materials.

Incineration has brought its own problems – the release of acid gases and the ash left by the burning process. However, modern plants now use lime filters in the chimney to control the problem. Use of these plants can be linked to the development of what is termed 'district heating' in UK cities such as Newcastle upon Tyne, Nottingham and Sheffield. In such systems, the incinerator acts as a huge central boiler that then supplies heat to a large area, such as a housing estate.

A number of developing technologies can produce energy from waste without using incineration. Some can efficiently convert the energy into liquid or gaseous fuels. These systems include:

- gasification (which produces combustible gas, hydrogen and synthetic fuels)
- pyrolysis (which produces combustible tar/bio-oil and a liquid fuel similar to diesel).

Disposal

The most common disposal methods, particularly in the UK, are landfill and, to a lesser extent, incineration. There are over 4000 landfill sites in the UK, and each year millions of tonnes of controlled waste (household, commercial and industrial) are disposed of there. Some waste from sewage sludge is also placed in landfill sites, along with waste from mining and quarrying.

As landfill waste decomposes, methane is released in considerable quantities. Methane is a strong greenhouse gas that contributes to global warming. Also, the leachate fluids formed from decomposing waste can permeate through the underlying and surrounding ground, polluting groundwater, which may be used for drinking water supplies.

Anaerobic digestion is another method of waste disposal. It is less common, but more sustainable than incineration or landfill. In this process, waste decomposes in an enclosed chamber, unlike in a landfill site. Digestion takes place in an oxygen-free environment where bacteria decompose waste by breaking down the molecules to form gaseous by-products (methane) and small quantities of solid residue. Anaerobic sewage plants produce significant quantities of methane, which can be burnt to generate electricity. Liquid and solid organic fertilisers are also formed with this method.

Waste transport

You can transport most business waste directly to an authorised waste management site or recycling facility. However, a business must register with their environmental regulator as a 'waste carrier' if they transport:

- construction and demolition waste produced by the business
- any waste produced by another business.

As waste carriers, the business must check that it holds the correct registration for the type of waste being transported. It must also complete a Waste Transfer Note (WTN) for every load of waste passed on or accepted. Copies of all WTNs must be held for at least two years.

Figure 2.04 gives guidelines for good waste disposal practice.

> **Did you know?**
>
> Generally, the burning of waste in the open is an environmentally unsound practice, and you should use less damaging options for waste disposal.

❏ Do not burn waste on site; find another method of disposal.

❏ Before allowing any waste hauler or contractor to remove a waste material from your site, ask where the material will be taken and ask for a copy of the waste management licence or evidence of exemption for that facility.

❏ Segregate the different types of waste that arise from your works. This will make it easier to supply an accurate description of the waste for waste transfer purposes.

❏ Minimise the quantity of waste you produce to save you money on raw materials and disposal costs.

❏ Label all waste skips – make it clear to everyone which waste types should be disposed of in each skip.

❏ Check if waste is hazardous or special waste before you transport it.

Figure 2.04 Checklist for good waste disposal practice

You must complete a consignment note whenever you or anyone else moves or transfers hazardous waste. Copies of consignment notes must be kept for at least three years.

If you transport hazardous/special waste you must:

* keep it separate from other wastes
* use sealed and clearly labelled containers
* check that it is transferred to a facility that is authorised to receive it.

The Waste Electrical and Electronic Equipment (WEEE) Regulations

Electrical and electronic waste (such as TVs, monitors and computers) has become the fastest growing waste in the UK, with nearly two million tonnes of it generated every year.

The Waste Electrical and Electronic Equipment (WEEE) Regulations aim to reduce the amount of this waste going to landfill and improve recovery and recycling rates. You need to comply with the WEEE Regulations if you:

* manufacture or import electrical or electronic equipment
* distribute electrical or electronic equipment
* generate any electrical or electronic waste
* collect electrical or electronic waste from your customers for treatment or disposal
* operate a waste treatment facility
* export electrical or electronic waste.

These Regulations apply to electrical and electronic equipment (EEE) in the categories listed below with a voltage of up to 1000 V for a.c. or up to 1500 V for d.c. You need to comply with the WEEE Regulations if you generate, handle or dispose of waste that falls under one of these ten categories.

1. Large household appliances
2. Small household appliances
3. IT and telecommunications equipment
4. Consumer equipment
5. Lighting equipment
6. Electrical and electronic tools
7. Toys, leisure and sports equipment

8. Medical devices

9. Monitoring and control equipment

10. Automatic dispensers.

All businesses that use EEE must comply with these Regulations. This includes all domestic or household EEE that you may use on your premises. For all non-household EEE, either the producer or end user is responsible for the disposal of the products.

You must obtain and keep proof that your waste EEE was given to a waste management business, and was subsequently treated and disposed of in an environmentally sound way.

The Packaging (Essential Requirements) (Amendment) Regulations 2009

One part of the waste process that is easy to forget is the **packaging** that helped deliver the items you wanted safely.

A lot of waste can be generated by packaging so these Regulations state that packaging should be:

- manufactured so that the packaging volume and weight are limited to the minimum adequate amount to maintain the necessary level of safety, hygiene and acceptance for the packed product and for the consumer

- designed, produced and commercialised in such a way as to allow for re-use or recovery, including recycling, and to minimise its impact on the environment when packaging waste or residues from packaging waste management operations are disposed of

- manufactured so that the presence of noxious and other hazardous substances and materials as constituents of the packaging material or of any of the packaging components is minimised, in emissions, ash or leachate when packaging or residues from management operations or packaging waste are incinerated or land-filled.

The Control of Noise at Work Regulations 2005

The Control of Noise at Work Regulations 2005 (the Noise Regulations) aim to ensure that workers' hearing is protected from excessive noise at their place of work, as this could cause them to lose their hearing or to suffer from **tinnitus**.

> **Key term**
>
> **Packaging** – 'all products, made of any materials of any nature, used for the containment, protection, handling, delivery and presentation of goods, from raw materials to processed goods, from the producer to the user or the consumer, including non-returnable items used for the same purposes' (official definition from the Regulations)

> **Key term**
>
> **Tinnitus** – a permanent ringing in the ears

People often experience temporary deafness after leaving a noisy place. Although hearing recovers within a few hours, you should never ignore these things: it is a sign that, if you continue to be exposed to the noise, your hearing could be permanently damaged.

Hearing loss is usually gradual, and happens because of prolonged exposure to noise. It may only be when damage caused by noise combines with hearing loss due to ageing that people realise how deaf they have become. This may mean their family complains about the television being too loud, they cannot keep up with conversations in a group, or they have trouble using the telephone. However, young people can be affected by noise just as badly as older people.

Noise can also be a safety hazard at work, interfering with communication and making warnings harder to hear. Employers are obliged by law to carry out risk assessments and identify measures to eliminate or reduce risks from exposure to noise, to protect the hearing of their employees.

As a rough guide, employers will probably need to do something about the noise if any of these questions gets a 'yes'.

- Is the noise intrusive (such as a busy street, a vacuum cleaner or a crowded restaurant) for most of the working day?
- Do employees have to raise their voices to carry out a normal conversation when about 2 metres apart for at least part of the day?
- Do employees use noisy powered tools or machinery for more than half an hour each day?
- Do you work in a noisy industry such as: construction, demolition or road repair; woodworking; plastics processing; engineering; textile manufacture; general fabrication; forging, pressing or stamping; paper or board making; canning or bottling; foundries?
- Are there noises due to impacts (such as hammering, drop-forging or pneumatic impact tools), explosive sources such as cartridge-operated tools or detonators, or guns?

Noise levels

The Noise Regulations require you to take specific action at certain action values, relating to:

- the levels of exposure to noise of your employees averaged over a working day or week
- the maximum noise (peak sound pressure) to which employees are exposed in a working day.

> **Remember**
>
> Bear in mind that permanent hearing damage can be caused immediately by a sudden, extremely loud, explosive noise, such as from a gun or cartridge-operated machine.

The values are:

- lower exposure action values:
  - daily or weekly exposure of 80 **dB**
  - peak sound pressure of 135 dB
- upper exposure action values:
  - daily or weekly exposure of 85 dB
  - peak sound pressure of 137 dB.

There are also levels of noise exposure that must not be exceeded:

- exposure limit values:
  - daily or weekly exposure of 87 dB
  - peak sound pressure of 140 dB.

These exposure limit values take account of any reduction in exposure provided by hearing protection.

Noise and statutory nuisance

As mentioned, Part 3 of the Environmental Protection Act 1990 (as amended) creates various statutory nuisances, including noise emitted from premises. The Act further defines statutory nuisance in relation to noise as being 'noise emitted from premises that is prejudicial to health or a nuisance'.

For action to be taken, the nuisance complained of must be, or be likely to be, 'prejudicial to people's health or interfere with a person's legitimate use and enjoyment of land'. This particularly applies to nuisance to neighbours in their homes and gardens.

Here are some examples of good practice concerning noise for any business.

- Establish whether your business might cause a nuisance to neighbours by checking noise, odours and other emissions near the boundary of your site during different operating conditions and at different times of the day. Take all reasonable steps to prevent or minimise a nuisance or a potential nuisance.
- Make sure there is a good level of housekeeping on your site and that your site manager and staff are aware of the need to avoid nuisances. Regularly check your site for any waste accumulations, evidence of vermin, noise or smell as applicable.

Key term

dB – the abbreviation for decibel, the unit in which noise level is measured

Remember

Where appropriate, use mains-generated electricity in preference to diesel generators. This will help you to reduce noise levels and to reduce the risk of pollution through fuel spillage.

Did you know?

A statutory nuisance could arise from the poor state of your premises or any noise, smoke, fumes, gases, dust, steam, smell, effluvia, the keeping of animals, deposits and accumulations of refuse and/or other material.

- Even if a complaint does not amount to a statutory nuisance you should consider if there are simple, practical things that you can do to keep the peace.

- Try to establish a good relationship with your neighbours, particularly in relation to transient events likely to affect them. Advise neighbours in advance if you believe that a particular operation such as building work or an installation process for new plant could cause an adverse effect. If neighbours are kept informed they will perceive the business as more considerate and are less likely to make a complaint.

- At noise-sensitive locations, monitor background noise before your works begin, and include in your method statement any actions you can take to reduce noise levels.

- Reduce noise levels outside or escaping your buildings by increasing insulation to the building fabric and keeping doors and windows closed.

- Make sure that any burglar alarms on your premises have a maintenance contract and a callout agreement.

- Avoid or minimise noisy activities, especially at night; pay particular attention to traffic movements, reversing sirens, deliveries, external public address systems and radios.

- Where practical, schedule or restrict noisy activities to the normal working day (for example, 0800 to 1800, Monday to Friday and 0800 to 1300 on Saturday).

- If noisy operations take place near site boundaries, relocate them if you can, perhaps further away, or make use of existing buildings/stockpiles/topography as noise barriers.

- Consider replacing any noisy equipment and take account of noise emissions when buying new or replacement equipment.

- Maintain fans and refrigeration equipment to help keep noise levels as low as possible.

- Do not have any bonfires; find other ways to re-use or recover wastes – see the Clean Air and Waste Management Guidelines.

- Keep abatement equipment, such as filters and cyclones, in good working order. Ensure boilers, especially oil or solid-fuel units, are operating efficiently and do not emit dark smoke during start-up.

The Building Regulations

The Building Regulations are made under powers provided in the Building Act 1984 and apply in England and Wales. The current edition is 'The Building Regulations 2000' (as amended), and most building projects are required to comply with them.

The Building Regulations exist to ensure the health and safety of people in and around all types of building (domestic, commercial and industrial) and to provide for energy conservation, plus access to and use of buildings.

These Regulations contain various sections dealing with definitions, procedures, and what is expected in terms of the technical performance of building work. They set out the notification procedures to follow when starting, carrying out, and completing building work and set out the requirements with which the individual aspects of building design and construction must comply in the interests of the health and safety of building users, of energy conservation and of access to and use of buildings.

These requirements are contained in a schedule (Schedule 1) to the Building Regulations and are grouped under fourteen 'Parts', as shown in Table 2.01.

The requirements in each Part set out the broad objectives or functions which the individual aspects of the building design and construction must set out to achieve. They are often referred to as 'functional requirements', and are expressed in terms of what is reasonable, adequate or appropriate.

Part	Area covered
A	Structure
B	Fire safety
C	Site preparation and resistance to contaminants and moisture
D	Toxic substances
E	Resistance to the passage of sound
F	Ventilation
G	Hygiene
H	Drainage and waste disposal
J	Combustion appliances and fuel storage systems
K	Protection from falling, collision and impact
L	Conservation of fuel and power
M	Access to and use of buildings
N	Glazing – safety in relation to impact, opening and cleaning
P	Electrical safety

Table 2.01 Parts of Schedule I

Building work is defined in Regulation 3 of the Building Regulations as covering:

- the erection or extension of a building
- the installation or extension of a service or fitting controlled under the Regulations
- an alteration project involving work that will temporarily or permanently affect the ongoing compliance of the building, service or fitting with the requirements relating to structure, fire, or access to and use of buildings
- the insertion of insulation into a cavity wall
- the underpinning of the foundations of a building.

The second and third points in the above list are of particular relevance to our industry as electrical installation is classed as a service.

Part P of the Building Regulations has been introduced to reduce the risk of death and injury caused by electricity or fires started by electrical faults. It sets the requirement that 'reasonable provision shall be made in the design, installation, inspection and testing of electrical installations to protect persons against fire or injury'.

This means that, unless the work is carried out by a Competent Person for Part P, it will have to be notified to the Local Building Control office (before work commences) so that they can inspect the work during construction and upon completion.

This will apply to all work carried out, either professionally or as DIY, except where:

- the proposed work is to be undertaken by a Competent Person for Part P authorised to self-certify
- the work is of a minor nature and is not in a kitchen or a special location.

Compliance with Part P of the Building Regulations is achieved by following safety rules and the applicable regulations in BS 7671, the IET On-Site Guide and associated Guidance Notes.

As other buildings benefit from the Electricity at Work Regulations 1989, Part P is restricted to fixed electrical installations in:

- dwellings
- combined dwellings and business premises such as shops with a common supply
- common access parts in blocks of flats
- shared amenities in flats (e.g. laundries, gyms etc).

Table 2.02 on page 136 shows some links between electrical installations and other Parts of the Building Regulations.

Did you know?

Complying with BS 7671 is **not** sufficient to guarantee compliance with the Building Regulations. The non-electrical builder's work that you carry out in connection with any electrical installation work must also comply with all the applicable requirements of the Building Regulations – not just Part P.

The Code for Sustainable Homes

Developed by government, the code is essentially a rating system for buildings, linked to the Energy Performance Certificate when houses are bought or sold. The Code aims to influence the house building industry to improve in terms of sustainability.

The Code is closely linked to the Building Regulations, in that the minimum standards for the compliance with the Code have been set above the requirements of Building Regulations. It is intended that the Code will signal the future direction of Building Regulations in relation to carbon emissions from, and energy use in, homes, providing greater regulatory certainty for the homebuilding industry.

Part	Area covered	Link to electrical installations
Part A	Structure	Depth of chases in walls and size of openings in joists or structural elements
Part B	Fire safety	Fire resistance of ceilings and walls including provision of fire alarm and detection systems
Part C	Site preparation and resistance to moisture	Moisture-resistance of openings or penetrations for cables through external walls
Part E	Resistance to the passage of sound	Penetrations through ceilings and walls
Part F	Ventilation	Ventilation rates for areas within a dwelling
Part L	Conservation of fuel and power	Energy-efficient lighting, effective controls including automatic controls
Part M	Access to and use of buildings	Heights of switches, socket outlets, etc.
Part P	Electrical safety	Fixed electrical installations in dwellings

Table 2.02 Links between Building Regulations and electrical installations

The Code shows the overall sustainability performance of a home by a scale of one to six 'stars'. One star is the entry level, which is above the level of the Building Regulations; six stars reflects an exemplar development in sustainability terms. Table 2.03 summarises the minimum standards that exist under the Code.

Progress check

1   Your site supervisor asks you to remove all the packaging from equipment and clear the work area. Most of the waste comprises cardboard, paper, plastic sheet, wooden pallets and metal strapping. How should the waste be disposed of and whose responsibility is it?

2   Your company has been engaged to replace all the old fluorescent lighting in an office block with new. You are made responsible for the clear up and disposal of all the old fluorescent fittings. Which set of regulations covers the disposal of such material and how do you ensure you have complied with them?

3   Noise in the workplace can cause workers some permanent problems.
    a. Which regulations identify these risks? What are those risks?
    b. What should be done to protect workers from the risks associated with noise?
    c. What maximum levels should not be exceeded?

4   An electrical company is hired to rewire a block of residential flats. The work includes the provision of fire alarm detection and low voltage energy efficient spotlights in the kitchen areas. It is the first contract for domestic work the company has undertaken.
    a. What options have they to notify work?
    b. What is the purpose of The Building Regulations 2000 as amended?
    c. Which parts of the Building Regulations would need to be consulted for this project?

Minimum standards		
Code Level	**Category**	**Minimum standard**
1(*)	**Energy CO$_2$** Percentage improvement over Target Emission Rate (TER) as determined by the 2006 Building Regulation Standards	10%
2(**)		18%
3(***)		25%
4(****)		44%
5(*****)		100%
6(******)		A 'zero carbon home' (heating, lighting, hot water and all other energy uses in the home)
1(*)	**Water** Internal potable (drinking) water consumption measured in litres per person per day (l/p/d)	120 l/p/d
2(**)		120 l/p/d
3(***)		105 l/p/d
4(****)		105 l/p/d
5(*****)		80 l/p/d
6(******)		80 l/p/d
1(*)	**Materials** Environmental impact of materials	At least three of the following five key elements of construction are specified to achieve a BRE Green Guide 2006 rating of at least D • Roof structure and finishes • External walls • Upper floor • Internal walls • Windows and doors
1(*)	**Surface water run-off** Surface water management	Ensure that peak and annual volumes of run-off will be no greater than the previous conditions of the development site
1(*)	**Waste** Site waste management	Ensure that there is a site waste management plan in operation, which requires the monitoring of waste on site and the setting of targets to promote resource efficiency
	Waste Household waste storage	Where there is adequate space for the containment of waste storage for each dwelling. This should allow for the greater (by volume) of: • EITHER accommodation of all external containers provided under the relevant Local Authority refuse collection/recycling scheme. Containers should not be stacked to facilitate ease of use. They should also be accessible to disabled people, particularly wheelchair users and those with a mobility impairment • OR at least 0.8m³ per dwelling for waste management as required by BS 5906 (Code of Practice for Storage and On-site Treatment for Solid Waste from Buildings)

Table 2.03 Minimum standards star system under the Code for Sustainable Homes

K2. Understand how work methods and procedures can reduce material waste and impact on the environment

It is projected that the world could see temperatures increased by 6°C by the end of the century if we don't begin to think and act sensibly regarding emissions. However, there are also other problems to consider, such as contamination of the land, air or water by dangerous substances, processes and wastes.

There are several areas you need to consider when choosing work methods and procedures to reduce material wastage. You will need to:

- report any hazards to the environment
- choose methods that can help to reduce material wastage
- use environmentally friendly materials, products and procedures that can be used in the installation and maintenance of electrotechnical systems.

Importance of reporting hazards to the environment

Additionally, if during your work you even think that something is having a negative effect on the environment (for example, fumes, noise or asbestos particles), inform your supervisor immediately and get it checked out.

The consequences to everyone, both fellow workers and passers-by, could be disastrous if you don't.

Reducing material wastage and using environmentally friendly materials, products and procedures

Using less energy can save companies and households money and 'greener' living and working are going to be key factors. As an overview of practical measures for householders, the UK government currently suggests the following options. Bear in mind how these could be applied to the world of work.

Saving energy and water

Burning fossil fuels to heat our homes or produce electricity releases carbon emissions, which cause climate change. The energy you use at home is likely to be your biggest contribution to climate change. Approximately 80 per cent of it goes on heating and hot water, so this is a good place to look for savings.

Water is a precious resource and we should use it wisely at all times because it's not as abundant as you might think. Using water, especially hot water, also uses energy and increases emissions of greenhouse gases contributing to climate change.

Here are some ways to save energy.

- **Turn down the thermostat** – turning your thermostat down by one degree could reduce carbon emissions and cut fuel bills by up to 10 per cent.

- **Look for the labels** – when buying products that use energy (this could be anything from light bulbs to fridge-freezers) help tackle climate change by looking for the Energy Saving Recommended label or European energy label rating of A or higher. The European energy label also tells you how much water appliances use, so you can choose a more efficient model.

- **Improve insulation** – more than half the heat lost in your home escapes through the walls and roof. Cavity wall insulation costs about £450, can take a couple of hours to install, and could save about £100 a year on fuel bills, as well as reducing your carbon footprint.

- **Install water efficient products** – low flush-volume toilets, water-efficient showerheads and aerating heads on washbasin taps help to reduce your water use significantly.

- **Fix dripping taps** – this and fitting a 'hippo' device in toilet cisterns are cheap ways of saving water. You can also collect rainwater in water butts and use it for watering your garden instead of a hose.

- **Fix leaking pipes** – every day more than 3.3 billion litres of treated water – 20 per cent of the nation's supply and 234 million litres a day more than a decade ago – are lost through leaking pipes in England and Wales. The water lost would meet the daily needs of 21.5 million people.

Did you know?

Most water is used for washing and toilet flushing, but it also includes drinking, cooking, car washing and watering the garden. We use almost 50 per cent more water than 25 years ago, partly because of the use of power showers and household appliances.

Find out

Use the Internet to research 'reducing energy' and see how many websites you can come up with. Try some of their ideas or do an audit as a group on energy conservation.

Find out

As a trial, work in groups to monitor your own household's daily water usage. Does the amount surprise you? What could you do to reduce it?

Figure 2.05 Exhaust fumes from traffic produce a large amount of pollution

Remember

Walking, cycling, or taking the bus or train will help reduce local air pollution and climate change effects, and help your personal fitness too.

Saving energy spent on transport

Travelling accounts for around a quarter of all the damage individuals do to the environment, including climate change effects. Individual car travel is responsible for the majority of these impacts, so if you're buying a car, look for the fuel efficiency label to choose the most efficient model.

Also try to reduce your car use, especially for short trips, which are the least energy efficient.

With flying, consider alternatives to taking a plane. If you do fly, you can offset your CO_2. You could consider options for reducing your travel, such as taking fewer, longer breaks instead of several short ones. Maybe you can find what you want closer to home, by taking a holiday in the UK or travelling to nearby countries by rail or sea.

Saving energy when eating and drinking

Producing, transporting and consuming food is responsible for nearly a fifth of our climate change effects. Some foods have a much bigger impact on the environment than others.

- **Look for the labels** – to help you choose food that has been produced with the aim of reducing the negative impact on wildlife and the environment.
- **Buy fresh, local and in season** – unprocessed or lightly processed food is likely to mean that less energy has been used in its production. Providing it has been produced and stored under similar conditions, choosing food that has travelled a shorter distance will help to reduce any congestion and transport emissions that contribute to climate change.
- **Reduce your food waste** – the average UK household spends £424 a year on food that goes in the bin. If this ends up in landfill, it produces methane, a greenhouse gas judged to be more than 20 times as powerful as carbon dioxide in causing climate change. Throwing less food away produces less methane and reduces other harmful environmental impacts from producing, packaging and transporting food.

Recycling and cutting waste

Reducing, re-using and recycling waste saves on raw materials and energy needed to make new paper, metal, glass and other items.

- **Re-use and repair and recycle more** – nearly two thirds of all household rubbish can be recycled. Most councils run doorstep recycling collections for paper, glass and plastics, often more. Local sites often accept many other things, from wood and shoes, to textiles and TVs.

- **Take a bag** – hang on to your shopping bags and take some with you when you next go to the supermarket.

Reducing energy use in the workplace

Buildings produce nearly half of the UK's carbon emissions. The way a building is constructed, insulated, heated, and ventilated, and the type of fuel used, all contribute to its carbon emissions. New measures to improve the energy performance of our buildings are being introduced by the government, including:

- the introduction of Energy Performance Certificates (**EPCs**) for all buildings, whether they are built, sold or rented out

- requiring public buildings to display EPCs

- requiring inspections for air conditioning systems

- giving advice and guidance for boiler users.

The selection of construction materials, air conditioning systems and boilers is the work of others, but it is clear that the selection of efficient materials and systems before installation and the maintenance of any existing systems will be key factors.

Planning work methods

Planning can make a big difference. A carefully planned installation can reduce the environmental impact in terms of the number of visits to site, consequent travel and welfare arrangements, material and equipment delivery arrangements, selection of material and equipment, in terms of both installation and maintenance factors and the removal or disposal of waste.

One trend in construction that is gathering momentum is the move away from traditional on-site construction to off-site (modular) construction. Prefabricated construction has been around for some time but, with modern technology at our disposal, the modular approach is felt to offer many advantages. Although definitive statistics are difficult to find, some well respected architects have said that some of the advantages are:

Remember

Even if the work of an electrician has little direct environmental impact, each of us should take every step we can to reduce our environmental impact.

Key term

EPC – a certificate that gives buildings a rating for energy efficiency, with ratings from 'A' to 'G', 'A' being the most energy-efficient and 'G' the least. The average to date is 'D'

Remember

Although electricians do not design systems, they may be involved in a system's installation or maintenance. Every visit to site has an environmental effect, so electricians have a responsibility to work in an environmentally friendly way.

- up to 60 per cent less energy required to produce a modular building compared to an equivalent traditionally built project
- up to 90 per cent fewer vehicle movements to site, reducing carbon emissions, congestion and disruption
- a potential reduction in on-site waste by up to 80 per cent.

The current Building Regulations set standards for the design and construction of most buildings, primarily to ensure the safety and health for people in or around those buildings, but also for energy conservation and access to and about buildings.

Part L – Conservation of fuel and power is relevant here and particularly Approved Document L2A. Part L states that 'responsible provision shall be made for the conservation of fuel and power in buildings by … providing and commissioning energy efficient fixed building services with effective controls'. Document L2A clarifies 'fixed building services' as including 'any part of or control associated with, fixed internal or external lighting, excluding emergency lighting'.

Document L2A also states 'lighting controls should be provided so as to avoid unnecessary lighting during times when daylight levels are adequate or when spaces are unoccupied … with automatically switched systems being subject to risk assessment'.

Part L also makes requirements for the type of lamps used in luminaires, as we still rely too heavily on the incandescent lamp and older low-efficiency fluorescent tubes (the incandescent lamp is a very inefficient lamp converting only about 5 per cent of the energy received into light). However, the advances in modern compact fluorescent tubes and LEDs should be considered.

'Knowledge is power' is an old expression, but in this case it is true and legislated for, as Part L states the owner of a building must be provided with sufficient information about the building, its fixed building services and any maintenance requirements, to enable the building to be operated so that it uses no more fuel and power than is reasonable under the circumstances.

Progress check

1 Briefly describe six areas where potential energy savings could be made in buildings.

2 What three considerations should be made when planning working methods and procedures to reduce material wastage?

3 Which part of the Building Regulations covers energy efficiency? Give three examples of those requirements.

Saving energy in an eco-friendly building

For the construction industry, the phrase 'eco-friendly' can be hard to define exactly: many things interact to reduce our impact on the environment. Table 2.04 shows some ways to apply energy saving for electrical installation within an eco-friendly building.

Materials
Use natural, sustainably managed renewable sources (for example there are timber building companies in Norway that have planting policies which mean they are growing more trees than they cut down to make their products)
Source materials near to the point of use, reducing transportation effects
Should be made using minimal processing or with added content such as chemicals
Make use of their natural insulation properties
Use non-toxic and non-hazardous to users or building occupants (e.g. paint fumes)
Use durable materials with low maintenance
Materials should have the capability to be recycled
'Measure twice, cut once' is a good rule to avoid wastage
Energy
Use natural light wherever possible
Use low-energy appliances wherever possible (e.g. 'A' rated washing machines)
Use local expertise and labour wherever possible, thus reducing the effects of transportation
Use renewable sources where possible
Water
Use rainwater harvesting (using rainwater for irrigation, vehicle washing or toilet flushing)
Use grey-water recycling (using water from baths, showers, etc. for toilet flushing)
Use low-volume flush toilets
Use aerated taps (these make the water spray)
Use instantaneous water heaters over sinks instead of heating large volumes of water
Lag hot water pipes to avoid losing heat
Only heat as much as you need in a kettle or boiler

Table 2.04 Methods of saving energy in eco-friendly buildings

K3. Understand how and where environmental technology systems can be applied

As well as using technology within the building as shown in Table 2.04, we also need to consider its application elsewhere. We know that conventional power generation can be very damaging to the environment, but there is a range of alternatives that come under the heading of micro-generation (small scale as used in individual households) and macro-generation (large-scale commercial) renewable energy sources. Some of these are discussed below.

Operating principles, applications and limitations

Solar photovoltaic

These systems work by converting solar radiation into electricity that can be used immediately, stored or even sold back to the electricity provider (with a grid-connected system). Correctly installed systems, even in the UK, could allow efficiently managed domestic properties to generate about half of their own electricity.

Virtually all photovoltaic devices are some type of photodiode, and in most photovoltaic applications the radiation is sunlight. This is why the devices are known as solar photovoltaic.

A photodiode is a p-n junction that responds to light. If we take a zero-biased p-n junction, sunlight falling on the material causes a voltage to develop across the device, leading to a current in the forward-biased direction. This is the 'photovoltaic effect'. Assemble many photodiodes together and you have a solar cell. Solar cells are commonly assembled into panels (solar panels) to produce direct-current electricity from sunlight.

In a grid-connected system, power generated is first used within the property to reduce electricity consumption then any surplus is exported to the grid. Various schemes and tariffs are available through electricity suppliers to give credits or payments for both generated and exported power. A standard power inverter then takes the direct current (d.c.) from the photovoltaic system and converts it to a.c. for use in the property.

However, you can't just plug any inverter into the UK grid. Try thinking of the problems faced by a car trying to join a busy motorway from a slip road (such as the right speed, right direction) and you have a good starting point.

The problem is that electricity in the two systems may not be 'in phase': the electricity from your inverter may be cycling up when the grid is cycling down. A 'grid-tied inverter' monitors the power from the grid and makes sure the power coming from the inverter stays in sync with the phase of the electricity from the grid.

The grid-tied inverter also differs from the stand-alone variety in that the control circuit has to be able to operate in the presence of the existing grid voltage and force the grid to accept power, instead of providing it. Because the grid is essentially a very low-impedance voltage source, the inverter must be able to act as a current source, only allowing the desired amount of current to be sent into the grid. This process requires close control of the inverter output voltage.

Generally, the inverter will control its bulk d.c. input voltage and use this to determine the output power level. The power level signal is then used to determine the output power, and the inverter output will be adjusted upward until this amount of power is delivered to the grid.

Because of the extra work involved in monitoring and conditioning the output power to match the power from the grid, a grid-tied inverter is more expensive than a more simple power inverter that cannot tie into the grid. Figure 2.06 shows a simple solar photovoltaic system.

Remember

Isolators are required in several locations in the installation. Remember also that PV installations are unusual in that they can be live from both sides, because as soon as the sun shines on a PV array it will start to produce power.

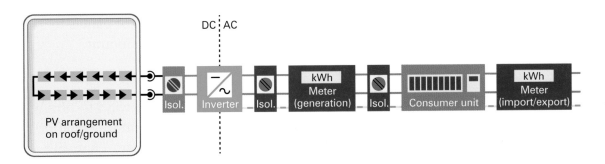

Figure 2.06 A simple power inverter system

Advantages

- Sunlight is free and solar power is pollution-free during use.
- PV installations can operate for many years, with little maintenance or intervention after their initial set-up.
- Solar electricity generation is economically superior where grid connection or fuel transport is difficult, costly or impossible.
- When grid connected, solar electric generation replaces some or all of the highest cost.
- Grid connected solar electricity can be used locally thus reducing transmission and distribution losses.

Disadvantages

- Photovoltaics are costly to install.
- While the modules are often guaranteed for about 20 years, the personal investment in a domestic system can be lost if the owner moves.
- You need enough space (roof or ground) for solar panels and the panels need a clear view of the southern sky. For houses, solar panels are usually located on the roof; solar panels can also be located on the ground. For maximum efficiency, it is best for the solar panels to be built to turn to face the sun throughout the day.
- Solar panels work most efficiently in parts of the world that get a lot of sun. The more sun, the more electricity. Less sun and cloudy days doesn't mean that you won't produce solar electricity, but you need a bigger array of panels to produce the same amount.

Wind energy generation

These systems rely on wind to turn a shaft linked to a generator, which in turn produces electricity. There are two common types of wind turbine used in homes in the UK:

- mast-mounted (2.5 kW to 6 kW)
- roof-mounted (1kW to 2 kW).

Figure 2.07 Wind power is one of many alternative power sources which may increase over the next few years

The greater the wind speed, the greater the power produced. As wind speed increases with height, large rotating blades are normally seen at the top of a tall supporting mast or tower. As wind speed and direction vary, this may not be the best option for domestic properties unless they are in remote locations. As with solar PV, any surplus electricity can be exported back to the grid.

In the UK, large-scale developments, known as **wind farms**, are becoming more popular, both on shore and off shore. The UK is the windiest country in Europe – so much so that we could power our country several times over using this free fuel. A modern 2.5 MW turbine located at a reasonable site will probably generate 6.5 million units of electricity each year, and that is enough to meet the annual needs of over 1400 households.

> **Key term**
>
> **wind farm** – a large number of wind turbines gathered in one location. There are currently on-shore and off-shore wind farms in the UK.

Advantages

- It is a plentiful energy source.
- It is a renewable energy that doesn't release any harmful carbon dioxide or other pollutants.
- The equipment is reasonably low maintenance and wind is free, so once you've paid for the initial installation your electricity costs will be reduced.
- It is useful in remote locations away from the grid.
- You can store electricity in batteries and use it when there is no wind.
- You can sell surplus electricity back to the grid.

Disadvantages

- You rely on wind to turn the turbine.
- Wind speed is variable.
- Small domestic turbines really need an exposed location without nearby obstacles that cause turbulence and affect air speed.
- Unlike solar PV, planning permission is required.

Micro (small scale) hydro generation

Although you may not realise it, people in the UK have used this concept for hundreds of years. Micro hydro generation is the process of using running water to turn something, on a small scale. Years ago, water was used to turn wheels in a mill, which in turn was used to grind flour. Today the system uses running or falling water to turn a turbine to produce electricity.

Useful power may be produced from even a small stream. For houses with no mains connection but with access to a micro hydro site, a good hydro system can generate a steady, more reliable electricity supply than any of the other renewable technologies.

Turbine technology is now available that can use quite small spring-fed streams for power, as long as the fall is sufficient.

Advantages

- It is environmentally friendly.
- No fuel deliveries are required.
- It is very good in rural areas.
- It is very low maintenance.

Disadvantages

- Total system costs can be high initially (£20,000), but often less than the cost of a grid connection.
- It requires sufficient flow of water.

Air and ground source heat pumps

Simply put, a ground source heat pump works by absorbing heat from the ground and raising its temperature. It is normally for use in the home. An air source heat pump does the same from the air.

The two types of pump work in the same way. The operating principles for the ground source version described are below and shown in Figure 2.08.

- Brine circulates in a closed loop known as the collector coil, which is buried in the ground and absorbs the heat energy from the ground.
- Inside the heat pump, the collector coil is wound around a heat exchanger (the evaporator). The lukewarm brine in the collector coil begins to warm the ice-cold refrigerant in the heat exchanger. This refrigerant has a very low boiling point, and the brine in the collector causes its temperature to rise by a few degrees. This is enough to cause it to boil and evaporate. Think of a boiling kettle which, as it boils, turns liquid into vapour.

Did you know?

Air and ground source heat pumps work using the same operating principles as a fridge.

- The evaporated refrigerant now moves into a compressor. Compression raises pressure, causing the refrigerant to rise in temperature (to about 50°C). It then passes into another heat exchanger (the condenser) and condenses. Think of the kettle again; the boiling steam vapour condenses back to liquid.

- Another closed loop (the distribution system out to the radiators) is wound around the condenser, causing the generated heat in the condenser to transfer from it to the closed loop feeding the radiators. From there it is sent out hot to the radiators, cools as it is passes through them and returns to be heated again by the condenser.

- The condensing refrigerant circulates from the heat exchanger and into an expansion valve that lowers the pressure, and the refrigerant becomes cold once again. The process then begins again when the circulating refrigerant meets the warm brine in the collector coil.

Advantages

- It has a low **carbon footprint**.
- No fuel deliveries are required.
- It can provide space heating and hot water.
- It can lower fuel bills, especially when measured against electric heating.
- It is often classed as a 'fit and forget' technology, as it needs so little maintenance.

Disadvantages

- It has high set-up costs.
- Ground source heat pumps produce a lower temperature heat than traditional boilers. It's essential that your home is insulated and well draught-proofed for the heating system to be effective. It could also make the system cheaper and smaller.
- Air source heat pumps tend to produce lower temperatures than ground source ones.

> **Remember**
>
> With the air source heat pump, the operating principle is exactly the same except that it absorbs heat from the air (even very cold air has heat within it). By contrast the ground source system makes use of the fact that, throughout the world, only a few metres under the ground surface, the temperature remains at a fairly constant 12°C, even in winter.

> **Key term**
>
> **Carbon footprint** – the total amount of greenhouse gases produced by an organisation, event, product or person

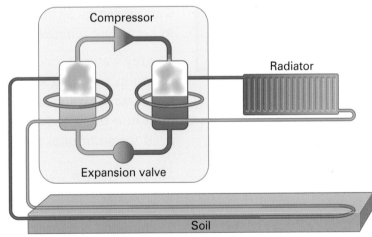

Figure 2.08 A ground source heat pump

Micro combined heat and power (Micro-CHP) unit

Although not strictly a renewable energy system, this system does have an energy-saving aspect to it. Sometimes called co-generation, micro-CHP involves having a domestic boiler that contains a condensing boiler to heat the home and provide hot water, but also includes something called a Stirling engine to generate electricity.

Invented by Scotsman Dr. Robert Stirling in 1816, these engines generate motion from heat without combustion. However, as you only generate electricity when the boiler is generating heat – and they only generate approximately 1.5 kW of electricity – the idea is to reduce your electricity bills, rather than replace your electricity supply.

Figure 2.09 A micro combined heat and power (Micro-CHP) unit

Did you know?

The micro-CHP concept can be operated on a larger scale, and there are currently trials underway at community level.

Grey water recycling

Grey water is waste water generated by activities in the home such as bathing, laundry and dish washing. Up to half of the water used in a house can end up as grey water.

Grey water does not include waste water from toilets, but with water reserves dwindling, it would seem sensible to put grey water to better use wherever possible.

As grey water is less contaminated than other waste water, common recycling sees it treated and then used for flushing toilets or for watering gardens. Some toilets now include a sink built into them to feed grey water straight to them without the need for additional pipework.

Advantages

- It can bring a reduction in water consumption without changing consumer behaviour.
- It is easy to install and maintenance free.
- It removes the need for complex water treatment.
- If properly designed, a grey water system can lower sewage costs.
- It can reduce ground water usage for irrigation.
- Less water will enter a city's sewage systems. This saves building new, or extending old, treatment plants.

Disadvantages

The main disadvantage is that the water is not suitable for drinking.

Rainwater harvesting

Rainwater harvesting is simply the gathering and storage of rainwater. Once stored, it can be used for any normal use of water, including drinking. Collecting rainwater can make an important contribution to drinking water. In some situations, rainwater may be the only available, or economical, water source.

Most rainwater systems are simple and inexpensive to build, as they will use the existing guttering and downpipes on a house. Water then flows via a filter to an underwater storage tank.

Water can be of such good quality that it may not need any form of treatment before being of drinking standard (potable). Even when the rainwater is not going to be used for drinking, it would is fine for flushing, washing and gardening – processes that make up half the water used in the average home.

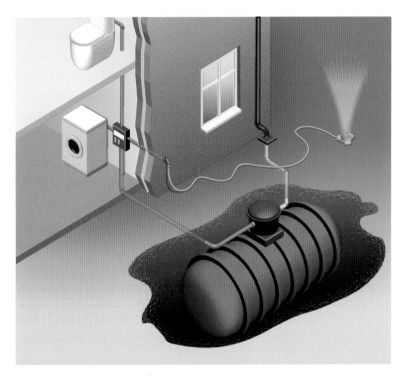

Figure 2.10 Rainwater harvesting system

Some commercial and domestic projects have also seen the introduction of rainwater harvesting systems as part of a fire protection system for the building, where the stored water can be used to feed the sprinkler system.

Advantages

- It is generally free (although some countries do charge).
- Large volumes of water are kept out of the storm-water management system, helping to reduce flooding risks.
- It is low maintenance.

Disadvantages

Treatment is often required to make the water reach a drinkable standard.

Biomass heating

Biomass is biological material (derived from living or recently living organisms) that can be converted into electricity or clean-burning fuels in an environmentally friendly and sustainable manner. It is accepted as a renewable replacement for fossil fuels as it can be replaced at the same rate as it used. The essential difference between biomass and fossil fuels is one of timescale.

Fossil fuels such as coal, oil and gas originally came from biological material, but that material absorbed CO_2 from the atmosphere many millions of years ago. Using these materials as fuel involves burning them, a process which leads to the carbon oxidising into carbon dioxide and the hydrogen becoming water (vapour). Unless these combustion products are captured and stored, they are usually released back into the atmosphere, resulting in increased concentrations in the atmosphere.

Biomass, on the other hand, takes carbon out of the atmosphere while it is growing, and returns it as it is burned. If it is managed on a sustainable basis, biomass is harvested as part of a constantly replenished crop. This maintains a closed carbon cycle, with no net increase in atmospheric CO_2 levels – hence it is called 'carbon neutral'.

Biomass energy is derived from five distinct energy sources:

- virgin wood from forestry or wood processing
- energy crops (high-yield crops grown specifically for energy applications)
- agricultural residues (residues from agriculture, harvesting or processing)
- food waste from food and drink manufacture, preparation and processing and consumer waste
- industrial waste and co-products from manufacturing and industrial processes.

Wood energy is derived both from direct use of wood as a fuel to provide heat (for example, in domestic wood-burning stoves) or in some cases as a replacement for fossil fuel in a power station. Crops such as corn and sugar cane can be fermented to produce ethanol, which can be used as transportation fuel or as an additive to petrol.

Did you know?

Landfill produces landfill gas, also known as biogas. This contains methane, which is very bad for the **ozone layer**, but which can be used for electrical generation by burning it as a fuel in a gas turbine or steam boiler. Compared to other hydrocarbon fuels, burning methane produces less carbon dioxide for each unit of heat released.

Biodiesel, another transportation fuel, can be produced from leftover food products like vegetable oils and animal fats. Biomass alcohol fuel, or ethanol, is derived primarily from sugarcane and corn and can be used directly as a fuel or as an additive to petrol.

Advantages

- Biomass production is carbon neutral – it produces no more carbon dioxide than it absorbs.
- The materials used to fuel biomass energy are otherwise sent to landfill, so it's an excellent green use for waste.
- Biomass energy could realistically produce up to 80 per cent of a home's energy needs.

Disadvantages

- Although biomass energy (like most alternative fuels) is more cost-effective than fossil fuels, there are both initial set-up costs and ongoing running costs.
- Storage space is required, and that requires ventilation and a dry environment.

Solar water heating

Solar water-heating systems use heat from the sun to work alongside a conventional water heater. They collect heat from the sun's radiation via a flat plate system or an evacuated glass-tube system before being stored in a hot-water cylinder. The system is normally not pressurised and often uses 'drain back' technology: the pipework slopes between items and so doesn't have water inside it once it is used, meaning there is no need for insulation.

When a solar water-heating and hot-water central heating system are used in conjunction, solar heat will either be concentrated inside a pre-heating tank that feeds into the tank heated by the central heating, or the solar heat exchanger will replace the lower heating element. The upper element will remain in place to provide for any heating that solar cannot provide.

However, we all normally use our central heating at night and in winter when solar gain is lower. Solar water heating for washing and bathing can often be a better application than central heating because supply and demand is better matched.

Did you know?

In many climates, a solar hot water system can provide up to 85 per cent of domestic hot water energy. In many northern European countries, combined hot water and space heating systems (solar combi-systems) are used to provide 15 to 25 per cent of home heating energy.

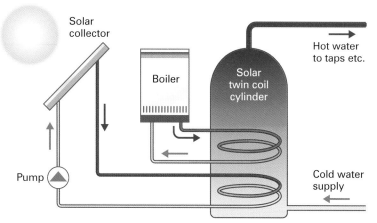

Figure 2.11 Solar water heating system

Advantages

- It is environmentally friendly.
- No fuel deliveries are needed.
- It is low maintenance.

Disadvantages

The main disadvantage is that the system has a relatively high set-up cost.

Voltage Management

Voltage in Europe was harmonised to be 230 V ± 10%, so many installations may experience supply voltages in excess of 240 V. BS 7671, Appendix 12 specifies a maximum value of voltage drop of 3% for lighting and 5% for all other uses, where the supply is directly from a public distribution system. On such installations, electrical equipment is running at a voltage much higher than the optimum voltage needed, resulting in excessive losses (optimum voltage will depend on the type of equipment). Voltage management is intended to reduce the voltage to this optimum level to reduce losses.

Before using this system, assess the implications. Supply voltages vary and the type of equipment used may have a bearing on the decision to use it. Installing a voltage regulator can help to ensure the voltage is maintained at the desired level and properly balanced to improve equipment efficiency and reliability.

Local Authority Building Control (LABC) requirements

Each local authority building control section has a duty to ensure that building works comply with building regulations, except those that fall under the remit of an approved inspector. They operate under an umbrella organisation known as the Local Authority Building Control.

Formed in 2005, the LABC's purpose is to ensure that buildings are healthy, safe, sustainable and accessible for all users and tenants whether they be domestic, commercial or public service.

The LABC also co-ordinates the technical application of the Building Regulations and influences new or revised regulations covering new technology, new building methods and environmental performance.

The ten main functions of the LABC are to:

- encourage innovation to produce energy-efficient and sustainable buildings
- support local, regional and national businesses
- educate and inform building professionals, contractors and tradespeople
- defend vulnerable communities and householders
- drive out dangerous cheats and rogue traders
- safeguard the investments of individuals and companies
- enhance access for disabled, sick, young and old people
- protect the community from dangerous structures
- provide life-saving advice to the emergency services
- ensure sports grounds and public venues are safe for crowds.

Consequently, if you are applying for building approval through the LABC services, you may be required to submit a Full Plans Application or give a Building Notice, depending on the scale of the work.

Full Plans Application

The Full Plans Application should contain plans and relevant construction details about the proposed building work. The procedure takes about five weeks. If plans comply with the Building Regulations, your application will be provided with a notice stating that they have been approved.

A conditional approval may be issued, which may require further plans to be given to the authority before proceeding with the building work. A full plans approval notice is valid for three years from the date of deposit of the plans, and building work must start within this period or the approval may lapse.

Building Notice

This does not require the submission of detailed full plans and normally applies to minor building work. After the building notice has been given and building control section informed of the intention to start work, the work will be inspected as it progresses.

The LABC will let you know if the work complies with the standards in the Building Regulations. Subject to the need to satisfy energy conservation requirements, a building notice is

> **Did you know?**
>
> If the LABC is not satisfied with an application, they may need the application to be amended or need more details provided to help them reach a satisfactory decision.

valid for three years from the date the notice was given and will lapse if no building work is carried out within this period.

Environmental Management Systems (EMS)

Concern for the environment and awareness of the need to improve management of resources is on the increase. The issue has found its way into the political arena and our everyday lives, and is creating pressure on businesses to demonstrate a commitment to minimising their impact on the environment.

An effective EMS certified to ISO 14001 or registered to EMAS (the European Eco-Management and Audit Scheme) can help an organisation to operate in a more cost-efficient and environmentally responsible manner by managing its activities while also complying with relevant environmental legislation and its own environmental policy. There are a number of key benefits associated with the implementation of a certified EMS as it:

- demonstrates conformance
- promotes management confidence
- improves management of environmental risk
- is an independent assessment
- shows compliance with legal and other requirements
- shows continual improvement
- shows a reduction in costs
- eases supply chain pressures.

Bearing in mind the previous environmental topics, employers may wish to become qualified to ISO 14001 for much the same reasons as the ISO 9001 series.

Progress check

1   Identify five environmental technology systems that may be used to reduce our energy requirements.

2   Briefly describe how a solar photovoltaic system works.

3   You have a client looking at having a solar PV system installed. As the work will involve the building, you check with your Local Authority Building Control regarding planning permission. What are the main functions of LABC? What may you have to submit for approval?

Getting ready for assessment

Environmental awareness is an increasingly important part of all stages of the construction process, from planning to final installation. You will always need to think about the environmental impact of your actions wherever you work and whatever your role is.

For this unit you will need to be familiar with:
- environmental legislation, working practices and principles which are relevant to work activities .
- work methods and procedures can reduce material wastage and impact on the environment
- how and where environmental technology systems can be applied

For each Learning Outcome, there are several skills you will need to acquire, so you must make sure you are familiar with the assessment criteria for each outcome. For example, for Learning Outcome 3 you will need to be able to describe the fundamental operating principles of a range of environmental technology systems and state their applications and limitations. You should also be able to state the Local Authority Building Control requirements which apply to the installation of these systems.

One of the focuses of environmental work is on sustainability and energy efficiency. It is important to understand these methods, and their benefits, when working. Although you may not be immediately working with all the technology systems you will encounter in this unit, you need to be familiar with how they operate so that you would be able to relate this information back into planning an environmentally friendly project.

It is important to read each question carefully and take your time. Try and complete both progress checks and multiple choice questions without assistance to see how much you have understood. Refer to the relevant pages in the book for subsequent checks. Always use correct terminology as used in BS7671. There are some simple tips to follow when writing answers to exam questions:

- **Explain briefly** – usually a sentence or two to cover the topic. The word to note is 'briefly' meaning do not ramble on. Keep to the point.
- **Identify** – refer to reference material, showing which the correct answers are.
- **List** – a simple bullet list is all that is required. An example could include, listing the installation tests required in the correct order.
- **Describe** – a reasonably detailed explanation to cover the subject in the question.

Before you start work, always remember to have a plan of action. You will need to know the clear sequence for working in order to make sure you not making any mistakes as you work and that you are working safely at all times.

Good luck!

CHECK YOUR KNOWLEDGE

1. Which of the following would **not** be covered by the WEEE Regulations?
   a) Large household appliances
   b) Medical devices
   c) Wooden stepladders
   d) Lighting equipment

2. What is the maximum noise exposure level that must not be exceeded?
   a) 100 dB
   b) 120 dB
   c) 140 dB
   d) 160 dB

3. Which of the following Parts of the Building Regulations 2000, amended, covers electrical safety?
   a) Part A
   b) Part B
   c) Part M
   d) Part P

4. The Sustainable Homes Code uses a star rating to show the most efficient buildings in terms of sustainability. Which, of the following star codings, is the highest possible rating?
   a) 1
   b) 2
   c) 4
   d) 6

5. You observe fumes that you consider will have a negative effect upon the environment from one of the site enclosures. Who should you inform?
   a) Electrician
   b) Site supervisor
   c) Site security
   d) Client

6. Solar photovoltaic systems use which of the following devices?
   a) Photo diode and transformer
   b) Diode and transformer
   c) Photo diode and inverter
   d) Diode and inverter

7. Which of the following statements related to wind energy generation is regarded as a **disadvantage**?
   a) Electricity generated can be sold back to the grid
   b) Planning permission is required
   c) Plentiful source of energy
   d) Wind energy generation does not release carbon dioxide

8. What is **not** included in grey water recycling?
   a) Dishwashers
   b) Showers
   c) Washing machine
   d) Rainwater

9. A measurement of noise on site from one operation reveals a level of 150 dB. What should be done?
   a) Ensure the noise level is reduced or removed
   b) Reduce the time operatives are exposed to the noise
   c) Issue all staff with ear plugs
   d) Carry on with the work

10. You are concerned that the mounting height of switches, marked on the site drawing, is incorrect. Which part of the Building Regulations would you check?
    a) Part B
    b) Part L
    c) Part M
    d) Part P

UNIT ELTK 03

Understanding practices and procedures for overseeing and organising the work environment

One of the key skills in all workplaces is the ability to share knowledge and communicate effectively with the people you work with. We need to be aware that everyone is different and will therefore see things differently from us. We can then use this knowledge to help remove misunderstanding or frustration and improve relationships to aid site progress and develop business.

This unit looks at the methods and systems used to communicate within the construction industry.

This unit will cover the following learning outcomes:

- understand the types of technical and functional information that is available for the installation of electrotechnical systems and equipment

- understand the procedures for supplying technical and functional information to relevant people

- understand the requirements for overseeing health and safety in the work environment

- understand the requirements for liaising with others when organising and overseeing work activities

- understand the requirements for organising and overseeing work programmes

- understand the requirements for organising the provision and storage of resources that are required for work activities.

K1. Understand the types of technical and functional information that is available for the installation of electrotechnical systems and equipment

There are many different people involved in an electrical installation and it's important that you can effectively communicate with your colleagues. This involves passing and receiving information.

Sources and interpretation of technical and functional information

The sorts of information you will come across in your work fall into two main categories:

- **functional information** – information such as user instructions, including when your professional expertise may be called on to advise a customer
- **technical information** – information such as installation and equipment specifications, manufacturer's data and instructions or test information.

There are hundreds of sources of such information and these would include consulting engineers, manufacturers, suppliers, trade associations, clients, libraries, book stores, the Internet or more formal sources such as the British Standards Institute (BSI).

Information provided by the manufacturer and the supplier

Almost all equipment will have the manufacturer's fitting instructions and other technical data or information sheets. You should read and understand these before fitting the item concerned.

Quite often on a project, the consulting engineer or client will have specified the use of particular products. For many of these products you will already be comfortable with their use and any requirements for installation, such as connecting a 13 A socket outlet.

However, you will also come across more complex products with many component parts, which you may be less familiar with. You will need to access information if you are to install these items correctly, so that they function properly.

The supplier may provide a certain level of information. However, in some cases, you may need to contact the manufacturer directly. Any information such as fitting or operating instructions should be handed over to the client on completion of the installation.

Once installation of the item is complete, any information should be kept safely in a central file so that they can be given to the customer in a handover manual when the project is completed. Sometimes you may need more details about an item. You can get this from the manufacturer's catalogue, datasheets or website, or by speaking to the manufacturer directly. It may be useful to talk to the contracts engineer to make sure that they know what is happening.

The sort of information that we need from suppliers and manufacturers might include:

- the materials used in constructing the product
- the types of component used in constructing the product
- general equipment details, such as operating characteristics, physical size and assembly instructions (or even supply characteristics from the REC)
- drawings and diagrams
- any special installation requirements
- delivery and storage details.

However, receiving information and interpreting it correctly are very different things.

Materials used in constructing the product

You need to be aware that certain materials are better suited to certain situations than others.

For example, if the manufacturer's information states they are supplying a brass cable gland for the single-core aluminium wire armoured cable that you have installed, alarm bells should be ringing, as there will be corrosion at the points of contact between the two dissimilar metals. This is called an electrolytic reaction between the two metals.

Remember

It is your responsibility to ensure that the installation is satisfactory.

Working life

You are checking the specification drawn up by the consulting engineer and notice they have specified the use of IP56 external luminaires. You are carrying out checks on the material that has been delivered on site.

- What information could you check to ascertain the luminaires are suitable for that purpose?
- What should happen to all information supplied with equipment?
- What should happen to information relating to equipment installed, once the contract is complete?

Types of component used in constructing the product

As with materials, certain components within a product may not suit the intended application.

For example, the manufacturer's data might tell you that the 60 A four-pole contactor that will be provided to control hall lighting in a leisure centre will have a 400 V coil fitted. However, the project specification might require the hall lighting to be controlled via a single-pole 230 V switch from the reception desk.

In such a circumstance the hall lighting will not work as designed so you would need to obtain the correct equipment.

General equipment details

With general details such as operating characteristics, physical size and assembly, you will need to take care how you interpret and apply the information.

For example, imagine that the manufacturer sends operating characteristics about the standard batten-style fluorescent lighting fittings that are to be provided.

You know that the lighting fittings are to be recessed into a 100 m long ceiling trough formed from plasterboard, which would give little ventilation and require components to operate at a higher operating temperature than normal. Are these high-temperature versions being provided by the manufacturer?

Drawings and diagrams

A technical diagram is simply a means of conveying information more easily or clearly than can be expressed in words. In the electrical industry, drawings and diagrams are used in different forms. However, those most frequently used by manufacturers and suppliers are block diagrams, circuit diagrams, schematic diagrams and assembly drawings.

Block diagrams

A block diagram can be used to relate information about a circuit without giving details of components or the manner in which they are connected. It is typically used for a higher level, less detailed description of overall concepts, rather than for understanding the finer details of a system.

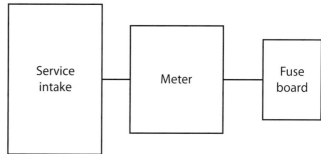

Figure 3.01 A block diagram

In block diagrams the various items are represented by a square or rectangle, each clearly labelled to indicate its purpose. This type of diagram shows the sequence of control for installations in its simplest form, as shown in Figure 3.01.

Circuit diagrams

A circuit diagram uses symbols to represent all the circuit components and shows how these are connected. The circuit diagram should be as clear as possible and should follow a logical progression route from supply to output. This sort of diagram is useful when testing a circuit and for

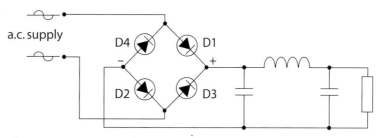

Figure 3.02 A circuit diagram

understanding how it works. However, in all other respects the circuit diagram cannot be regarded as a direct source of information.

Figure 3.02 shows a circuit diagram in relation to a rectification circuit. The shape of the diagram does not represent the physical outline of the circuit; it has no dimensions; and the symbols used do not bear any resemblance to the components they represent. The symbols used are BS EN 60617 circuit diagram symbols.

Schematic diagrams (working diagrams)

Schematic diagrams are similar in concept to the circuit diagram. They do not show you how to wire components, but they do show you how the circuit is intended to work. They tend to be used for larger, more complicated electrical diagrams. Figure 3.03 shows a schematic for the control circuit of an automatic star/delta starter.

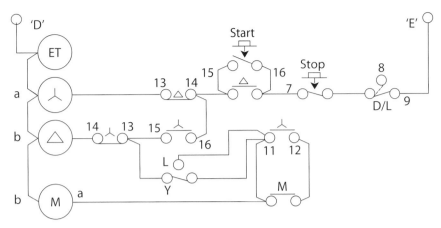

Figure 3.03 A schematic diagram

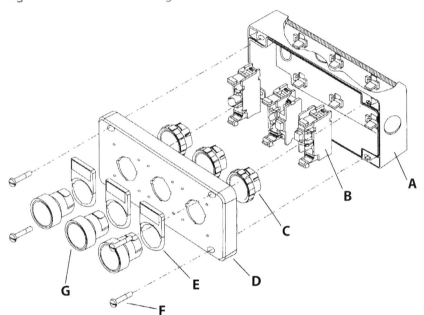

A Enclosure base with built in contact block clips

B Contact blocks/lamp holders

C Locking ring

D Enclosure lid

E Legend plate

F Captive screws (after screw in) loose in enclosure on delivery

G Actuators and lens cap

Figure 3.04 An assembly drawing

Assembly drawings

Assembly drawings show how the individual parts (or modules) of a product fit together. They normally contain scale drawings of all the components shown in their correct position relative to each other, with some overall dimensions.

Where there are internal components, these are shown as a section (a vertical plane cut through the object, in the same way as a floor plan is a horizontal section viewed from the top). Each component is listed and described on the drawing. Figure 3.04 shows the assembly drawing for a push-button enclosure.

Special installation requirements

A simple example of this may concern the installation of a purpose-made lighting fitting where, because of either its weight or its design, special fixing arrangements may be required on site.

Delivery and storage details

In the case of fragile or bulky items, manufacturers may have extended delivery dates and the equipment may require special storage arrangements on site.

Measuring and test instruments

When carrying out either measurement/testing of an installation or fault finding/rectification of an installation or equipment, you need to understand the operating requirements and implications of using various measuring and test instruments. This could involve something as simple as ensuring that the instrument complies with GS.38 in terms of its leads.

Other considerations regarding instruments are covered in the Health and safety (ELTK 01), Electrical principles (ELTK 08) and Testing (ELTK 06) units of this qualification.

Information provided by your employer

Most information needed during an installation will flow to site from your employer. It may originate from outside parties, such as the client, consulting engineer or main contractor, and those outside parties may require information from site. Your employer will normally act as the focus to gather and disseminate such information.

Here are some of the typical documents and information provided by employers on site.

Layout drawings

Layout drawings show where everything must go. They are usually prepared by an electrical consulting engineer with responsibility for the project, based on the architect's drawing.

Specifications

A classic example of how specifications work is a client requiring a particular finish on electrical accessories, for example satin chrome. Normally the consulting engineer will have two references to this in the specification as follows:

- All electrical accessories will be of satin chrome finish (general specification)
- Mounting heights of all accessories to be as detailed in Part M of the building regulations (specific detailed area of specification).

In other words one section informs a contractor for estimating purposes, that is quantities and type, while the other specifies how these are installed and to what standard. All specifications need careful scrutiny, to ensure the full requirements have been fully understood and interpreted.

The electrical specification adds to the layout drawings, by telling you the height of electrical work, how things are to be wired in and what systems you need to use. These specifications are also prepared by the consulting engineer.

In its most basic generic format, an electrical specification comprises two sections.

- Section 1 – the general specification – tends to give general information about installation circumstances, such as wiring systems, enclosures and equipment. The section gives generic requirements applicable to any project issued by the particular consulting engineer. It gives a general description of all the consultant's expectations and requirements for any installation circumstances.

- Section 2 – the particular specification – gives the specific details of the project. For example, it may say that, for this job, the installation will be carried out in galvanised steel conduit. Unless there are specific conditions applicable to the project, you need to see what the consulting engineer's general requirements are for installing galvanised steel conduit, set out in Section 1 requirements specific to the particular installation that you are about to begin working on.

In reality, the electrical contractor will have to read both sections to understand the full requirements of the installation.

The Case study gives an example.

Case study

Here is an extract from an electrical specification for the refurbishment of the main warehouse within a university. It shows what the consultant has written about the proposed fire alarm installation.

Fire alarms

SYSTEM OF WIRING

The wiring of the fire alarm system shall be carried out using MICC cable (600 volt grade) as manufactured by Pyrotenax.

The electrical contractor shall allow additional protection for cables to be provided at any point where it could be subject to mechanical damage.

You may think this extract is quite clear and easy to follow. However, within the general specification the consultant goes on to say this about MICC cables:

- Cables shall be made with ring-type glands with screw-on pot-type seals utilising cold plastic compound and neoprene sleeving and applied in the manner recommended by the manufacturer of the cable.

- Where ambient temperature is likely to exceed 150°C, medium temperature seals shall be used and completed with LSF (Low Smoke and Fume) shrouds and all fixing saddles shall be LSF (Low Smoke and Fume) coated or nylon throughout.

- Where cables pass through wall and floors, they shall be protected by galvanised conduit bushed at both ends. Cables rising on external or exposed wall surfaces shall be protected by galvanised conduit to a height of 2 metres above ground level. The conduit shall be suitably bushed and sealed with MICC sealing compound.

- Where MICC cables terminate into electric motors or similar equipment liable to vibration or movement, a vibration loop of a single coil of the cable shall be made before final connection.

- Where cables do not terminated in a conduit box with spouts, they shall be fitted with a coupling and brass bush.

- The cores of the cables shall be identified by means of coloured sleeves in accordance with BS 7671: 2008, Requirements for Electrical Installations Wiring Regulations Seventeenth Edition and BS 3858:1992.

- The whole of the system of cable and fittings shall provide both electrical and mechanical continuity and shall be efficiently earthed. Where the cable is cut, kinked or badly bent, shows signs of inferior workmanship, or an insulation reading less than 'infinity' is obtained, the Electrical Services Installer shall replace the defective cable or seals. All unmade ends of MICC cables on site shall be sealed to prevent the ingress of moisture.

- Wherever possible, the cable shall be covered with LSF (Low Smoke and Fume) oversheath.

As you can see, you must read the two sections of the specification together to fully comply with the installation requirements.

The layout drawings, which are scaled drawings based on the architect's drawings of the building, show the required position of all equipment, metering and control gear to be installed in plan view, using standard IEC.60617 location symbols.

Figure 3.05 shows the intended lighting installation within the university warehouse.

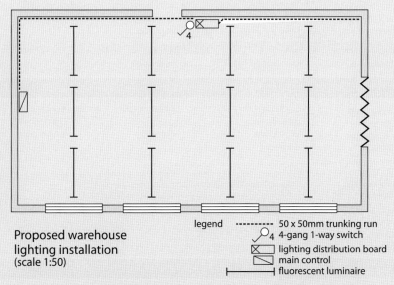

Proposed warehouse
lighting installation
(scale 1:50)

legend   `----------` 50 x 50mm trunking run
4-gang 1-way switch
lighting distribution board
main control
fluorescent luminaire

Figure 3.05 Intended lighting arrangement in a university warehouse

The layout diagram shows where things go, but does not specify the height of equipment such as switches or sockets or the type and manufacturer of such equipment.

This information will normally be contained in the specification.

Be aware that there are no hard and fast rules when it comes to drawings and specifications. Some consulting engineers include details of equipment on the layout drawing, while others issue a specification with a clear and detailed 'specific' section that illustrates their total requirements for the project. Consulting engineers generally use IEC 60617 location symbols to show the location of systems and equipment.

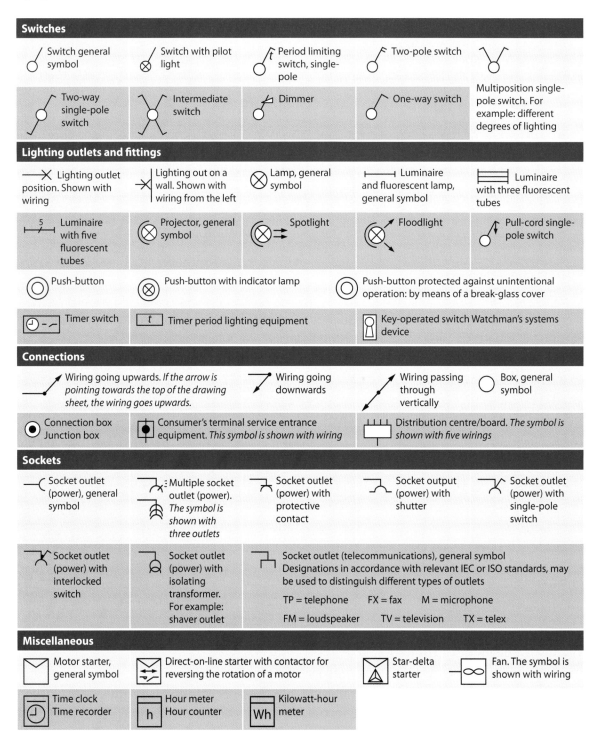

Figure 3.06 IEC.60617 lists the standard symbols for use in installation drawings

Wiring diagrams

Wiring diagrams show the physical layout, with a pictorial version of the components and connections in the actual circuit.

Wiring diagrams can carry specific information about the wiring or connection of components. As a rule, they do not use circuit or location symbols, although you will come across some that do.

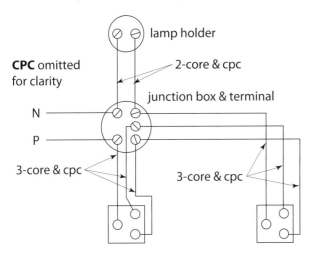

Figure 3.07 A wiring diagram

Record drawings

Record drawings, sometimes called 'as fitted' drawings, are used to record what has been fitted. This means completing a set of installation layout drawings and marking on them the routes of the installed wiring systems and enclosures, as well as showing any amended locations of equipment.

To help you complete these drawings accurately, you should gather information as you go along throughout the project, depending on the size of the installation. The completed drawings should form part of the information given to the client on handover of the project, as they will help with any maintenance or alterations to the installation.

Records and certificates for inspection, testing and completion

So far you have looked at information issued at the start of or during an electrical installation. However, you will also be expected to interpret information contained in documents that come into existence towards the end or completion of the installation.

Remember

These documents also act as relevant information for the Health and Safety File, a requirement of the CDM Regulations (see page 20) to ensure the safety of anyone involved in any future construction work on that project.

BS 7671 requires that the appropriate documentation required by 514.9, Part 6 and Part 7 be provided for every electrical installation. This includes electrical installation certificates, minor electrical works certificates and periodic inspection reports. These documents should exist for the entire life of the installation and will help to establish whether it is deteriorating in any way.

You will cover these documents in full in the Inspection and testing section of Unit ELTK 06, which is covered in Book B.

Interpreting information on the operation and control of equipment

Interpretation of information may also involve combining it with other information (such as the general and specific parts of specifications) or comparing it with another set of information (such as comparing two sets of test results to see whether an installation is deteriorating).

Another variation is to receive verbal or written information in the form of a request and then being able to interpret it and apply it to equipment: for example, setting up the controls of a central heating system to meet the requirements of the house owner.

All of these information types could be relevant when you are either setting up equipment for the first time or making alterations.

When interpreting information about equipment, you will need to understand:

- its operation
- its controls
- its settings
- how to make adjustments.

It can be relatively simple to install something without understanding why and how it works. However, it becomes a lot more difficult to alter settings or find faults without that information.

Thermostat operating systems

Did you know?

Although bi-metallic thermostats are still manufactured, you are more likely to find electronic versions on newer systems.

A thermostat is nothing more than a switch, but one that automatically reacts to changes in temperature. However, there are varied operating systems and you will need to establish which one is being used. There are two main types of thermostat operating system.

Bi-metallic strip

With a bi-metallic strip, temperature change causes mechanical movement. The strip is made up of two joined strips of different metals, rolled into a coil. The two metals have different co-efficient of expansion when coiled, and expand at different rates as they are heated. These different expansions force the strip to bend one way if heated, and the other way if cooled.

As the room heats up or cools down, the metal reacts to the change in temperature. Once the thermostat reaches a specific set level in either direction, it sends a signal to the central heating boiler to switch it on or off. These strips are low-cost, but slow to react to temperature change.

Electronic thermostat

An electronic thermostat uses a device known as a **thermistor** to sense room temperature, along with a **micro-controller** that measures the resistance change and converts that number to a temperature reading. Once the temperature in the room moves above or below the set temperature, the thermostat sends an electric signal to the boiler to switch it on or off.

Dial-operated device

With a dial-operated device, the rotating dial has temperature markings that increase in increments of 5°. A raised arrow makes it easy to see what the required temperature has been set to. To adjust the temperature to a different setting, you simply rotate the dial until the temperature required is aligned with the raised arrow.

Digital electronic thermostat

With a digital device the information is presented clearly on a liquid crystal display (LCD). Its default setting is to show the current room temperature. Buttons on the front of this particular model allow the temperature to be raised or lowered; the LCD changes to show this before returning to the default setting once the new temperature has been confirmed.

> **Key terms**
>
> **Thermistor** – a resistor with an electrical resistance that changes with temperature
> **Micro-controller** – a micro-computer on a single chip

Figure 3.08 Dial-operated thermostat

Figure 3.09 Digital electronic thermostat

Case study

With winter approaching, you are asked to go to a children's nursery and investigate the central heating system: following some building work, the heating seems to come on and then stay on far longer than required. The building work was the construction of a new double door into the outside play area.

The nursery staff show you where the central heating boiler and room thermostat are. The room thermostat hasn't been moved and is now next to the new doors leading outside to the play area.

What's gone wrong?

You've been shown the thermostat and it is now adjacent to the double doors leading outside. You can see that it is a digital thermostat with its display showing a room temperature of 20°C. On examination you can see the staff have set the required temperature at 22°C.

Unfortunately, as the thermostat is now next to the new doors it is in the coldest part of the room. This means the temperature around it rarely reaches 22°C. It senses this lower temperature and therefore has the boiler switched on most of the time.

Site requirements and procedures

All the information we have looked at in this unit is based around your actual work as an electrician. However there are also several more generic topics affecting an electrical installation that will both generate and require information. These include:

- health and safety requirements
- services provision
- ventilation provision
- waste disposal procedures
- equipment and material storage
- access for personnel.

The first four of these areas have already been covered in Units ELTK 01 and ELTK 02. Look back at pages 7–10 and 121–29 now to remind yourself of the key points. Equipment and material storage was covered in ELTK 01, on page 84.

Here we will look at the final point: access for personnel.

Access for personnel

During normal on-site work activities it is inevitable that sites will receive visitors, from delivery drivers to architects and clients. This may be less of a problem in areas such as factories, office blocks or hospitals, where a construction site may be located within the boundary of these organisations, as they probably already have a visitor facility. The same can also be said for larger construction sites.

At sites without designated facilities, the motivation and co-operation of all site personnel is key. Irrespective of the quality of the facilities available for receiving visitors, the reasons for putting into practice a 'visitor procedure' remain the same:

- to meet with health and safety requirements
- to maintain site security
- to project a professional approach for the company
- to establish and maintain good client relationships.

Generally speaking, there are some simple points of good practice when receiving visitors.

- Be polite and courteous, responding only to requests that are within your authority.
- Check their identity and the reason for their visit.
- Brief them on site safety and issue them with PPE if necessary.
- Ask them to complete the site visitors' book, as shown in Figure 3.10.

Remember

General access requirements were covered in unit ELTK 01. Take a look back at pages 101–114 now to remind yourself of the main points.

Date	Visitor's name	Company	To see	Time in	Time out	I.D. checked	Badge number	H&S briefed	Visitor's signature
Enter date	PRINT visitor's name	Enter name of company or organisation visitor is representing	Name of person to be visited	Time visitor is booked into the site	Time visitor is booked out of the site	Type of ID used – e.g. student card, letterhead	Number of visitor pass or badge issued	Enter 'Yes' when briefed and PPE issued, if required	Ask visitor to sign here

Figure 3.10 Site visitors' book

Progress check

1   You have a heater to connect on site and have concerns its position could present a hazard due to surface temperature. Where could you check the equipment's technical details and whether it is suitable for the position it is installed?

2   Explain the difference between a wiring diagram and a circuit diagram.

3   List the main items a site visitors' book should record.

K2. Understand the procedures for supplying technical and functional information to relevant people

Functional and technical information will be required by a variety of people throughout the duration of any construction project. Before you can understand the procedures involved, you need to understand the parties involved, the relationship between them and their varying activities and levels of responsibility.

The construction industry

The construction industry is one of the biggest industries in the UK and the electrical contracting industry is just one sector within it. Organisations in the construction industry range from sole traders (for example, a jobbing builder) to large multinational companies employing thousands of workers.

The work done by these companies is very varied, but we can broadly think of it under three headings:

- **building and structural engineering** – which covers the construction and installation of services for buildings such as factories, offices, shops, hospitals, schools, leisure centres and, of course, houses
- **civil engineering** – which involves the construction and installation of services for large structures such as bridges, roads, motorways, docks, harbours and mines
- **maintenance** – which covers the repair, refurbishment and restoration of existing buildings and structures.

Larger construction companies will be able to undertake work in all these areas, but smaller companies may specialise in one area.

Nearly every project, large or small, will involve a variety of different trades, such as bricklaying, plastering, plumbing and joining, as well as electrical work. The ability to work and communicate well with others, as well as establishing good professional relationships with colleagues and people in other trades, is crucial to the successful completion of a project.

The electrotechnical industry

Most people are unaware of the vast range of activities and occupations that make up the electrical industry. They usually think of an electrician as someone who installs lights and sockets in their house.

Did you know?

Only about half the people working in the construction industry are skilled craft operatives such as electricians; the other half are management, technical and clerical staff.

In reality, during their career an electrician could be involved with the installation, maintenance and repair of electrical services (both inside and outside) associated with buildings and structures such as houses, hospitals, schools, factories, car parks, leisure centres and shops. There are also specialist areas that call for additional knowledge and training, such as motor repair, street lighting, alarm systems or panel building.

Electricians may be employed within the electrical contracting industry, but they may also be employed directly by organisations needing their skills, such as refineries or factories.

The electrical contracting industry structure

Industry bodies

There are a number of organisations you should be aware of within the industry. You have already looked at the main health and safety organisation, the Health and Safety Executive (HSE), in Unit ELTK 01 (see page 26), but the following bodies play important roles too.

The Electrical Contractors' Association (ECA)

The ECA represents the interests of electrical installation companies in England, Wales and Northern Ireland and is the major association in the electrical installation industry. Founded in 1901, it has over 2000 member companies, from small traders with only a few employees to large multinational organisations.

The ECA aims to ensure all electrical installation work is carried out to the highest standards by properly qualified staff. Firms that wish to become members of the ECA must demonstrate that they have procedures, staff and systems of the highest calibre.

NICEIC

The NICEIC is an accredited certification body that protects users of electricity against the hazards of unsafe and unsound electrical installations. It is the industry's independent electrical safety regulatory body and is not a trade association.

The NICEIC maintains a roll of approved contractors who meet the council's rules relating to enrolment and national technical safety standards, including BS 7671 (IET Wiring Regulations). The roll is published annually and is regularly updated on the NICEIC website so that consumers and specifiers can select contractors who are technically competent.

Find out

Who is your local NICEIC Inspector, and where are their offices?

The Council also employs inspecting engineers, who make annual visits to approved contractors to assess their technical capability and to inspect samples of their work to the National Standard of BS 7671. As of early 2011 there were 74 area engineers working for the NICEIC.

Unite the union

For many years workers within the electrical installation industry have enjoyed good labour relationships and co-operation with their employers, largely due to the EETPU (the main trade union at the time) and the ECA. Since 2007 Unite, the UK's biggest trade union, has continued these excellent relationships. Unite has over 2 million members, and represents a large number of other trades and industries.

If you are a Unite member, they promise to protect your rights, health, safety and well-being at work, negotiating on your behalf with employers and the UK and European governments to get a fair deal for you at work.

The Joint Industry Board (JIB)

The Joint Industry Board for the Electrical Contracting Industry came into existence in 1968 through an agreement between the ECA and the Electrical, Electronic, Telecommunications and Plumbing Union (EETPU). Effectively the industrial relations arm of the industry, the JIB has as its main responsibility the agreement of national working conditions and wage rates.

SummitSkills

SummitSkills is the Sector Skills Council for the building services engineering sector, representing the electrotechnical, heating, ventilating, air-conditioning, refrigeration and plumbing industries.

SummitSkills has been created for employers to identify skills shortages and deliver action plans to address them. The organisation will also provide careers information for all industries within the sector, as well as dealing with all training standards and policy matters previously handled by the former National Training Organisations (NTOs).

The Institution of Lighting Engineers (ILE)

The Institution of Lighting Engineers is the UK's most influential professional lighting association, dedicated solely to excellence in lighting. Since its foundation, the ILE has evolved to include lighting designers, architects, consultants and engineers among its

2500-strong membership. The ILE's key purpose is to promote excellence in all forms of lighting. This includes interior, exterior, sports, road, flood, emergency, tunnel, security and festive lighting as well as design and consultancy services.

The Institution is a registered charity, a limited company and a licensed body of the Engineering Council.

The Institution of Engineering and Technology (IET)

Formerly the IEE, the IET was founded in 2006 and is now the largest professional engineering society in Europe, with a worldwide membership of just over 375,000. As well as setting standards of qualifications for professional electrical, electronics, software, systems and manufacturing engineers, the IET prepares regulations for the safety of electrical installations for buildings. The IET Electrical Wiring Regulations (Requirements For Electrical Installations: BS 7671) has become the standard for the UK and many other countries.

Employer structure

There are about 21,000 electrical contracting companies registered in the UK dealing with many tasks, including:

- handling initial enquiries
- estimating costs
- issuing quotations
- dealing with suppliers and sub-contractors
- supervising contracts
- carrying out the installation
- financial control of a project
- final settlement of accounts.

These companies range from one-person organisations to large multinational contractors, but the majority employ fewer than 10 people. The structure of electrical companies also varies considerably between firms, depending on the number of employees and the type and size of the business. You can think of them broadly in terms of two groups: small firms and large firms.

Small firms

In a small company, the main tasks are often the responsibility of one person, and there is only a narrow range of people at each management level. This type of company structure is known as a vertical structure. The principal advantage is that lines of communication are short: everyone knows who is responsible for different aspects of a project. When someone from outside – another tradesperson, contractor or customer – needs any information, they know who to talk to, and are not passed from one department to another.

However, if the person dealing with the project is ill, is on holiday, or leaves the company, vital information can be missing and it can be difficult for someone else to continue handling the project smoothly.

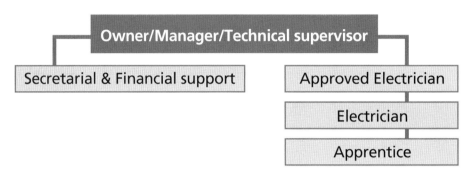

Figure 3.11 Small firm organisational flow chart

Large firms

In larger companies, the tasks are allocated to a range of people: for example, estimators who handle the initial enquiry and produce an estimate, and contracts engineers who see that the work is done. This type of structure is known as a horizontal structure as there are more people at each management level. The advantage is that individuals become specialists in a particular task and can work more efficiently.

Companies of this size often belong to a trade organisation such as the Electrical Contractors' Association (ECA) and, because of their structure, can offer good career development prospects.

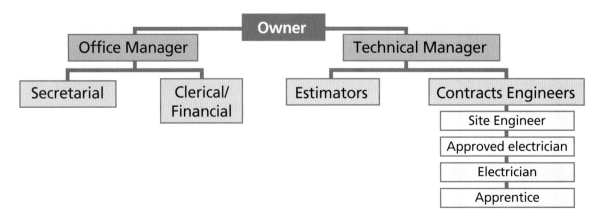

Figure 3.12 Large firm organisational flow chart

Project roles and responsibilities

Many different people are involved in a construction project, from its initial design through to construction and completion. As an electrician, you will be dealing with many of these people, and it is important that you understand their job function and how they fit into the overall project.

Here you will look at the people involved in the three main stages of a typical construction project: design, tendering and construction.

Key term

Bill of quantities – a list of all the materials required, their specification and the quantities needed; contractors who are tendering for the project use this information to prepare their estimates

The design stage

Client	Person or organisation that wants the work done and is paying for it • Specifies the purpose of the building • Usually gives an idea of number of rooms, size, design, etc., and any specific wants • May also give an idea of the price they are willing to pay for the work
Architect	Designs the appearance and construction of the building so that it fulfils its proper function • Advises the client on the practicality of their wishes • Ideally provides a design solution that satisfies the client and also complies with the appropriate rules and regulations for the type of building • For small projects, may draw up a complete plan • For larger, more complex buildings, consults specialist design engineers about technical details
Consulting engineer (Design engineer)	Acts on behalf of the architect, advising on and designing specific services such as electrical installation, heating and ventilation • Creates a design that satisfies the client and architect, the supply company and regulations • Ensures that cable sizes have been calculated properly, that the capacities of any cable trunking and conduit are adequate, and that protective devices are rated correctly • Produces drawings, schedules and specifications for the project that will be sent out to the companies tendering for the contract • Answers any questions that may arise from this • Once the contract has been placed, produces additional drawings to show any amendments • Acts as a link between the client, the main contractor and the electrical contractor
Quantity surveyor (QS)	Manages and controls costs for a building project • Responsible for taking the plans and preparing an initial **bill of quantities** • During construction, monitors the actual quantities used, and also checks on claims for additional work and materials
Clerk of works	Checks that the quality of the materials, equipment and workmanship used on the project meet the standards laid down in the specification and drawings (on big contracts there may be several clerks of work, each responsible for one aspect, such as electrical, heating or ventilation) • Effectively employed by the client • Inspects the job at different stages • Checks any tests carried out • May also be given the authority by the architect to sign day worksheets and to issue Architect's Instructions for alterations or additional work

Table 3.01 People involved at the design stage

Did you know?

Larger electrical contractors, and in particular those that are multi-disciplinary, often offer a design service for customers, which eliminates the need to use an external architect and makes things simpler (and possibly cheaper) for the client. However, if a dispute arises over design aspects of the job, the client no longer has anyone to arbitrate.

The architect, consulting engineer, quantity surveyor and clerks of works are traditionally part of the architect's design team. However, installations may not always be tackled like this.

For example, on a small job, the client may approach an electrical contractor directly and ask it to carry out the work. The client will provide a few basic details and requirements, and expects the electrician to ensure that the installation is properly designed and carried out.

The tendering stage

Tendering is the process by which a contractor bids for contracts. The contractor works through the drawings and specifications issued by the consulting engineer and submits in writing a total cost for carrying out the work, including materials, tools, equipment and labour.

This is done in competition with other firms who want to carry out the work.

Most invitations to tender (also known as enquiries to submit a tender) have strict guidelines about the information to be supplied and a fixed deadline by which the tender must be received.

The estimator

The estimator's task is to calculate the total cost that will be given in the tender. At the start of the tendering process, a consulting engineer or building contractor usually issues an invitation to tender. This contains various documents, such as:

- a covering letter, giving a broad description of the work, including start and finish dates
- the form of contract that will be applicable to the project (for example, whether it is to be a fixed or a fluctuating price)
- drawings and specifications for the project
- a tender submission document that must be used
- a day work schedule.

A fixed price contract is exactly what it says. The contractor agrees to complete the job for the price quoted in the tender, even if material or labour costs go up before the project is completed.

A fluctuating price contract always has a time limit, so that the contractor is not held to a fixed price if the project is delayed. A fluctuating price contract allows the contractor to claim back the difference between costs included at the time of estimate and actual costs incurred at the time of installation.

When the invitation to tender is received, the estimator will check with their management (sometimes the contracts manager) to see whether the company wishes to tender on the basis of the submitted documents. If it does, the estimator reads the specifications carefully to understand the requirements.

As you read on page 165, most specifications have two sections and these need to be read together. Using this information and the scaled drawings, the estimator calculates the amount of materials and labour required to complete the job within the time specified by the client. This information is then recorded on a 'take-off' sheet (see Figure 3.13).

Item	Qty	Description	Init cost	Discount	Material cost £	Hours to install	Hourly rate £	Total labour £
1	200 m	20 mm galvanised conduit	1.2/m	0	240.00	70	9.50	665.00
2	30	Earthing couplings	.24	0	7.20	0	0	0
3	30	20 mm std. brass bushes	.10	0	3.00	0	0	0
4	150	20 mm distance saddles	.48	0	72.00	0	0	0
5	150	1.5 x 8 brass screw and plugs	.03	0	4.50	0	0	0

Figure 3.13 Take-off sheet

Nowadays this work is normally done using dedicated computer software. The final tender is based on the results of these calculations.

The construction stage

The contract

Before any work starts, **contracts** must be agreed and entered into by all the firms involved, as any failure to comply with the details of the contract by either party could result in a court

Key term

Contract – a legally binding agreement between two or more parties

action and heavy financial damages. This might happen if the contractor does not complete the work or uses sub-standard materials, or if the client does not pay.

Contracts do not have to be made in writing. A verbal agreement, even one made on the telephone, can constitute a legal contract. However, most companies use written contracts that cover all aspects of the terms and conditions of the work to be carried out.

Several conditions must be met for a contract to be binding:

- An offer must be made that is clear, concise and understandable to the customer.
- The customer must accept the offer, and the contractor must receive this acceptance. The acceptance must be unqualified (i.e. with no additional conditions). Up to this point, there is no agreement or obligation binding either side. The contractor is free to withdraw the offer, and the customer can reject it.
- There must be a 'consideration' on both sides. This shows what each party is agreeing to do for the other.

There are several reasons why a contract may not be made. Here are just some of them.

- **Withdrawal** The contractor can withdraw the offer at any time until the offer is accepted. The contractor must notify the customer of the withdrawal.
- **Lapse due to time** Most offers put a time limit on acceptance. After this time, the offer expires and the contractor is under no obligation, even if the customer later accepts the offer.
- **Rejection** The customer can simply reject the contractor's offer; no reason has to be given. If the customer asks the contractor to submit a second offer (for additional work, or simply to lower the price), the contractor is under no legal obligation to quote again. Each quote is self-contained: the terms or conditions for a previous quote do not automatically apply.
- **Death of contractor** If the contractor dies before the offer is accepted, the customer must be notified; otherwise the customer could agree (within the offer time limit) and the contract would become valid.

A **breach of contract** occurs when one of the parties does not fulfil the terms of the contract: for example, if the contractor does not perform the work to the specification in the offer (it is the contractor's responsibility to ensure that the installation is in complete compliance with the specification) or if the customer refuses to pay for the work.

Contract law is very complex, and it cannot be covered fully on this course. To minimise the risks, you can use a standard form of contract. The Joint Contracts Tribunal (JCT) Standard Form of Contract (normally for projects of a complex nature or in excess of 12 months' duration) or the Intermediate Form of Contract are typical of contracts used in the industry. In line with understanding your responsibilities, if you are involved in any kind of contract, it is always advisable to seek professional legal assistance before making or accepting an offer.

Once contracts are agreed, you move into the construction stage, where construction and installation work begin and many more people become involved, each with their own particular roles and responsibilities. Table 3.02 shows the roles and responsibilities of those most commonly involved.

Main contractor	<ul><li>Usually the builders, because they have the bulk of the work to carry out</li><li>Has the contract for the whole project</li><li>Employs sub-contractors to carry out different parts of the work</li><li>In refurbishment projects, where the amount of building work is small, the electrical contractor could be the main contractor</li><li>Paying and co-ordinating sub-contractors</li></ul>
Nominated sub-contractors	<ul><li>Named (nominated) specifically in the contract by the client or architect to carry out certain work</li><li>Must be used by the main contractor</li><li>Normally have to prepare a competitive tender</li><li>Sub-contractors will include electrical installation companies</li></ul>
Non-nominated sub-contractors	<ul><li>Companies chosen by the main contractor, rather than being specified by the client</li><li>Their contract is with the main contractor</li></ul>
Nominated supplier	<ul><li>Supplier chosen by the architect or consulting engineer to supply specific equipment required for the project</li><li>Main contractor must use these suppliers</li></ul>
Non-nominated supplier	<ul><li>Selected by main contractor or sub-contractors</li><li>For electrical supplies, this will be a wholesaler selected by the sub-contractor who can provide the materials needed for the project</li></ul>
Contracts manager	<ul><li>Oversees the work of the contracts engineers</li><li>May also be responsible during tendering for deciding whether a tender is to be submitted and the costs and rates to be used in it</li></ul>

Table 3.02 People involved in the construction stage

Contracts engineer	• Employed by the electrical contractor to manage all aspects of the contract and installation through to completion • Responsible for planning labour levels, ordering and organising materials required • Ensures the contract is completed within the contract timescales and on budget • Liaises with suppliers to ensure planned delivery dates and builders' work programme are acceptable • May negotiate preferential discounts with suppliers • Attends site meetings
Project engineer	*Role definitions vary within the industry but generally the role is similar to that of a contracts engineer* • Responsible for day-to-day management of on-site operations relative to a specific project • Often based on the site
Site supervisor	• Contractor's representative on site • Oversees normal day-to-day operations on site • Experienced in electrical installation work – normally an Approved Electrician • Responsible for the supervision of the approved electricians, apprentices and labourers • Uses the drawings and specification to direct the day-to-day aspects of the installation • Liaises with contracts engineer to ensure that the installation is as the estimator originally planned it • Ensures materials are available on site when required • Liaises with the contracts engineer where plans are changed or amended to ensure additional costs and labour/materials are acceptable and quoted for
Electricians, apprentices and labourers	• The people who actually carry out the installation work • Work to the supervisor's instructions
Electrical fitter	• Usually someone with mechanical experience • Involved in varied work including panel building and panel wiring and the maintenance and servicing of equipment
Electrical technician	*Job definition varies from company to company* • Can involve carrying out surveys of electrical systems, updating electrical drawings and maintaining records, obtaining costs, and assisting in the inspection, commissioning, testing and maintenance of electrical systems and services • May also be involved in recommending corrective action to solve electrical problems
Service manager	*Similar role to contracts manager (and in some cases the roles are combined) but focuses on customer satisfaction rather than contractual obligations* • Monitors the quality of the service delivered under contract • Checks that contract targets (e.g. performance, cost and quality) are met • Ensures customer remains fully satisfied with the service received
Maintenance manager	*Once the building has been completed* • Keeps installed electrotechnical plant working efficiently • May issue specifications and organise contracts for a programme of routine and preventive maintenance • Responsible for fixing faults and breakdowns • Ensures legal requirements are met • Carries out maintenance audits

Table 3.02 People involved in the construction stage (cont.)

Limits of responsibility for supplying technical and functional information

As you can see from Table 3.02, there can be many people involved with an electrical installation and their needs in terms of information may change daily, depending on the circumstances.

As an apprentice, you may be required to give information to, or receive information from, any of them. However, when doing so, you must be aware of your level of responsibility. This will involve recognising when a limit to that responsibility has been reached.

As an example, if the main contractor approaches you and asks where the main switchboard panel will be located and when it will be delivered to site, you may well be in possession of the correct information. However, it is more likely that you know the 'where' part, but not the 'when'.

If you do not know the answer, either fully or partially, then do not guess. Just say that you don't know, but that you'll ask your supervisor to get back to them with the information.

And if you say you are going to do something, then do it!

Working life

Your company is working on a large construction site and acts as the main contractors. You have been asked to ensure the relevant people have been invited to a site meeting to discuss the preliminary stages of the electrical installation and how the works programme will affect other trades.

- Which personnel would most likely be invited?
- What are their responsibilities on a large contract?
- Who would normally ensure materials and equipment are programmed to arrive in time for installation to avoid delays.

Your level of responsibility may also be reflected when handing over the completed installation, which you will look at in the next section.

Organisation policies and procedures for the handover and demonstration of electrotechnical systems, products and equipment

At the handover stage, it is usual to issue the client with a handover manual that contains documents such as 'as fitted' drawings, specifications, maintenance/manufacturer's instructions and guarantees or warranties.

Remember

Usually more than one copy of the handover manual is given to the client or other interested parties. Your employer should also keep copies for future reference.

It is also usual to walk the client around the completed installation and show them how various key components work, so that they can operate them correctly and safely. Your supervisor may ask you to take part in this activity. For example, your supervisor might ask you to show the client how the central heating system works. To do this properly you would need to:

- know the operating principles yourself
- know the location of the various components in the central heating system
- ensure that the client has understood your demonstration and explanation, possibly by asking them to explain or demonstrate back to you.

Depending on the size of an installation, these activities may be co-ordinated to take place during a formal handover meeting, where the outcome of a successful handover is recorded and documented for all relevant parties. Such meetings may typically involve:

- closing out any defects or amendments
- final inspection of the installation by the client
- demonstrating key features of the installation to the client to ensure it functions as specified
- issue of the handover manual(s)
- establishment of contacts and procedures throughout the **defect liability period**.

Some projects also include a Handover Summary Report, which can be used to summarise and clarify any changes to scope, budget, schedule or quality, as well as detailing who agreed the changes and why they were felt to be necessary.

Appropriateness of different customer relations methods and procedures

Who is your customer?

At first, you'll probably think of the person who wants the work done and is paying for it – but does this mean you can ignore everyone else you will meet in your job? A wider definition of a customer is: 'Anyone who has a need or expectation of you.'

Using this definition, almost everyone you work with becomes your customer, and you will, in turn, be theirs. They may be architects, consultants, clients, other tradespeople or a member of the public. They may ask you to do some complex task, or simply

---

Did you know?

Occasionally part of the handover is to invite everyone involved in the project to celebrate the project's completion. For everyone involved, this is perceived as a strong motivational tool for future work activities.

Key term

Defect liability period – a contracted period of time and coverage, during which the installer is responsible for the repair of any defective items

Remember

Be polite to everyone you come into contact with during your work projects, not just the main 'customer'.

ask you a question. Whatever the case, they will certainly expect a good, polite response.

There may be certain procedures to follow: for example, all architect enquiries may have to be directed via the site engineer. However, treating everyone as a valuable customer and always trying to give them your best service will bring you many benefits. Dealing with people is an important part of your job. It is never wise to upset people, if only because it may cause problems later for you or your firm.

Table 3.03 gives are some general points to consider that improve customer relations.

DO	DON'T
<ul><li>be honest</li><li>be neat and tidy in your personal appearance, and look after your personal hygiene</li><li>learn how to put people at ease, and be pleasant and cheerful</li><li>show enthusiasm for the job</li><li>try to maintain friendly relationships with customers, but don't get overfamiliar</li><li>know your job and do it well – good knowledge of the installation and keeping to relevant standards gives the customer confidence in you and your company</li><li>explain what you are going to do, and how long it will take</li><li>if you are not sure about something – ask!</li></ul>	<ul><li>'bad-mouth' your employer</li><li>use company property and materials to do favours for others</li><li>speak for your employer when you have no authority to do so</li><li>use bad language</li><li>smoke on customer premises</li><li>gossip about the customer or anyone else</li><li>tell lies – the customer will find out eventually if they are being misled or ripped off</li><li>assume that you know what your employer wants without bothering to ask</li></ul>
When working in someone's home or office:	
<ul><li>ensure you protect their property. Use dust covers, and ask them to remove objects that might get damaged</li><li>make sure that you understand exactly what the client expects</li><li>if you have recommendations for improvements or alterations, take time to discuss these with the client and explain any technical information</li></ul>	<ul><li>work with pets or small children around – ask for them to be kept well away from the working area</li><li>use hazardous substances without informing the customer. Take the correct precautions and respect any instructions you get from the customer</li></ul>

Table 3.03 Some dos and don'ts to improve customer relations

Always try to provide customers with answers to any questions they have about your work. You might be asked:

- Is this the right product for the job?
- Will it cost a lot to buy and install?
- Will it do what I need?
- How reliable is it?
- Will I be able to use it?
- How easy is it to repair?
- How long does the guarantee last?
- Will you be finished on time?

Before answering the question, try to understand why the customer is asking it: what do they really want to know? Are they worried that they cannot afford the installation? Have they booked a holiday that starts just after you are scheduled to finish?

If you don't know the answer, don't guess. Promise to find out – and then do so! When the job is completed, make sure the client fully understands how to use the installation, and leave behind any manufacturers' user guides or installation manuals. Invite the customer to contact you if there are any problems in the future.

If you follow this advice you will have excellent relationships with your customers. In the unlikely event that a dispute arises, you will both need to seek the help of an independent mediator to settle your differences.

> **Remember**
>
> When you talk with customers you are representing not just yourself, but also your company. People will judge the company by the way you behave. If you do well, your company could get more work from the client; on the other hand, if you don't, contracts may be lost and you may be out of a job.

> **Working life**
>
> Ellie is an apprentice electrician who always cleans up any mess she has made when working on site. She is working with Joe, the senior electrician. While installing an extra socket outlet in an old people's home, Joe has left a mess in the kitchen. He has also burnt the kitchen work surface. Ellie is tasked with clearing up after the work has finished.
>
> How could this situation be avoided? You will need to think about the different working practices that may not have been used here. What steps could Ellie take to explain the situation to the client? What should happen to Joe when he gets back to the company premises? Remember, Joe and Ellie are both representing the company while they work and when they communicate with the client.

Methods of providing technical and functional information

Earlier in this unit we looked at some the different types of information, such as project specifications and site plans (see pages 165–6). There are many sources of such information, such as consulting engineers, manufacturers, suppliers and libraries.

Later in this unit we will look in more detail at the importance of ensuring that information is accurate and complete and that it is communicated clearly with copies kept. Here we will look at the methods used to communicate and store information.

There are two main methods of transmitting information: verbal and written communication.

Verbal information

Information requested verbally can be received either face to face or by telephone. The person requesting the information is usually looking to gain an instant response to an immediate need, irrespective of the importance of that need. You do this every day with other people in normal conversation.

There are some disadvantages with verbal communication. It makes the assumption that people have correctly heard what is being said and that they then understand it correctly.

In construction, not understanding something could prove to be very costly. This means that, whenever there are contractual responsibilities such as with installation projects, people tend not to rely on verbal communication. Instead they seek written confirmation of what has been agreed, leaving a permanent record that can be referred to later if necessary.

Written information

There are several ways of viewing, sharing or storing written information, and different methods are best suited to different purposes. Here are some of the most common.

Printed materials

This is still the most common form of storage for general information: for example, books, newspapers, leaflets, data sheets and catalogues. Their major advantage is that they need no special equipment for reading. Increasingly, many of these items are also available in the other formats described in this unit. This is particularly true of technical information, where electronic methods have the advantage of lower costs and increased delivery speed.

Did you know?

If you are involved in a dispute, it is vital to have a written statement of what was agreed at the start of the work. This will be your 'insurance', demonstrating that you have carried out what was requested.

Remember

Electronic methods are cheaper and faster than ordinary printed methods.

Fax machines

A fax machine enables written or printed material to be sent to another fax machine over a normal telephone line. Email has greatly reduced the use of fax but it is still useful for sending handwritten documents and printed text without a computer and is quite fast. Many computers and 'all in one' printers now include the software to fully replicate the system, so standalone fax machines are less frequently used.

Microforms

Microforms (also known as microfiches) store large amounts of information such as text, documents and photographs using photographic techniques. Their main advantage is long life, and therefore they are popular for storing information that needs to be preserved for many years or even centuries, for example in libraries or newspaper offices.

Figure 3.14 A microform reader

Microforms are based on polyester film, and have a life expectancy of about 500 years when stored correctly. Microfilm is a length of 16 mm or 25 mm film; microfiche is a larger piece of photographic film (around 150 mm × 100 mm).

Both formats store information as tiny pages, and each type can hold hundreds of pages on one microform. Special optical reading devices are used to read the images. They have a temperature-controlled light source and a magnifying device to show the images on a screen similar to a computer display.

CD/DVD/USB device

CDs and DVDs are among the most popular storage media today. Information – text, images, film, music and documents – can be encoded into a digital format and stored on the disc. The data can be read using a computer with suitable drive and software, or on a stand-alone device. The discs are low-cost, and any information that exists as an electronic file can easily be transferred to a disc.

Perhaps even more popular now is the use of USB memory sticks or similar devices. These can store data in a digital format and are physically very small, typically about 50 mm long. They can be used on computers without a CD/DVD drive and can therefore easily transfer information between machines.

> **Did you know?**
>
> Videotape is a polyester film that stores television signals magnetically. The main format is VHS. Video recorders are not now widely available as the technology has been replaced by DVD and Blu-Ray, but there are still many of them out there.

Figure 3.15 USB Storage device

Table 3.04 shows the storage capacity of these formats.

CD	700 MB
DVD	4.7 GB (Single layer)
USB	16 GB

Table 3.04 Storage capacity of media devices

Information can be encoded using a number of different formats, each having certain advantages for different applications. Some of these are more common than others. To read the information, the reading device must be equipped with software to handle it. Some typical file types (denoted by their normal file-name suffix) are shown in Table 3.05.

Images	jpg, bmp, tiff, pcx, gif, png
Movies	wmv, mpg, mov
Text	txt
Documents	doc, pdf
Music	wav, mp3

Table 3.05 File types

Electronic file servers

The storage methods described so far need the reader to have their own personal copy of the information, whether it is a book, CD or video; in effect, there is one copy per reader (although one copy may be shared among several readers, similar to a library book).

The increasing use of electronic networking to connect computers together means that a single copy of information can be stored and accessed by many users at the same time.

Figure 3.16 Electronic file server

Key term

Server – a computer being used only to store information

The information is held on a file **server**. In the case of a home or small company network, the file server could be similar in size to a PC and located in the premises; in the case of large organisations, a file server could be the size of a room and located anywhere in the world, provided that it is connected to the network.

The Internet is the ultimate example of this. The information is stored in digital format, using the same formats as a CD or DVD. Access to the information can be limited to certain individuals or groups, or may be open to anyone.

The Internet

The Internet, sometimes called simply 'the Net', is a worldwide system of computer networks. In practice, the Internet is really a network of networks, accessible to millions of people across the world, giving access to information from other computers. The Internet is used to transmit voice, radio and video as well as data, particularly through email. More recently, Internet telephony hardware and software has made it possible to use the Internet for normal voice conversations and video conferencing.

Using the World Wide Web (often abbreviated as 'www', or simply 'the Web'), you have access to billions of pages of information.

Did you know?

Web pages are coded using HyperText MarkUp Language (HTML). This enables the same source document to be read using any compatible browser, although the appearance of a particular web page may vary slightly.

Web browsing is done using web-browser software, of which Internet Explorer®, Firefox®, Chrome® and Safari® are currently the most popular. Most companies now have websites that allow you to browse their products, services or technical information.

Email

Email is now one of the most popular methods of communicating, in the business world and elsewhere. You can transmit simple and complex messages to almost any country in the world in seconds, and you can also attach other files, such as documents, images and movies.

To send and receive email messages you generally need a computer running a software application called a 'mail client', such as Outlook®, a connection to the Internet (via modem or broadband) and an account with an Internet service provider (ISP).

Printers

Sometimes we need to have a 'hard copy' of the information provided by the information storage. Many manufacturers have now combined the roles of a printer, scanner, copier and fax into what is now commonly referred to as an 'all in one' printer.

The two most common technologies used for printing are inkjet and laser. Inkjet printers use ink cartridges and can produce excellent results, depending on the paper used. If the machine is not used for some time, the ink may dry and clog the cartridge, giving a poor image. Laser printers produce the best images, but are generally more expensive to buy.

Post

Post is a method for transmitting information in envelopes or parcels, through a postal service and then delivered to destinations around the world.

Sometimes and perhaps unkindly referred to as 'snail mail' because of the delay between dispatch of a letter and its receipt (as opposed to the 'immediate' dispatch and delivery of its electronic equivalent, email), some may view this system as outdated. However, it definitely has its uses.

For example, it is normal for manufacturers to despatch small parts directly to an electrical contractor using a parcel delivered by post. Equally in terms of authenticity, it is perceived that an actual document with an actual signature currently holds more sway than an electronic document and electronic signature.

Whatever method is used to provide information, you should always ensure that it is clear, complete, accurate and provided in a professional and courteous way. Always keep a copy for your own records.

Working life

A client is visiting site and advises you a piece of equipment is of the wrong type and has been installed in the wrong place. You are convinced this is not the case and state you will report this to your supervisor, who will check and come back to the client with the answers.

- What documentation could be referenced to check the client's claims?
- How should the information be presented to the client?
- What should be done with the answers, following discussion with the client?

Progress check

1. During a progress check with your supervisor, they advise you to consider joining the union and looking at the grading requirements for electricians. Where would you make those enquiries?

2. What are the main functions of a design engineer?

3. On completion of a contract, it is normal to have a handover meeting with a client. What are the usual areas to be covered?

4. Identify four dos and don'ts, relating to good customer relations, that all employees on site should be aware of particularly if visitors appear on site.

K3. Understand the requirements for overseeing health and safety in the work environment

Everyone on site has health and safety responsibilities. Always check working conditions are safe before work begins. Ensuring the proposed work will not put others at risk requires planning and organisation.

The procedures and requirements for health and safety have been covered in Unit ELTK 01 on pages 1–118. This section will give a brief overview of the key points relevant to this unit. The key to achieving safe working conditions is to ensure health and safety issues are planned, organised, controlled, monitored and reviewed.

Remember that planning has to consider changes to the site as it develops – from welfare arrangements at the set-up, through installation to **snagging** work and the dismantling of site cabins at the end of the contract. The person supervising the work must be adequately trained, experienced and able to manage health and safety procedures on site. The same applies to your staff on site and the supervisor should be aware of any training workers have received and their experience of safe working practice before setting them to work.

Risk assessment and method statements aim to prevent some of the risk from hazards. Risk assessments were covered in depth on pages 40–47. Before work starts:

- consider if there are any hazards you can avoid altogether
- decide which risks need to be controlled along with the best ways of controlling them
- having decided what needs to be done, make sure it happens
- ensure everyone is properly trained and competent, and that they have the equipment they need
- make sure that agreed work methods are put into practice.

Once work has started, continually monitor all situations as circumstances can change quickly during an installation.

Keep all storage areas secure and tidy, whether in an agreed storage area or on the site itself. Try to plan deliveries to keep the amount of materials on site to a minimum. See page 84 for more information on safe storage.

Key term

Snagging – a list of omissions, normally prepared by the Consulting Engineer, that require correction before an installation can be classed as complete

Remember

Always be aware that the activities of other trades could have a significant impact on your health and safety.

K4. Understand the requirements for liaising with others when organising and overseeing work activities

Successful projects require good teamwork. But who is in the team?

In one sense, everyone working on the project is in the team. However, it is more likely that each individual contractor sees their own group of people as being the team, and that their team will be working with the other trades' teams to complete the project.

Before building a team, it is vital that you are familiar with the best methods for communicating with the people you work with. Communication is a vital part of all teamwork and poor communication can have a serious impact on the team.

Promoting good relationships with fellow workers

There are generally three ways in which people interact with each other:

- competing against each other to see who is best
- working as individuals toward a target without paying attention to others
- working as a team towards common targets paying attention to others in the team.

During an installation it is possible all three ways of interacting are in operation – individuals may be paid bonuses for their individual performance, but the whole team may get paid bonus for overall project completion.

A site where everyone works together to finish the job is much happier and more productive than one where people are at loggerheads. Figure 3.17 shows a checklist of the things you can do to help make things run smoothly.

✔ co-operate with other trades – it's always better than conflict

✔ be patient and tolerant with others

✔ attend site meetings regularly – this helps liaison with other trades

✔ keep to the agreed work programme

✔ do your work in a professional manner

✔ finish your work on time; don't hold others up if you can help it

✔ respond cheerfully to reasonable requests from colleagues

✔ don't leave the site for long periods of time

✔ don't borrow tools and materials unless it is necessary, and return them promptly and in good condition if you do

✔ tell your employer if you have personal or work difficulties – don't be too proud

✔ take good care of your and others' property

✔ keep noise down, especially from your radio

✔ show respect for everyone on site – make an effort to learn their names

✔ make sure everyone, including visitors to the site, has the right PPE

✔ report any breakdown in discipline or disputes between co-contractors promptly to the site supervisor

✔ keep a current edition of the Wiring Regulations or the Amicus guide book with you on site

✔ always do your best to answer questions from visitors or other tradespeople

✔ never play practical jokes on colleagues (for example hiding tools, lunch boxes, car keys). This can cause bad feeling and may result in injury or accident.

Figure 3.17 Checklist for good relationships with fellow workers

Communicating with a purpose

When you communicate with someone, you usually have a clear purpose in mind. Here you will look at four of the main reasons you will have to communicate as an electrician.

Motivation

Motivation can be described as happening when common aims, purposes and values are shared between individuals and teams. However, motivation will often take a different form for different people. The case study gives one example.

Case study

A BBC radio journalist once interviewed a worker in a biscuit factory.

Interviewer: How long have you worked here?
Worker: Since I left school, oh, about 15 years ago.
Interviewer: What do you do?
Worker: I take packets of biscuits off the conveyor belt and put them into cardboard boxes.
Interviewer: Have you always done the same job?
Worker: Yes.
Interviewer: Do you enjoy it?
Worker: Oh yes, it's great; everyone is so nice and friendly and we have a good laugh.
Interviewer: Really? Don't you find it a bit boring?
Worker: Oh gosh no, sometimes they change the biscuits ...

The point is that you should never assume that things that motivate one person will motivate someone else. This means that motivational methods need to be varied too, from inspirational speeches, quotes and poems through to team-building games and activities. Team workshops and meetings can also prove motivational.

Playing 'games' can enable people to experience achievement in a new way, and that experience of success tends to lead to higher motivation. This is one reason why Outward Bound courses and paintballing are successful, as they create an environment that allows people to achieve outside of their normal work environment, as individuals and as teams.

When people play games, socialise or compete in teams they learn about each other, communicate and see each other from a different perspective. Mutual respect can grow out of these activities.

Role-play exercises do not have the best reputation but, used correctly, they can be another highly effective motivational tool. Role play can aid instruction by getting people to be practically involved, and can encourage co-operation, as those taking part get to see situations and issues from perspectives other than their own.

Instruction

There are various methods available for instructing. To choose the right one, you will need to consider what the instruction is for and the learning style of the person being instructed.

Did you know?

People also often enjoy non-work related activities, especially if managers are seen to take part alongside everyone else.

Remember

Strong, positive images also encourage motivation, confidence and belief.

There are many different learning styles, but for this industry most learners could be said to be:

- **visual** – learn through seeing (thinking in pictures, visual aids such as slides, diagrams, hand-outs, etc.)
- **auditory** – learn through listening (lectures, discussions, tapes, etc.)
- **reading** – learn through reading and writing
- **kinaesthetic or tactile** – learn through experience, moving, touching and doing (active exploration of the workplace, projects, experiments, etc.)

Some of the most common instruction methods you will come across are:

- **tutor-led** – tutor presenting information and interacting by frequent questioning and periodic summaries or logical points of development
- **demonstration** – observation of a procedure, technique, or operation, which shows you how to do something or how something works
- **practice** – repeated performance of previously learned actions, sequences, operations, or procedures
- **independent** – independently learning and practising, perhaps at home, with advice on offer if required.

To cater for the mixture of learning styles, instructors should try to use a range of learning methods.

Monitoring

All learning needs to be monitored, in order to:

- provide an effective means of measuring the progress of an individual or project
- establish levels of skill and understanding
- regularly assess the achievement of technical, financial and economic goals
- determine whether any corrective actions or training are required
- assist in defining new or modified performance techniques or measures.

Here are some key monitoring techniques.

- **Direct observation** An observer watches the real-time performance of an individual as they undertake work or

> **Remember**
>
> Many people believe in the old phrase 'practice makes perfect'.

visually inspects overall project progress. This can establish if an individual has a training need or the project needs a process changing. It can be very effective: the personal approach is often appreciated by the individual.

- **Written examination** An individual's knowledge is checked via an independent examination. Many courses, such as the 17th Edition Wiring Regulations, are assessed like this.

- **Interview** One or more people are interviewed to establish an outcome. This could be to check individual knowledge or receive verbal updates to establish progress.

- **Reports or other written documents** These are used to collect information from a variety of sources about specific circumstances or project progress.

To achieve the overall aim of a successfully completed installation, it is an advantage if staff understand, without suspicion, why monitoring is taking place. When this understanding exists, self-monitoring by staff can be introduced. This is very useful to achieving targets.

Co-operation

The aim of motivating, instructing and monitoring is to help bring about co-operation between individuals and, as a result, aid the development of the team.

Team development

The Form-Storm-Norm-Perform model gives useful insights into how any team develops, whether the team is just the squad of electricians, or everyone working on the site. As a team develops, relationships between team members shift, and the team leader must change their leadership style.

Find out

As you read through this team development model, try to apply it to teams you have been involved with. It could be your team at work, but could also be in your social life – a gang of mates, perhaps, or a sports team. What stages did the team go through? How do you think the team may develop in the future?

The four stages

The four stages of team development

1. Forming
This is the first stage, when the team has just come together. The members probably don't know one another very well, and individual roles and responsibilities are unclear. There is little agreement between members about what the team is trying to do. Some will feel confused and won't know what they should be doing. At this stage the team relies heavily on the team leader for guidance and direction. The leader must provide lots of answers about the team's purpose and objectives and relationships with groups outside the team.

2. Storming
During this stage, team members jockey for position as they try to find themselves a role in the team. The leader might receive challenges to their authority from other team members. The team's purpose becomes clearer, but there remains plenty of uncertainty, and decisions are hard to achieve because members will argue a lot. Small groups or factions may form, and there may be power struggles. The team needs to be focused on its goals to avoid being broken up by relationship and emotional issues. Some compromises will be needed to make any progress. The leader has to become less bossy and more of a coach.

3. Norming
This is a more peaceful stage, when team members generally reach agreements easily. Roles and responsibilities are clear and accepted by all, and the team works together. Members develop ways and styles of working by discussion and agreement together. Big decisions are made by the whole team or the team leader, but smaller decisions are left to individuals or small groups inside the team. The team may also enjoy fun and social activities together. The leader now acts to guide the team gently and enable it to do its job, and has no need to enforce decisions. The team may share some leadership roles.

4. Performing
In this final stage, the team knows clearly what it is doing and why. It has a shared vision and needs little or no input from the leader. If disagreements occur they are tackled positively by the team itself. The team works together towards achieving the goal, and also copes with relationship, style and process issues along the way. Team members look after each other. The leader's role is to delegate and oversee tasks and projects, and there is no need for instruction or assistance, except for individual personal development.

Remember

If the installation team is to succeed, then everyone must be of the correct ability level to handle the tasks that they are asked to do.

Determining the competence of operatives

On page 184 you saw how important it is that the person supervising work is adequately trained and experienced and can manage the project. To do so they must be aware of the competence levels of the project team, including any training they have received and their experience and working practices, before setting them to work.

There are several ways to check the competence of an individual.

Checking competency cards

Competency cards are a practical tool. They generally have individual identification and competency level on one side of a card and the behaviours and levels printed on the opposite side. There are two cards that are widely used in your industry: CSCS and ECS cards.

Construction Skills Certification Scheme (CSCS)

The CSCS was set up to help the construction industry improve quality and reduce accidents. CSCS cards are increasingly demanded as proof of occupational competence by contractors, public and private clients and others, and they cover hundreds of occupations. With a variety of grades available, the objective is to hold the Gold Card – Skilled Worker. People qualify for this card if they achieve an NVQ Level 3 and have passed the Construction Skills health and safety test.

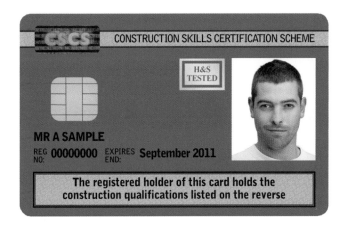

Figure 3.18 CSCS Card

Electrotechnical Certification Scheme (ECS)

It is now almost impossible to gain access to a construction site without proof of identification, your competence and qualification levels. For this reason, the electrotechnical industry combines the requirements of the CSCS card with a sector-specific card known as the ECS card.

ECS is the sole ID and competence card scheme for electrotechnical operatives in the UK. Holding an ECS card means an individual can prove their identity, qualified status and occupation when working on site. It also proves that you have passed a health and safety assessment. The ECS is affiliated to the CSCS and endorsed and supported by UK Part P Competent Person Schemes.

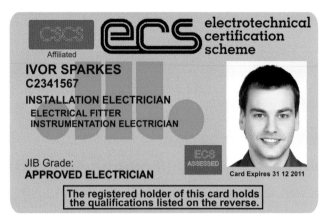

Figure 3.19 ECS Installation Electrician card

To be eligible for an ECS Installation Electrician card, applicants must either:

- have successfully completed an NVQ Level 3 Approved Apprenticeship

or

- hold the full City and Guilds 2360, 2351 or 2330 qualifications

and

- have NVQ Level 3 in Installation and commissioning and;
- hold an up-to-date Health and Safety Certificate or recognised health and safety qualification and;
- hold a formal BS 7671 qualification in the current edition of the wiring regulations.

Checking technical qualifications

It is important to understand what type of technical certificate you are looking at and what you hope to gain from checking it. In essence, there are two types: those that certificate knowledge only and those that certificate knowledge and practical performance.

Working life

In response to an advert for someone to carry out electrical inspection and testing, an electrician approaches a contracting company to seek employment. The electrician presents the following certificates for consideration:

- C&G 2330 Level 2 certificate
- C&G Unit 301 Certificate of Unit Credit
- EAL BS 7671 certificate.

What would you do if you were the manager? You will need to think about what level of practical experience these certificates prove.

In the Working life, the problem is that the certificates the electrician has presented are all knowledge-only certificates, so you have no proof of the person's practical ability. You need to be aware that a Certificate of Unit Credit is not a full certificate – it is recording success in a part of a certificate. The individual has not fully met the industry knowledge requirements.

There may be a number of valid reasons for the above scenario, and the individual may actually have all of the knowledge and practical experience needed for the job. However, as a manager, you would be expected to dig deeper before you could safely make an offer of employment. Here are some other things you might want to check.

Written references

Although the source of the reference needs to be established as trustworthy, a reference from a previous employer can help establish whether someone has relevant experience, as well as their attitude and skill to deal with work.

Monitoring of performance

If an employer remains uncertain of someone's ability, they can employ them for a period of time to assess their suitability. This 'probation period' usually lasts between one and three months and allows an employer to see that an applicant fits in with the company ethos and has the abilities needed for the job, as well as being able to form satisfactory relationships with colleagues. Normally this period will involve trusted members of staff monitoring the applicant's progress and reporting back to management.

As an employee, a probation period requires you to work under modified terms of employment. For example, should you choose to leave the company (or for that matter, if the firm decides to fire you), your notice period will be reduced. Other employers may not allow you your full holiday entitlement until this initial term has elapsed.

Communication

Earlier in this unit, we looked at the different people you may interact with throughout an installation project. The level and type of communication will depend on the person involved and the level of responsibility that you hold. The key people you may need to communicate with include:

- customers
- clients
- site managers
- major and sub-contractors
- other services
- the public.

Accepting our earlier points about not exceeding your level of responsibility, all of the ideas we have looked at so far in this learning outcome involve communication. Look back at page 185 for more information on your likely role in communicating with these groups.

Communication is about much more than just speaking or writing. You communicate an enormous amount by how you look, the gestures and facial expressions you use, and the way you behave. Even something as simple as a smile can make a big difference to your communication.

The benefits of good communication are huge. Good attitude, appearance and behaviour will immediately put your customers at ease and earn you respect from others. Clearly conveyed thoughts and ideas, when backed up with good working practices and procedures, will invariably improve productivity, increase profitability and produce satisfied customers.

Speaking, writing, appearance, attitude and behaviour all combine to make up your personal 'communication package'. It affects your relationships with everyone you deal with – client, client representative, architect, main contractor, other site trades, your supervisor and your work colleagues.

This section will look at some of the methods for communicating with others.

Letter and report writing

Good written communication skills are vital for avoiding confusion, especially in an industry that relies heavily on documentation. Writing good letters and reports is a part of the communication process. Letters can be used to request information (such as delivery dates from a wholesaler, or progress updates from a builder) or inform people of situations.

Reports can be used to:

- summarise investigations into causes and effects of problems or trends and recommend solutions (for example, why there has been an increase in absenteeism, or why a piece of work is below standard)
- provide statistical or financial summaries (such as end-of-week materials costs)
- record decisions made at meetings
- supply information for legal purposes (such as accident reports)
- monitor progress (for example, of negotiations, construction work or the implementation of a new system)
- look into the feasibility of introducing new procedures, processes or products, or changing company policy.

The writing process

Writing isn't too difficult if you break it down into steps and then focus on one step at a time. You can tackle almost any writing project if you follow this method.

Step 1: Have a clear idea about the overall purpose of your letter or report. Think about exactly what you want to say, and why. Most people get stuck because they have not thought carefully about what they are trying to achieve.

Step 2: Gather all the information you will need.

Step 3: Plan the logical order for presenting your ideas. Write a list of headings, and check that each one follows on from the previous one. If it is a report, the first paragraph should be a summary of the rest of the document; if you want your reader to take some action, say so in your last paragraph. Don't start writing in detail until you have completed this step.

> **Remember**
>
> Try to use simple language and don't try to impress the reader with your huge vocabulary – it will only frustrate or annoy them. Don't use jargon or abbreviations unless you are sure your reader is familiar with them.

Step 4: Now that you know what each section or paragraph will contain, go ahead and write each one. Then read it through again and, if necessary, edit what you have written. Finally, present it as a properly formatted and typed document.

Hints and tips for good writing

Avoid 'wordy' and complicated sentences

Read this sentence:

> 'I found out that I should take an investigative look at the new plant room in order to establish a prospective plan to help us re-evaluate the installation methods and techniques that we intend to use for the duration of the task.'

It could have been written as:

> 'I will be checking our proposed installation in the new plant room to see if there is a better way of doing things.'

Much easier to understand, isn't it?

Communication always involves two people: the writer/speaker and the reader/listener. You should think about your reader and write for them, clearly conveying your ideas so that the reader can easily understand them. For example, you wouldn't expect to write the same for a five-year-old as you would for a college lecturer.

Make your writing interesting

Don't let your document become boring and repetitive. Use a **thesaurus** to find alternative words to use.

For example, instead of the word 'wherewithal' in the sentence 'I have the wherewithal to pay the bill', the thesaurus suggests other words that you could use such as 'means', 'resources', or 'ability'.

Check your work before sending it

Everyone makes mistakes. Checking your written work (known as proofreading) is an important aspect of writing. Pay attention to grammar, spelling and punctuation.

If you are using word-processing software to write your letter or report, it will often indicate when something might be wrong, so get into the habit of finding out what the problem is and correcting it. However, always remember that the software cannot detect every mistake.

Key term

Thesaurus – a type of dictionary that lists words with similar meaning

Did you know?

You may find a thesaurus function on your computer.

Remember

If you have any doubts about what you have written, read it out loud to see how it sounds.

For example, the following sentence has several errors, but they won't all show up on the computer screen (try typing it in!).

> Two many electricians were note iced too bee erecting the too extract fans in 2 the to staff toilets.

A dictionary can help: if you aren't sure how to spell a word, look it up. If possible, don't proofread something immediately after you have written it. You are more likely to spot any mistakes and find better ways of saying things if you go back to it later.

Report writing

Reports usually describe a problem or an investigation. Their purpose is to help someone, or a team, to make a decision, by setting out all the relevant facts and perhaps making some recommendations for action. Some reports will present a solution to a problem; others may simply list historical and factual data.

Many companies will have a standard structure or format for reports and you should use this whenever you can. However, a lot depends on the subject and type of the report: there is no merit in slavishly following a format if it makes it difficult to understand what the report is about.

As with any writing, it is important to know clearly what you are trying to say, and then to put that into a style and format that will be acceptable to the reader.

Style, although hard to define, can make a big difference to the success of a report. A good style can help to convince the reader of the merits of the report and its recommendations. A bad style may put the reader off, even if the content is perfectly OK.

Report structure

Companies may have a required format, but they will probably include most or all of the elements described below. For a business report or memo:

- basic identification data (who it is to, who is writing it, the date, subject, reference number)
- summary – the project or problem and the purpose of the report
- background – the history of the issue being reported
- relevant data – the evidence you have gathered
- conclusions and recommendations.

Remember

Always proofread material before sending it out.

Find out

Find some examples of reports and look through them. What features do they have in common? What sort of information do they contain? How are they presented? Who are they aimed at? Use them to help complete your own report about an issue you may have encountered in a work environment.

When writing a more technical report, consider organising it under the following headings, in the order shown:

- Title page
- Acknowledgements
- History of any changes
- Contents list
- Report summary
- Introduction
- Technical chapters
- Conclusions and recommendations
- References
- Appendices (these contain specific, additional information).

Whatever style and format you use in your report, the principles of good writing in Table 3.06 still apply.

Be clear	• The reader must be able to understand the report easily. • Explain symbols, tables or diagrams. • Never assume that your reader has prior knowledge of the issues. • Don't use jargon or abbreviations unless you know that the reader will be familiar with them.
Be concise	• Don't waffle. • Be succinct without making the report hard to understand. • Think about what the reader wants to know, i.e. evidence and conclusions, not a vivid description of the valiant effort you made to compile them.
Be logical	• Ensure you have a beginning, middle and end. • There should be a sensible flow between sections, chapters, paragraphs and sentences. • Avoid jumping randomly from idea to idea. • Take out unnecessary distractions such as pretty graphics that don't add any extra information.
Be accurate and objective	• Base your report on honest facts – an inaccurate report is at best pointless and at worst harmful. • Remember that someone may make an important decision based on your recommendations. • If you have to make assumptions, make it clear that you have done so. • Be tactful. What you say may contradict others or what the reader thinks. • If you feel very strongly about something, don't send the report until you have had some time to think about it, and make sure it contains only facts. • Don't use reports as a sounding-off platform. It is very easy to destroy a relationship by sending an angry, ill-conceived letter or report.

Table 3.06 Basic rules for good report writing

Organisational procedures for completing documentation

Now that you know a bit more about writing, you can look at the various types of documentation in use on a site and how you can apply these communication skills to them.

We have already seen how technical information is recorded on diagrams and drawings, so that everyone involved in a project during installation, and in the future, knows what to do and what has been done. There is plenty of other information people need to communicate to others in the workplace, and various types of documents are used for this. Information about manufacturer's data and service manuals can be found on pages 160–61.

Job sheets

Job sheets give detailed and accurate information about a job to be done. Electrical contracting companies issue them to their electricians. They will include:

- the customer's name and address
- a clear description of the work to be carried out
- any special instructions or special conditions (such as needing to pick up special tools or materials).

Sometimes extra work is done which is not included in the job sheet. In this case it is recorded on a day worksheet so that the customer can be charged for it.

Job Sheet	**Evan Dimmer** Electrical contractors
Customer	Dave Wilkins
Address	2 The Avenue Townsville Droopshire
Work to be carried out	Install 1 x additional 1200 mm fitting to rear of garage
Special conditions/instructions	Exact location to be specified by client

Figure 3.20 A typical job sheet

Find out

When might you need to use a Variation Order?

Variation Orders

A Variation Order (VO) is issued or 'raised' when the work done varies from the original work agreed in the contract or listed in the job sheet. If this situation arises, it is important for the site electrician to tell their supervisor immediately. A VO can then be made out to enable the new work to be done without breaking any of the terms of the contract. The purpose of the VO is to record the agreement of the client (or the consulting engineer representing the client) for the extra work to be done, as well as any alteration that this will make to the cost and completion date of the project.

Day-work sheets

Work done outside the original scope of the contract, perhaps as a result of a VO initiated by the architect, engineer or main contractor, is known as day work. When the work is completed, the electrician or supervisor fills out a day-work sheet and gets a signature of approval from the appropriate client representative. Day work is normally charged at higher rates than the work covered by the main contract, and these charges are usually quoted on the initial tender. Typical day-work charges are:

- labour – normal rates plus 130 per cent
- materials – normal costs plus 25 per cent
- plant – normal rate plus 10 per cent.

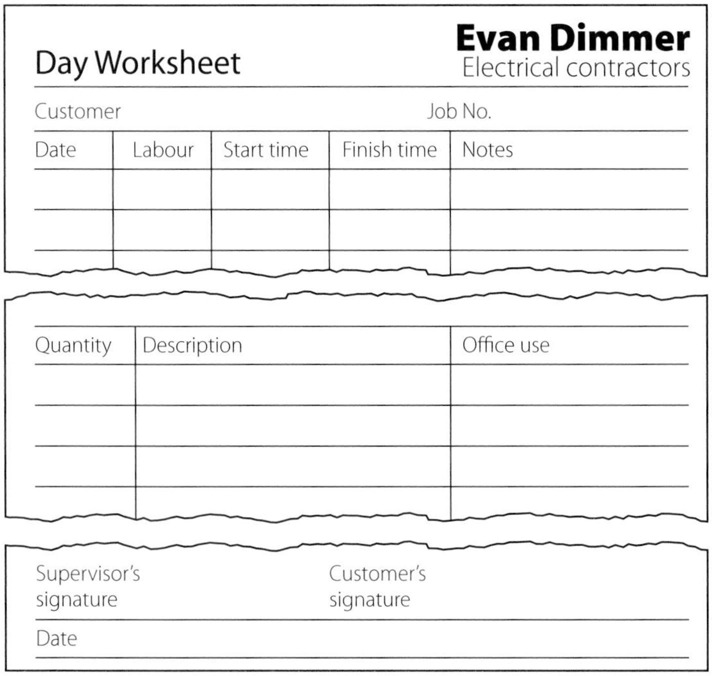

Disputes over day work can easily arise, so it is important that the installation team on site records any extra time, plant and materials used when doing day work.

Time sheets

Time sheets are very important to you and your company, because they are a permanent record of the labour on a site. Time sheets include details of:

- each job
- travelling time
- overtime
- expenses.

Day Worksheet

Evan Dimmer
Electrical contractors

Customer				Job No.
Date	Labour	Start time	Finish time	Notes

Quantity	Description		Office use

Supervisor's signature   Customer's signature

Date

Figure 3.21 A typical day-work sheet

This information allows the company to track its costs on a project and also to make up your wages. If you work on several sites during a week, you may need to fill in a separate time sheet for each job.

Remember

Accurate time sheets are essential to make sure you get paid correctly and that your customer is charged the right amount for the job.

Time Sheet

Evan Dimmer
Electrical contractors

Employee | Project/site

Date	Job No.	Start time	Finish time	Total time	Travel time	Expenses
Mon						
Tue						
Wed						
Thu						
Fri						
Sat						
Sun						
Totals						

Employee's signature | Supervisor's signature

Date

Figure 3.22 A typical time sheet

Purchase orders

Before a supplier will dispatch any materials or equipment, they will require a written purchase order. This will include details of the material, the quantity required, and sometimes the manufacturer; it may also specify a delivery date and place. In many cases the initial order is made on the telephone or via email or the Internet, and a written confirmation is sent immediately afterwards. The company keeps a copy of the original order in case there are any problems.

Usually the purchasing department sends out these orders, but sometimes an order is raised directly from site when there is a need for immediate action.

Delivery notes

Delivery notes are usually forms with several copies that record the delivery of materials and equipment to the site. All materials delivered by a third party directly to the site will arrive with a delivery note. As the company representative on site, this is the form you are most likely to deal with.

The delivery note should give the following information:

- the name of the supplier
- to whom the materials are being sent
- a list of the type, quantity and description of materials that are being delivered to the site in this particular load
- the time period allowed for claims for damage.

When materials arrive on site you should make sure that they are unloaded and stored correctly and should check each item against the delivery note and for obvious signs of damage.

If everything is OK, sign the note. If not, note any missing or rejected items on the delivery note, then you and the delivery driver should both sign it. Store your copy of the note safely.

Check the materials thoroughly for damage within the time given stated in the delivery note (usually three days), and inform the supplier immediately if there are any problems.

A delivery won't always contain all the materials listed in the purchase order. Sometimes the material is not all needed on site at the same time and it is delivered in several loads. This helps to reduce the need for on-site storage and minimises the risk of damage or loss.

Incomplete deliveries may also occur if the supplier is out of stock, or if some of the order is coming direct from the manufacturer. Linked to the delivery note, a completion order records the fact that all the material on an original purchase order has been delivered.

Delivery note		**A. POWERS** *Electrical wholesalers*
Order No.		Date
Delivery address 2 The Avenue Townsville Droopshire		Invoice address Evan Dimmer Electrical Contractors
Description	Quantity	Catalogue No.
Thorn PP 1200 mm fit fitting	1	
1.5 mm T/E cable	50 m	
Comments		
Date and time of receiving goods		
Name of recipient		Signed

Figure 3.23 A typical delivery note

Site reports, memos and minutes of meetings

The site foreman, supervisor or engineer in charge usually compiles reports for companies. Site reports contain details of work progress, defects, problems and delays. Sometimes other reports will be made about specific problems or incidents.

A memo is usually a short document sent to a relevant person about a single issue: for example, a problem installing a piece of equipment, or materials not being delivered on time.

Although perhaps more to do with the site engineer or supervisor, no installation project will be complete without a site meeting. Normally chaired by the main contractor, with representatives from the consultants and all related contractors present, these meetings tend to try to establish whether a project is progressing as hoped, as well as attempting to solve any identified problems.

INTERNAL MEMO

To: D Boss, Contracts Engineer

From: A Foreman, Site Supervisor

Project: The New Hospital

Following the arrival of the new essential services generator on site, we have found that it is too big for the entrance to the existing generator house. I have spoken to the Main Contractor, and we believe that a section of roof could be removed easily and the generator craned into position.

Please advise.

A Foreman

Figure 3.24 Memo

A record of each meeting is then sent to all relevant parties. These 'minutes' should be checked for accuracy and retained, as often there will be actions for you to comply with. Should any difficulty arise as a project progresses, these minutes can prove a useful tool to establish the situation at a given moment in time.

Progress check

1   Your supervisor needs to complete a risk assessment before you start work on site. What should they consider including in the risk assessment?

2   Identify the four key principles in building a team.

3   Identify the key people you may need to communicate with during work on site.

4   As materials begin to arrive on site you are tasked to check them into the storage area established on site. What should you check?

K5. Understand the requirements for organising and overseeing work programmes

To understand work programmes, you first need to understand the sequence of events that lead to their preparation.

How a work programme is developed

As a contractor on site, your presence will be the result of your employer having successfully tendered for the work. This will have meant your company estimator working with documentation, drawings and specifications.

In our industry, the estimator uses the layout drawings and specification provided by the consulting engineer during the tendering stage. The electrician will also need the same layout drawings during the installation stage – and there might be hundreds of them, especially if it is a large project like a school or hospital.

The next section looks at layout drawings and how to interpret them.

Layout drawings and their interpretation

Layout and assembly drawings give information about physical objects, such as the floor layout in a building, or a mechanical object.

If you were to make the drawing the same size as the object, most drawings would obviously be too big to handle. To make the drawing a sensible size, **scaled drawings** are used.

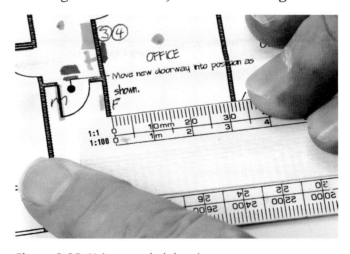

Figure 3.25 Using a scaled drawing

In these drawings a ratio scale is used: for every unit of measurement on the real object, a specific, much smaller measurement is used on the drawing.

For example, on most construction projects, the scale used to show the floor layout of a building is 1:100. So 10 mm on the drawing represents something that is 100 times bigger in reality (on this scale 1 metre).

1 metre = 1000 mm, so 1 metre is represented as 1000/100 mm = 10 mm.

The ratio scale makes the drawings easy to use. To find a measurement on the actual object, you measure the distance on the drawing and multiply it by the scale. Going back to the example earlier, you would measure 10 mm on the drawing, multiply by 100, and get 1000 mm – so you would know that the real measurement was 1 metre.

It doesn't matter what unit of measurement you choose, because you are simply going to multiply it by a number (the scale).

The drawing scale is chosen to make the drawing a reasonable size, according to its purpose. Although a scale of 1:100 may be fine for the layout of a building, it would be impractical for a road map, because you would only be able to get a few kilometres on each sheet. A scale of 1:500,000 (1 cm = 5 km) would be better.

In the same way, an assembly drawing for a wristwatch would be too small to read if we used 1:100; a better scale might be 20:1 (20 mm on the drawing represents 1 mm on the actual watch).

In using the scaled layout drawings to produce the successful tender, the estimator is taking information from the specification and drawings, planning an electrical installation in line with those requirements and then measuring what materials are involved and establishing the likely time taken to install them.

Did you know?

You may have built model aeroplanes from a kit. Quite often these are described as 1:32 scale: in other words, every part in the model is 32 times smaller than the real thing.

Remember

As well as compiling a materials list, you will need to work out what wiring is required to control the installation and feed the lighting distribution board.

Case study

Look again at the university warehouse you were working on earlier in this unit (see pages 166–7). The layout drawing shows the layout of the building, the location of specific services and how some items are to be installed and connected. As it is to scale (1:50), you can measure it to find the actual dimensions of the building and prepare a materials list for the job. You can also scale up positions shown on the drawing and mark them for real inside the building itself. However, what the drawing doesn't tell you is anything about the fitments or how they should be installed. This information will be in the specification, so you'll need a copy of that too. Opposite is an extract from the specification, which tells us what we need to know.

Step 1: Count up all the major pieces of equipment needed

Looking at the plan and referring to the specification, we can see that we need:

- 12 Tamlite TM58 fittings, plus tubes
- 1 MK Metalclad Plus 4-Gang one-way surface switch with aluminium front plate
- 1 MEM three-way SPN surface-mounted DB (distribution board).

Step 2: Decide the best runs of conduit and trunking

A logical way would be to install the 50 x 50 mm trunking as shown on the drawing and then install separate conduits from the trunking up to each row of lights and then along each row of lights, fixing the conduit to the roof structure.

Step 3: Calculate lengths of trunking, conduit and cable required

You measure these from the drawing and calculate the actual distance using the scale provided (1:50). You will also need to know the height of the building, which is not shown on the layout diagram.

The warehouse lighting and sub-main installation will be completed in PVC single-core cable contained within galvanised steel conduit and galvanised steel trunking where required. All fluorescent lighting fittings will be 1700 mm x 58 W of type Tamlite TM58 and suspended on chain from back boxes. All light switches will be of MK type Metalclad Plus with an aluminium front plate. The installation will be fed from the new lighting distribution board (MEM), which will be fed from the existing MCB DB located in the warehouse.

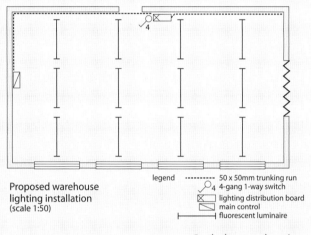

Proposed warehouse lighting installation (scale 1:50)

legend ---------- 50 x 50mm trunking run
4-gang 1-way switch
lighting distribution board
main control
fluorescent luminaire

Scale layout drawing

Step 4: Include accessories, fixings, etc.

You will need to allow back boxes and hook-and-chain arrangements at each luminaire, and to order sufficient fixings (screws, bolts, saddles for conduit, etc.) of various types for the installation.

Step 5: Consider special access equipment

Do you need special ladders, scaffolding, etc.? It's no use having the materials if you can't get to the right places.

Armed with this information the company now has an idea of the basic resources that will be needed to complete the electrical installation – but there is more to it than that.

The estimator will also have to consider the regulatory requirements that affect the planned installation. Legislation such as the HASAWA, CDM and the EAWR all need to be considered, along with BS 7671, if the installation is to be completed safely as well as being electrically and mechanically sound.

Looking the warehouse example, you know you will be installing fluorescent lighting fittings at high level at, therefore a mobile scaffold tower looks like the best option for this. However, apart from the costs involved, you need to consider access arrangements, storage facilities, the ability of your staff to use the scaffold and whether you can use it when there are other trades working on the project, possibly in the same area at the same time.

Industry standards relevant to installation of electrotechnical systems and equipment

There is a great deal of legislation that employers need to be familiar with in order to successfully plan and run their businesses. Some of these have been covered earlier in Unit ELTK 01.

Within England and Wales, the law regarding employment both protects and imposes obligations on employees, during their employment and after it ends. The law sets certain minimum rights and an employer cannot give you less than what the law stipulates. The principal rights and obligations imposed on employers and employees arise from three sources:

- common law, which governs any contract of employment between employer and employee and includes the body of law created by historical practice and decisions
- UK legislation
- European legislation and judgements from the European Court of Justice (ECJ).

UK employment law has been heavily influenced by European law, particularly in the areas of equal pay and equal treatment; many of our statutory minimum rights began their life in European legislation. Here are just some of the most relevant pieces of legislation.

The Employment Rights Act 1996, Employment Acts 2002 & 2008

Subject to certain qualifications, employees have a number of statutory minimum rights (such as the right to a minimum wage). The main vehicle for employment legislation is the Employment Rights Act 1996 – Chapter 18. If you did not agree certain matters at the time of commencing employment, your legal rights will apply automatically. The Employment Rights Act 1996 deals with many matters such as:

- right to statement of employment
- right to pay statement
- minimum pay

- minimum holidays
- maximum working hours
- right to maternity/paternity leave.

The Employment Act 2002 amended the 1996 Act to make provision for statutory rights to paternity and adoption leave and pay. The Employment Act 2008 makes provision for the resolutions of employment dispute including compensation for financial loss, enforcement of minimum wage and of offences under the Employment Agencies Act 1973 and the right of Trade Unions to expel members due to membership of political parties.

The Sex Discrimination Act

The Sex Discrimination Act 1975 makes discrimination unlawful on the grounds of sex and marital status and, to a certain degree, gender reassignment. The Act originated out of the Equal Treatment Directive, which made provisions for equality between men and women in terms of access to employment, vocational training, promotion and other terms and conditions of work.

The Equal Opportunities Commission (EOC) has since published a Code of Practice. While this is not a legally binding document, it does give guidance on best practice in the promotion of equality of opportunity in employment, and failure to follow it may be taken into account by the courts.

The Sex Discrimination Act was amended to ensure compliance with the Equal Treatment Directive, with all changes being effective from April 2008. The definition of harassment is extended so that if, for example, a male supervisor makes disparaging comments about women, it is no longer a defence to show that he makes similar comments about men. In addition if someone witnesses sexual harassment of a colleague, they can bring a claim of harassment themselves if they felt it made their work environment intimidating.

Employer liability has also been extended to make organisations liable if they haven't taken reasonable steps to prevent harassment by a third party such as a visitor or customer.

Employment Relations Act 1999 & 2004

The 1999 Act is based on the measures proposed in the White Paper: Fairness at Work (1998), which was part of the Government's programme to replace the notion of conflict between employers and employees with the promotion of partnership.

As such it comprises changes to the law on trade union membership, to prevent discrimination by omission and the blacklisting of people on grounds of trade union membership or activities; new rights and changes in family-related employment rights, aimed at making it easier for workers to balance the demands of work and the family and a new right for workers to be accompanied in certain disciplinary and grievance hearings.

The Employment Relations Act 2004 is mainly concerned with collective labour law and trade union rights. It implements the findings of the review of the Employment Relations Act 1999, announced by the Secretary of State in July 2002, with measures to tackle the intimidation of workers during recognition and de-recognition ballots and provisions to increase the protections against the dismissal of employees taking official, lawfully organised industrial action.

The Human Rights Act 1998

The Human Rights Act 1998 covers many different types of discrimination – including some not covered by other discrimination laws. However, it can be used only when one of the other 'articles' (the specific principles) of the Act applies, such as the right to 'respect for private and family life'.

Rights under this Act can only be used against a public authority (such as the police or a local council) and not a private company. However, court decisions on discrimination will generally have to take into account what the Human Rights Act says.

The main articles within this Act are: right to life, prohibition of torture, prohibition of slavery and forced labour, right to liberty and security, right to a fair trial, no punishment without law, right to respect for private and family life, freedom of thought, conscience and religion, freedom of expression, freedom of assembly and association, right to marry, prohibition of discrimination, restrictions on political activity of aliens, prohibition of abuse of rights, limitation on use of restrictions on rights.

The Race Relations Act 1976 and Amendment Act 2000

When originally passed, the Race Relations Act 1976 made it unlawful to discriminate on racial grounds in relation to employment, training and education, the provision of goods, facilities and services, and certain other specified activities. The 1976 Act applied to race discrimination by public authorities in these areas, but not all functions of public authorities were covered.

The 1976 Act also made employers vicariously (explicitly) liable for acts of race discrimination committed by their employees in the course of their employment, subject to a defence that the employer took all reasonable steps to prevent the employee discriminating.

The Commission for Racial Equality (CRE) proposed that the Act should be extended to all public services and that vicarious liability should be extended to the police. The main purposes of the 2000 Act were to:

- extend further the 1976 Act in relation to public authorities, thus outlawing race discrimination in functions not previously covered
- place a duty on specified public authorities to work towards the elimination of unlawful discrimination and promote equality of opportunity and good relations between persons of different racial groups
- make Chief Officers of police vicariously liable for acts of race discrimination by police officers
- amend the exemption under the 1976 Act for acts done for the purposes of safeguarding national security.

The Race Relations Act 1976 (Amendment) Regulations 2003

The Race Relations (Amendment) Regulations 2003 modify the Race Relations Act 1976.

- Indirect discrimination on grounds of race, ethnic origin or national origin is extended to cover informal as well as formal practices.
- The concept of a 'Genuine Occupational Requirement' is introduced for situations where having a particular ethnic or national origin is a genuine requirement for the employment in question.
- The definition of discriminatory practices is extended to cover those who put particular groups at a disadvantage, rather than only those where there is proof that a disadvantage has been experienced.
- The Act is extended to give protection even after a relationship (such as employment in an organisation, or tenancy under a landlord) has finished.
- The burden of proof is shifted, meaning an alleged discriminator (such as an employer or landlord) has to prove that he or she did not commit unlawful discrimination once an initial case is made.

Equality Act 2010

From 1 October 2010, the Equality Act replaced most of the Disability Discrimination Act (DDA). However, the Disability Equality Duty in the DDA continues to apply. The Equality Act 2010 aims to protect disabled people and prevent disability discrimination. It provides legal rights for disabled people in the areas of:

- employment
- education
- access to goods, services and facilities including larger private clubs and land-based transport services
- buying and renting land or property
- functions of public bodies, such as the issuing of licences.

The Equality Act also provides rights for people not to be directly discriminated against or harassed because they have an association with a disabled person. This can apply to a carer or parent of a disabled person. Also people must not be directly discriminated against or harassed because they are wrongly perceived to be disabled.

Protection from Harassment Act (PHA) 1997

Harassment is defined as any form of unwanted and unwelcome behaviour (ranging from mildly unpleasant remarks to physical violence) that causes alarm or distress by a course of conduct on more than one occasion (note that it doesn't need to be the same course of conduct).

The PHA is the main criminal legislation dealing with harassment, including stalking, racial or religious motivation and certain types of anti-social behaviour such as playing loud music. Significantly, the PHA gives emphasis to the target's perception of the harassment rather than the perpetrator's alleged intent.

Equality Act 2006

This amends the Sex Discrimination Act and places a statutory general duty on employers when carrying out their functions to have due regard to the need to eliminate unlawful discrimination and harassment, and also to promote equality of opportunity between men and women.

Employment Equality (Religion or Belief) Regulations 2003

These regulations make it unlawful to discriminate against, harass or victimise workers because of religion or religious or similar philosophical belief. They are applicable to vocational training and all aspects of employment, recruitment and training.

Employment Equality (Age) Regulations 2006

The Employment Equality (Age) Regulations 2006 is a piece of legislation that prohibits employers from unreasonably discriminating against employees on grounds of age.

Data Protection Act 1998

Information about people can also be subject to abuse, so the Information Commissioner enforces and oversees the Data Protection Act 1998 and the Freedom of Information Act 2000. The Commissioner is a UK independent supervisory authority reporting directly to the UK Parliament. It has an international role as well as a national one.

The principles put in place by the Data Protection Act 1998 aim to ensure that information is handled properly. Data must be:

- fairly and lawfully processed
- processed for limited purposes
- accurate, adequate, relevant and not excessive
- not kept for longer than is necessary
- processed in line with your rights
- secure and not transferred to other countries without adequate protection.

'Data controllers' have to keep to these principles by law.

Racial and Religious Hatred Act 2006

The Racial and Religious Hatred Act 2006 makes inciting hatred against a person on the grounds of their religion an offence in England and Wales. The House of Lords passed amendments to the Bill that effectively limit the legislation to 'a person who uses threatening words or behaviour, or displays any written material which is threatening... if they intend thereby to stir up religious hatred.' This removes the abusive and insulting concept, and requires the intention – rather than just the possibility – of stirring up religious hatred.

Child Protection and Safeguarding

Safeguarding means keeping children, young persons and vulnerable adults safe from harm (physical or psychological distress or injury) and abuse (physical, sexual or emotional), abuse or injury. The United Nations Convention on the Rights of the Child 1989 set out minimum standards for all children up to the age of 18. Underpinned by the Children's Act 2004, 'Every Child Matters 2004' and 'Working Together to Safeguard Children 2006' are government initiatives that try to address concerns about child protection.

'Every Child Matters' sets out a framework that allows the building of services around the needs of children. It highlights the following key outcomes for the well-being of children and young people:

- stay safe
- be healthy
- enjoy and achieve
- make a positive contribution
- achieve economic well-being.

'Working Together to Safeguard Children 2006' states all agencies and individuals should aim to proactively safeguard and promote the welfare of children. This means the need for action to protect them is reduced.

Discrimination and victimisation

Under current law, there are two main forms of discrimination: direct and indirect.

Direct discrimination

Direct discrimination occurs when someone is treated less favourably because of their sex, race or **disability**. With regard to employment, this could happen if an employer treats a job applicant or existing employee less favourably on the grounds of their sex, race or disability. In law, the applicable test is if someone would have been treated differently, or more favourably, had it not been for their sex, etc. This definition is expected to continue under future legislation.

An example of direct discrimination could be a woman of superior qualifications and experience being denied promotion in favour of a less experienced and less qualified man.

> **Did you know?**
>
> An employer cannot argue that it was not their intention to discriminate; the law only considers the end effect.

> **Key term**
>
> **Disability** – a physical or mental impairment that has a substantial and long-term effect on a person's ability to carry out normal everyday activities

Indirect discrimination

Indirect discrimination occurs where the effect of certain requirements, conditions or practices imposed by an employer disproportionately disadvantage one group more than another. Courts tend to consider three factors:

- the number of people from a racial group or of one sex that can meet the job criteria is considerably smaller than the rest of the population

- the criteria cannot actually be justified by the employer as being a real requirement of the job, so an applicant who could not meet the criteria could still do the job as well as anyone else

- because the person cannot comply with these criteria they have actually suffered in some way (this may seem obvious, but a person cannot complain unless they have lost out in some way).

With cases of indirect discrimination, employers may argue that there may be discrimination, but that it is actually required for the job. As an example, one individual claimed indirect discrimination on religious grounds against his employer as he was requested to shave off his beard. The court agreed that discrimination had been applied, but as the employer was a factory involved in food preparation, the particular case was rejected on the grounds of hygiene.

Disability discrimination

Disability discrimination relies on the same basic principles, but the complainant must be treated less favourably due to their disability. A person has a disability if they have a physical or mental impairment that has a substantial and long-term effect on their ability to carry out normal everyday activities.

Victimisation

This is where an employee is singled out for using their workplace complaints procedures or exercising their legal rights: for example, making a complaint of discrimination or giving evidence and information on behalf of another employee who has brought proceedings for discrimination.

Positive discrimination and positive action

Positive discrimination occurs when someone is selected to do a job purely on the basis of their gender or race, not on their ability to do the job. This is illegal under the Sex Discrimination Act and the Race Relations Act and is generally unlawful other than for what are called '**Genuine Occupational Requirements**'.

Positive action is activity to increase the numbers of men, women or minority ethnic groups in a workforce where they have been shown to be under-represented. This may be in proportion to the total employed by the employer or in relation to the profile of the local population.

An example of positive action might be carefully targeted advertising and courses to develop the careers of those from under-represented groups who are already employed by an organisation. Positive action is legal and is designed to help employers achieve a more balanced workforce.

Key term

Genuine Occupational Requirements – where an employer can demonstrate that there is a genuine identified need for someone of specific race or gender to the exclusion of others – for example a film company needs an Indian actor for a film set in India, or a modelling agency needs a woman to model female clothes

Planning work allocations, operative duties and co-ordination with other services and personnel

If we return again to the work programme, we now know that we need to consider the estimators' quantities and times, health and safety, employment legislation and site access.

However, as an employer, there are still more things to consider before we can commit to a plan. Some of the key points would be:

- the ability of your staff (Are they skilled in the areas of work that the installation requires?)
- the availability of your staff (Are they working on another site or on holiday?)
- the availability of material and equipment (Does specialist equipment need to have been ordered?)
- when will the site be available, safe and secure enough to receive labour and materials?
- what are the work plans of the other contractors involved and how do they affect your plan?
- is the weather likely to have an adverse effect?

Remember

As a worker, you will need to identify your own responsibilities within the work programme. If you are uncertain about your role and responsibility, then check with your manager.

An additional stage to the contract has been identified. This involves fitting of powered showers in several areas. Variation Orders have been issued to cover the altered work. You clearly see the impact on the program, as the work involves both the plumbing contractor and your company installing wiring to the showers.

- What considerations need to be checked for the operatives working upon these circuits?
- Identify the items you will need to check before committing to any plan of work.
- How could the plumbers' work affect your program?

Consequences of not completing work on time or meeting requirements of the programme of work

Finally, before you put your plan down on paper, you need to consider what would be the consequences of not meeting the plan. Invariably production of a work programme is a contractual requirement and, once it is issued, failure to comply with it constitutes a breach of contract that can have serious cost implications if someone feels they have incurred damages.

You should remember that all a client wants is their project completed to specification. That said, if something goes wrong, isn't operating properly or is delayed, the client will want a solution.

General damages (also known as unliquidated damages) are one of the most important remedies for a breach of contract, but require that the non-defaulting party proves that it has incurred actual loss as a result of the breach.

A liquidated damages clause in the contract avoids this requirement, as the non-defaulting party only needs to prove that a breach has occurred and the money calculation is based on an estimate of the costs resulting from that breach. This establishes some predictability involving costs so that, when planning, you can balance the cost of your anticipated performance against the cost involved should you breach the contract.

Some typical problems and their consequence are shown in Table 3.07.

Problem	Consequences
Not completing work in time	The effect of this may only be to create a minor inconvenience. However, if your delay affects others, you run the risk that all parties make a claim against you.
	Some contracts where loss of business can be reasonably predicted have a liquidated damages clause in excess of £1 million per week if the project isn't completed on time.
Not meeting the requirements of the work programme	The work programme will be drawn up to make the best use of time, skills and resources. If you do not meet the requirements of the programme in some way, you could be responsible for adding costs, delaying work or even the failure of the structure.
Using incorrect materials and equipment	If you use, without gaining approval, incorrect or unspecified materials or equipment you could face the costs (both financial and time) of having to replace them with the correct thing.
	If that delays project progress, it could lead to not completing work on time.
Not installing materials and equipment as required	The effect here could be just a low-quality piece of work, or a job where elements do not function properly. However, the outcome could be worse: rectifying faults could be very inconvenient and costly and, at worst, very dangerous.

Table 3.07 Typical problems and their consequences

Producing and illustrating work programmes

In addition to drawings and specifications, charts and reports are two other methods of showing and communicating technical information and data. In order to prepare a work programme and then monitor progress against it, you must be able to interpret the data contained in a chart. The charts you will see in this unit can be created using Excel®, a spreadsheet programme that can easily suit this purpose.

Charts

Charts can often make information easier to understand and allow the user to see clearly what they need to know. The most popular chart used within construction work is the bar chart. When it shows activities against time, it is sometimes referred to as a Gantt chart, after its inventor, Robert Gantt.

Figure 3.26 is a bar chart that shows several activities and when they are due to happen. The chart helps the supervisor keep an eye on how the contract is actually progressing when compared to the original plan. Main contractors often use this sort of bar chart to show when individual trades should be on site at any time during the contract.

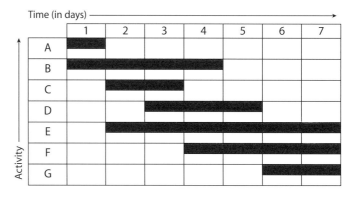

Figure 3.26 Bar Chart 1

Looking at Figure 3.26 you can see that:

- Activity A should take one day
- Activity B starts on the same day, and lasts four days
- Activity C lasts two days, but doesn't start until the second day, and so on.

Bar charts can show additional information by adding colours, codes and symbols.

Figure 3.27 gives a further example of a bar chart. Here the activities are the same as before, but the actual progress against each one is shown by the shaded blue area beneath the original bars. The chart shows progress to the end of Day 3. It is easy to see which activities have been completed, and which ones are lagging behind.

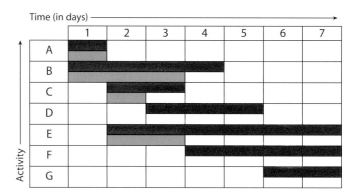

Figure 3.27 Bar Chart 2

Critical path analysis

In larger projects, many tasks must be completed before the project is finished. Not all of these activities can be done at the same time, and some can't begin until others are completed. You need a way of working out the best way to organise the project efficiently, and critical path analysis (CPA) is one solution.

Critical path networks (CPNs) are diagrams that represent each task and how they relate to one another. The critical path is the sequence of activities that fix the duration of a project; if you know the time needed for each activity, you can calculate the overall project completion time. CPA helps you to see what happens if a task is delayed unexpectedly.

Figure 3.28 represents an activity (A), which lasts for two days. The circles are 'events'; they have no duration, but represent the time at which the activity starts or finishes. The arrow represents the activity, and always goes from left to right. It starts and ends with an event.

Figure 3.28 Critical path diagram

The first step in constructing a critical path network is to list all the activities, what must be done before they can start (the sequence) and the expected time needed for each.

You then build the network by linking the activities from left to right at their start and end events. One rule of CPA is that no two activities can begin and end on the same two events. To explain this, we'll use an example.

Activity	Activities that need to be done before this activity
A	None
B	None
C	A and B
D	B

Table 3.08 Sequence of activities for a critical path analysis

In Table 3.08 you have four activities (A, B, C and D) to complete.

Activities A and B can run at the same time, but C cannot begin until Activities A and B have been completed, and Activity D cannot begin until Activity B has been completed.

As the first two activities (A and B) can begin at the same time, the temptation would be to draw them as shown in Figure 3.29.

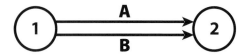

Figure 3.29 Timing of Activities A and B

However, to comply with the rule that no two activities can start and end on the same events, we introduce a 'dummy activity'. Shown as a dotted line, a dummy activity doesn't take any time. This means that the correct drawing becomes as shown in Figure 3.30.

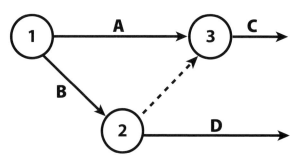

Figure 3.30 Introducing a dummy activity

This drawing shows that Activity C can only start at Event 3, when both A and B are complete.

Now you will see a more detailed example, where the information is shown as both a bar chart and as a critical path network.

This is the information we have about the activities.

- Activity A starts on Day 1 and lasts two days.
- Activity B starts on Day 1 and lasts three days.
- Activity C can only begin once Activity B is complete and lasts 3 days.
- Activity D can only begin once Activity C is complete and lasts 3 days.
- Activity E cannot start until Activity A is complete; it will take 5 days.
- Activity F can start at any time and lasts 2 days.

The question is how long will it take to complete the project? Figure 3.31 shows the information represented as a bar chart.

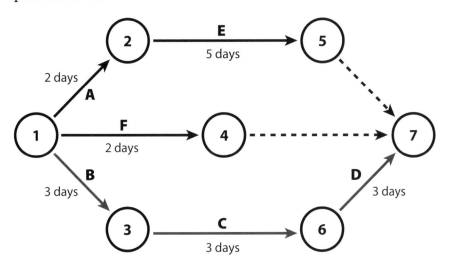

Figure 3.31 Activities shown as a bar chart

You can see that the project will finish at the end of Day 9. Figure 3.32 shows the same information represented by a critical path network.

Figure 3.32 Critical path network

This may need a little careful study to see how it all fits together.

The minimum time for project completion is taken as the longest time path through the network. In this example it is through points 1, 3, 6 and 7, giving us a project time of nine days.

In a CPN, the shortest time it will take to complete a project is found by taking the longest time path through the network. If you follow the red line through events 1, 3, 6 and 7 in the diagram, you can see that the project lasts for nine days.

You may feel that the bar chart is easier to understand, but it can't readily show when several activities have to be completed before another can begin. Using the critical path method gives you greater control over establishing when activities may start. Bar charts, spreadsheets and critical path networks will help you to see which activities affect your work and how it fits in with everyone else's, so you can:

- plan which areas to work in
- see when you can start each activity
- make sure you have the correct materials and equipment ready at the right time
- avoid causing delays to others and to the overall contract.

> **Remember**
>
> Both the bar chart and the critical path network are management planning tools; their purpose is to help a contract run smoothly.

Project management and completing a work plan

You should now be able to see that many different skills and activities are needed to complete a successful project. Every project is unique, but a lot of tasks are common to them all. You need to have a sensible plan for pulling all these together. This is called a work plan.

Typically, a work plan will include:

- checking the drawings, instructions and specifications
- identifying the tasks to be done
- checking that the work area and environment are suitable and safe at all times
- using scaled drawings, listing the tools, materials and equipment needed for the project and making sure that they are available when required
- establishing what skills are required
- allocating work in line with skills, experience and ability
- establishing responsibility levels for staff
- creating a logical sequence for the work activities
- co-ordinating with other contractors
- managing the installation process cost effectively (for example, by using time sheets and delivery notes to track materials and time used against estimated costs, without waste or incurring costs)
- ensuring compliance with the work programme
- making sure the worksite and installation comply with all appropriate legislation
- inspecting, testing and commissioning the installation
- ensuring the project has been completed to specified requirements and handing over to the client.

Depending on the size of the project, one person or many may have the responsibility for devising and monitoring the work plan. On a large project,

Figure 3.33 Teamwork and communication is an important part of project management

contract managers and engineers, project engineers, safety officers and site supervisors may all be involved.

For a small project, many of these tasks (if not all) may fall to the electrician on site. This might include producing drawings and specifications.

All the tools you have covered in this unit – including communication skills, bar charts, time sheets, critical path networks, requisitions, delivery records, day-work sheets, letters and reports and Variation Orders – will help you to complete a project successfully.

Rescheduling work

Some jobs may have their start dates delayed. This could be caused by a number of factors, such as an overrun on a previous contract. If this happens, it would be important to review the overall project and study the critical path to see if any changes can be made in the project management or plan for the work to make up the time. An important part of project management is to see if there is some way of keeping the project on target.

Working life

Your company is involved in a major refurbishment programme to an office block. The following form part of the contract:

- New distribution cables to each floor
- New lighting and power circuits in each office
- New luminaires to all office areas
- New data circuits and computers to install

- Replacement showers in washrooms
- Redecorating of all areas
- New workstations to be fitted for computers
- Replacement pipework in washrooms

1. Consider the tasks that are required to be done and show on a plan a suitable order for them to be completed.

2. How would you maintain electrical safety if the new pipework was to be completed after all the electrical work was finished?

3. What would your company need to do if the computing workstations were delayed from the manufacturers?

Progress check

1   The Employment Rights Act 1996 gives employees a number of rights. What are the main rights?

2   Explain briefly what a critical path network is and what we can ascertain from it.

3   The client asks you to alter the position of a socket outlet which has been wired and fixed in situ. What should you do and what document should be issued to cover the alteration? What is the purpose of this document?

K6. Understand requirements for provision and storage of resources that are required for work activities

By now you should have gathered an insight into what needs to be considered when planning an installation.

The key factors you need to remember are:

- interpreting the installation specification and work programme to identify the exact resource requirements for the work to be undertaken

- interpreting the schedule to confirm that the materials available are correct, fit for purpose, in the correct quantity and suitable for the work to be completed efficiently and to cost

- using the correct storage and transportation requirements for all work, materials, tools and equipment.

This material was all covered earlier in this unit and in Unit ELTK 01.

Working life

In the refurbishment task in the Working Life on page 233 you and a colleague are tasked with ensuring all materials are securely stored on site. This is to ensure they are readily available and readily accessible when required. All equipment is delivered at ground level.

- Explain what you need to consider to ensure everyone is fully supplied with materials at the correct time in the programme.

- Will any specialist equipment be required to move the stored material to the points of installation?

- How will you store the equipment materials to ensure breakages are kept to a minimum?

Progress check

Using the information provided in this unit and Units ELTK 01 and 02, try to answer the following questions:

1 How would you establish whether you have the right materials, whether they are fit for purpose, of the right quantity and available to you?

2 You are setting up on site and are awaiting the delivery of all of the materials and equipment needed. The equipment will include specialist glass lighting fittings, mineral-insulated cable and some purpose-made control panels that contain electronic components. You also will be using a mobile scaffold tower. The site is 50 miles from your employer's premises. Describe:
   - what arrangements need to be made to have everything delivered
   - once delivered, how you intend to transport them to your storage area
   - your considerations for storing everything.

3 List all of the resources that you think could be necessary on this project.

Getting ready for assessment

A large part of your working life will be working, and communicating with, other people and using information gained from this and other sources to plan your work successfully. These skills are vital, not only for those in management positions, but for anyone who needs to plan and carry out their work effectively.

For this unit you will need to be familiar with:

- types of technical and functional information that are available for the installation of electrotechnical systems and equipment
- procedures for supplying functional and technical information to relevant people
- requirements for overseeing health and safety in the work environment
- requirements for liaising with others when organising and overseeing work activities
- requirements for organising and overseeing work programmes
- requirements for organising the provision and storage of resources that are required for work activities

For each Learning Outcome, there are several skills you will need to acquire, so you must make sure you are familiar with the assessment criteria for each outcome. For example for Learning Outcome 4 you will need to describe the techniques used to motivate, instruct, monitor and co-operate with other workers. You will also need to know how to determine the competence of operatives and specify their roles. For working with others you will need to be able to identify the appropriate methods for communicating with others and specify procedures for re-scheduling work to co-ordinate with changing work conditions and work of other trades. Finally you will need to be able to clarify procedures for completing documentation required during work operations.

It is important to read each question carefully and take your time. Try and complete both progress checks and multiple choice questions, without assistance, to see how much you have understood. Refer to the relevant pages in the book for subsequent checks. Always use correct terminology as used in BS 7671. There are some simple tips to follow when writing answers to exam questions:

- **Explain briefly** – usually a sentence or two to cover the topic. The word to note is 'briefly' meaning do not ramble on. Keep to the point.
- **Identify** – refer to reference material, showing which the correct answers are.
- **List** – a simple bullet list is all that is required. An example could include, listing the installation tests required in the correct order.
- **Describe** – a reasonably detailed explanation to cover the subject in the question.

English skills are an important part of this unit, as they will help you to communicate clearly and effectively. You will also need to use your maths skills to help you complete plans and schedules. Before you start work, always remember to have a plan of action. You will need to know the clear sequence for working in order to make sure you not making any mistakes as you work and that you are working safely at all times.

Good luck!

CHECK YOUR KNOWLEDGE

1. Which of the following is a drawing that uses symbols to represent circuit components?
   a) Assembly drawing
   b) Layout drawing
   c) Block diagram
   d) Circuit diagram

2. Which one of the following would **not** be included in a site visitors' book?
   a) Date
   b) Visitor's name
   c) Visitor's car colour
   d) Time out

3. Identify the trade association from the list below.
   a) JIB
   b) ECA
   c) UNITE
   d) NICEIC

4. The main function of the electrician on a large construction site is to?
   a) Work to the supervisor's orders
   b) Alter circuits as the client verbally requests
   c) Pay and co-ordinate sub-contractors
   d) Oversee the work of contracts engineers

5. To maintain good customer relations on site, you should?
   a) Use derogatory terms about your employer
   b) Deliberately tell lies when questioned as to project progress
   c) Be honest
   d) Gossip about the customer

6. Site records would **not** be transmitted by using?
   a) Microforms
   b) CDs
   c) Emails
   d) Telephone

7. When checking the technical competence of a new employee what would you look for?
   a) Type of PPE
   b) ECS card
   c) First-aid certificate
   d) Certificate of unit credit

8. Which of the following documents is used for alterations to the original contract?
   a) Day work sheet
   b) Variation Order
   c) Time sheet
   d) Purchase order

9. Under the Data Protection Act 1998 data must?
   a) Be shared with all suppliers
   b) Be accessible to all on site workers
   c) Not be kept longer than is necessary
   d) Be transferred to other countries

10. Critical path networks allow site managers to see?
    a) How much the contract will cost
    b) The length of time tasks take
    c) The profitability of the contract
    d) The number of suppliers utilised

Understanding practices and procedures for preparing and installing systems and electrotechnical equipment

This unit will introduce a wide range of practical skills required to become a qualified electrician. You need to have practical skills that will allow you to understand wiring systems and equipment and how to install them safely and efficiently.

This unit will cover the following learning outcomes:

- understand procedures, practices, statutory and non-statutory requirements relative to preparing and installing

- understand procedures for checking the work location before commencing work activities

- understand requirements for safe isolation of circuits and complete systems

- understand types, applications and limitations of wiring systems and associated equipment

- understand procedures for selecting and using tools, equipment and fixings for wiring systems, enclosures and equipment

- understand practices and procedures for installing wiring systems, enclosures and equipment

- understand the regulatory requirements that apply to the installation of wiring systems, enclosures and equipment.

K1. Understand procedures, practices, statutory and non-statutory requirements relative to preparing and installing

In Unit ELTK 01 you saw that both employers and employees have legal duties to establish and maintain a safe working environment. Before carrying out any work on site, you need to have an understanding of the installation requirements. To do this you will need to use some of the documents covered in ELTK 03, including site drawings, wiring diagrams, specifications and relevant technical data, such as manufacturer's instructions. You will also need to use statutory documentation, such as HASAWA and EAWR.

Once you understand the installation requirements, you should undertake a risk assessment (see page 40) to ensure:

- safe access and egress to, from and around the work area
- all work area hazards to property, personnel and livestock are eliminated or controlled.

Once risk assessments have been completed, you need to make sure their findings are implemented before any work begins.

You will also be required to use relevant British Standards. Perhaps the most important being BS 7671 – Requirements for Electrical Installations, more commonly referred to as the IET Wiring Regulations.

BS 7671 applies to the design, erection and verification of electrical installations. You will explore it in more detail later in this unit.

Did you know?

Although not a statutory document itself, BS 7671 can be used in a court of law as evidence of compliance with a statutory document such as the EAWR or The Building Regulations.

K2. Understand procedures for checking the work location before commencing work activities

Planning a work programme and the considerations needed to do so were covered in Unit ELTK 03. As a reminder, here are some of the things you will need to do to prepare before starting work.

- Interpret drawings and specifications to produce accurate material and equipment requisites.
- Identify and select material and equipment that meets the specification (where not specified).
- Identify suitable installation, fitting and fixing methods.
- Identify where co-ordination with other trades is required.
- Confirm that the site is ready for the installation, including making sure there are suitable storage facilities.
- Confirm that all tools, equipment and instruments are ready and fit for purpose.
- Identify suitable access or lifting equipment, if required.

For an existing property, you will need to check for any pre-existing damage, and establish clearly with the client whether this needs to be fixed to allow you to do your job. The main areas to look for pre-existing damage are:

- wall and floor fabric
- equipment and components
- building decor and floor finishes.

You should also consider the possible damage that could be done to a client's property during an installation. The possibility of causing damage is relative to:

- the type of building (new or existing)
- the type of wiring systems/enclosures being used
- the contractual arrangements in place.

Case study

Example 1

You are asked to provide an electrical installation in a newly constructed school. Your installation is specified as being in a concealed steel conduit within plastered walls and trunking in ceiling voids.

Most of your installation will take place as the building develops and ultimately walls will be plastered. This is the responsibility of the main contractor.

However, it is possible that without due care and attention walls, floors and ceilings could be damaged as the project develops. For example, you may need to use a mobile scaffold tower in the school gym to install high-level lighting fittings and therefore you will be expected to protect the wooden floor.

Example 2

For this project you are required to use concealed MICC cable to install a fire alarm system in a 200-year-old stately home. The contract states that you are responsible for all repairs to the building structure to maintain its appearance. This will mean carefully selecting cable routes and being aware of the type of decorative plaster, paints and materials used throughout the building. There are several areas that can be accessed by the public and time limits on when the work can be carried out.

Figure 4.01 A stately home can present challenges

Obviously in a contract where you are responsible for all repair work, it also pays to have established what the condition of the building was before you arrive on site. Otherwise you may find yourself carrying out more restoration work than expected.

With the stately home example, it is normally a contractual requirement that the prospective contractor is deemed to have arranged to visit the site, inspected the property and then allowed in the estimate for all restorative work costs. Equally, it will normally state in the contract that having pointed this out, no claim may then be made by a contractor for lack of knowledge.

For the stately home, such an initial visit should mean:

- walking around the property
- looking at the intended location of the equipment
- planning your intended cable routes to accommodate these locations
- establishing alternatives if necessary
- marking your routes on the location drawings
- establishing what protective measures may be needed for the building or items within it
- noting any items that may need to be removed and stored
- establishing any special requirements for such removal, transport and storage
- taking photographs or video footage of areas affected by your installation
- establishing any special decorative techniques or finishes that may be required.

After the visit, the estimator will research the costs where relevant and include them within the quotation.

In terms of protective measures the estimator will consider whether walls and floors must be protected with a covering material, such as wood or sheeting; whether areas need to have restricted access and require the use of barriers; or whether items need to be carefully removed and stored. The requirements for this will change with each project, but the principle remains the same.

However, the estimator should also include a covering letter that states any damage noted during this initial inspection, including evidence of this and, where possible, any alternatives that may avoid further damaging the property.

If the contractor is awarded the contract, it would also be sensible for them to carry out another such visit immediately before starting work to see whether conditions have changed within the property, and to continually monitor.

K3. Understand requirements for safe isolation of circuits and complete systems

Why safe isolation is important

Safe electrical isolation procedures are important.

Each year about 20 people die from electric shock or electric burns at work and about 30 die from electrical accidents in the home. Many more people have had an electric shock but without any lasting injury – but more often than not they have been lucky.

Electric shock is not the only hazard. When electrical arcing occurs, the heat generated can be intense and can cause deep, slow healing burns. There are hundreds of serious burn accidents each year arising from unsafe working practices. The intense ultraviolet radiation from an electric arc can also cause damage to the eyes (this is why welders wear special goggles).

Another problem is that arcing, overheating and, in some cases, electrical leakage currents can cause fire or explosion by igniting flammable materials, which in turn can cause death, injury, damage to equipment and property, as well as considerable financial loss.

Most electrical accidents occur because people are working on or near equipment that is:

- thought to be dead but which is live
- known to be live, but those involved do not have adequate training, have not taken the right precautions, or do not have the appropriate equipment.

Procedures for completing safe isolation

Safe working practices

Safe working practices rely on clearly thought-out systems of work, carried through by trained, competent people who are aware of their own limitations.

To plan and execute electrical work safely, there should be adequate information available about the electrical system and the work to be done. In the case of a newly constructed electrical system (or newly installed equipment), there should be drawings and schedules relating to the design that are updated, if necessary, by the people carrying out the installation.

Records in the form of drawings and schedules should be kept for all but the most basic of installations. In the case of old installations where records may be poor, some measures should be taken to improve the records for the installation, such as a combination of surveying, testing and labelling of the installation. All equipment should be labelled as necessary for it, and its function, to be properly identified. However, when checking records before working on an installation, it is unwise to rely solely on one source of information, such as a label.

Remember that, with the best will and planning in the world, circumstances can change: for example, you may be carrying out inspection and testing, but discover a fault. You must always be monitoring such changes and be prepared to deal with them.

The establishment of a safe working system has three stages:

- identification
- disconnection
- secure isolation.

Identification

There is an old phrase that says 'To assume makes an ass of you and me (ass – u –me).' In terms of electricity, to assume also runs the risk of killing you.

Never assume that labelling is correct and that you can start work without having first proved that the equipment or circuit is dead. Check where equipment is fed from and bear in mind the presence of time-controlled or sensor-activated equipment. For example, outside security lights generally have a constant supply present at the light fitting; the light may not be lit because it hasn't sensed activity, not because the supply is missing.

In some special cases, such as with underground cables, special cable-locating instruments may be necessary.

> **Remember**
>
> A loss of supply to a whole building does not constitute safe isolation. Apart from the fact the Distribution Network Operator (DNO) may restore the supply without any warning, some installations also have back-up generators that could suddenly kick in.

Disconnection

Disconnect the equipment from every source of electrical energy before working on, or near, any part that has been live or is likely to be live. To ensure safety after disconnection, follow the procedure for secure isolation.

Secure isolation

This topic is fully covered in Unit ELTK 01 (see page 70).

Here are some points based on information from The Electricity Safety Council about low-voltage installations that also take into account the most relevant requirements of the Electricity at Work Regulations (EAWR) 1989.

Isolation

Regulation 12 of the EAWR states that where necessary to prevent danger, suitable means (including, where appropriate, methods of identifying circuits) shall be available for a) cutting off the supply of electrical energy to any electrical equipment and b) the isolation of any electrical equipment. In Regulation 12, 'isolation' is defined as the disconnection and separation of the electrical equipment from every source of electrical energy in such a way that this disconnection and separation is secure.

Security

The security aspect is reinforced in Regulation 13, which states that precautions need to be taken regarding equipment that has been made dead. This includes securing the means of disconnection in the OFF position, putting a warning notice or label at the point of disconnection, and proving 'dead' at the point of work with an approved voltage indicator.

'Dead' working

To comply with Regulation 14 of the EAWR, 'dead' working should be seen as the normal method of carrying out work on electrical equipment or circuits, with live work only being carried out in particular circumstances where it is unreasonable to work 'dead'. With particular reference to fault finding, most people are aware that certain activities require the circuit to be live. However, danger is always present when fault finding and you must take precautions to ensure safety and prevent injury.

Working near live conductors

It is also noticeable that there can be commercial pressure to carry out work on, or near, live conductors, especially in areas such as banking and in high-cost manufacturing premises or retail premises that operate 24 hours a day. However, the EAWR still applies, and live working should only be carried out when justified using the risk assessment criteria explained in HSE document HSG85.

Finally, Regulation 16 requires that no one shall engage in work with electricity unless they are trained and competent to do so.

To ensure compliance with Regulations 12 and 13 of the EAWR, there are certain working principles that you must follow.

- The correct point of isolation must be be identified.
- An appropriate means of isolation must be used.
- The point of isolation should ideally be under the control of the person who is carrying out the work on the isolated conductors.
- Warning notices should also be applied at the point(s) of isolation.
- Conductors must be proved 'dead' at the point of work before they can be touched.
- The supply cannot be re-energised, knowingly or unknowingly, while the work is in progress.

Remember

You may need to notify people of the intent to isolate and then put appropriate measures in place. For example, you might need to isolate a circuit to let you repair a light fitting in a dark, public corridor or over a lathe in a factory, or you might need to isolate a supply feeding a building's security and CCTV systems.

Progress check

1 Before commencing work on site you are obliged to take account of safety. What should you do first?

2 Identify the general preparations to be done before starting work.

3 What are the essential steps in the establishment of a safe working system?

K4. Understand types, applications, and limitations of wiring systems and associated equipment

Sometimes also referred to as a containment system, the phrase 'wiring enclosure' refers to systems such as conduit, trunking, cable tray, cable basket, busbar and ladder systems. The phrase 'wiring system' generally means cables such as PVC/PVC, MICC and SWA.

Constructional features, applications, advantages and limitations

Conductors and insulators

The correct selection of cable for an electrical installation is very important. To ensure you get the right sort of cable, you need to consider:

- conductor material
- conductor size
- insulation
- wiring system
- environmental conditions.

Wiring systems are covered in the next section (page 254).

Conductor material

Copper and aluminium

The choice generally is between copper and aluminium. Copper has better conductivity for a given cross-sectional area and is preferable, but its cost has risen over the years. Aluminium conductors are now sometimes preferred for the medium and larger range of cables, but all cables smaller than $16\,mm^2$ cross-sectional area (CSA) must have copper conductors.

Table 4.01 shows the main advantages/disadvantages of the two materials.

Table 4.02 shows other cable conductor materials available.

Conductor	Advantages	Disadvantages
Copper	• easier to joint and terminate • smaller cross-sectional area for given current rating	• more costly • heavier
Aluminium	• cheaper • lighter	• bulkier for given current rating • not recommended for use in hazardous areas

Table 4.01 Advantages and disadvantages of copper and aluminium as conductor materials

Conductor material	Application
Cadmium copper	Has a greater tensile strength for use with overhead lines
Steel-reinforced aluminium	Used for very long spans on overhead lines
Silver	Used where extremely good conductivity is required. However, it is extremely expensive
Copper-clad (copper-sheathed aluminium)	A conductor types that has some of the advantages of both copper and aluminium but is difficult to terminate

Table 4.02 Other cable conductor materials

Whatever the choice of conductor material, the conductors themselves will usually be either stranded or solid. Solid conductors are easier and cheaper to manufacture, but they are harder to install because they are not very pliable.

Stranded conductors are made up of individual strands that are brought together in set numbers. These provide a certain number of strands (such as 3, 7, 19 or 37). With the exception of the three-strand conductor, all have a central strand surrounded by the other strands to make up the overall conductor.

Conductor size

Many factors affect the choice of size of conductor and these will be discussed in greater detail in Unit ELTK 04a (covered in Book B). However, here are some of the factors you need to consider.

Load and future development

The current the cable is expected to carry can be found from the load, taking into account its possible future development, such as change in use of premises, extensions or additions.

Ambient temperature

The hotter the surrounding area, the less current the cable is permitted to carry.

Grouping

If a cable is run with other cables, its current-carrying capacity must be reduced.

Type of protection

Special factors must be used when BS 3036 (semi-enclosed) fuses are employed.

Thermal insulation

If cables are placed in thermal insulation, de-rating factors must be applied.

Voltage drop

The length of circuit, the current it carries and the CSA of the conductor will combine to affect the voltage drop. Appendix 12 of BS 7671 currently states that the maximum voltage drop between installation origin and load for an LV installation fed from the public supply must not exceed 3 per cent for lighting and 5 per cent for other uses.

Insulation

To insulate the conductors of a cable from each other and to insulate the conductors from any surrounding metalwork, materials with extremely good insulating properties must be used. However, cables can be installed in a variety of different situations and you must take care that the type of insulation selected is suitable for that particular situation.

Thermosetting and thermoplastic

BS 7671 has changed the classification of materials away from the main ingredient to a generic system based on the main properties of the material, where it is felt that the terms 'thermoplastic' and 'thermosetting' linked with the operating temperature of the cable will give a similar method of material classification. For example, XLPE 'cross linked polyethylene' is neither a plastic or rubber material, but is a thermosetting material rated at 90°C.

To help users, where the terms PVC and rubber are replaced in BS 7671, the old terms appear in brackets, e.g. 'thermoplastic (PVC)' or 'thermosetting (rubber)'.

Insulation types

Here are some of the more commonly found types of cable insulation and their properties.

PVC

PVC or polyvinyl chloride is a thermoplastic polymer. It is a good insulator, as well as being flexible and cheap, easy to work with and easy to install. However, thermoplastic polymers such as PVC do not stand up to extremes of heat and cold. BS 7671 recommends that ordinary PVC cables should not constantly be used in temperatures above 60°C or below 0°C.

XLPE

XLPE or cross-linked polyethylene is a thermosetting compound. The cable has a high softening temperature, small heat distortion and high mechanical strength under high temperature. XLPE is also lighter in weight than its PVC counterpart.

At high voltage there is a potential of failure because of 'treeing', which is when moisture penetrates the cable and causes a reaction that breaks down the insulation. Heat-shrink tubing, which also provides stress control, will aid this situation.

Paper

Dry paper is a surprising but excellent insulator that loses its insulating properties if it becomes wet. Dry paper is **hygroscopic** (it absorbs moisture), so it must be sealed to make sure that there is no contact with the air. Paper-insulated cables are sheathed with impervious materials, lead being the most common.

PILC (paper-insulated lead-covered) has been traditionally used for supply systems, with the paper insulation being impregnated with oil or non-draining compound to improve its long-term performance. Cables of this kind need special jointing methods so that the insulation stays sealed.

Magnesium oxide

Magnesium oxide is the white, powdered substance used as an insulator in mineral-insulated cables. This form of insulation is also hygroscopic and must be protected from damp with special seals. Mineral-insulated cables can withstand very high temperatures and, being metal-sheathed, can also withstand a high degree of mechanical damage.

> **Safety tip**
>
> Take care when burning off PVC insulation (to salvage the copper) because the fumes produced are toxic.

> **Key term**
>
> **Hygroscopic** – tending to attract and absorb water

Synthetic rubbers

Synthetic rubbers, such as butyl rubber, will withstand high temperatures much better than PVC. They are normally used for the flexible final connection to items such as immersion heaters, storage heaters and boiler house equipment.

Insulation colours

Since April 2004, all new installations in the UK and across Europe have to use cables with conductors that comply with Table 51 of BS 7671 (see Table 4.03).

Function	Colour
Protective conductors	Green and yellow
Functional earthing conductor	Cream
a.c. power circuit[1] Phase of single-phase circuit Neutral of single- or three-phase circuit Phase 1 of three-phase a.c. circuit Phase 2 of three-phase a.c. circuit Phase 3 of three-phase a.c. circuit	 Brown Blue Brown Black Grey
Two-wire unearthed d.c. power circuit Positive of two-wire circuit Negative of two-wire circuit	 Brown Grey
Two-wire earthed d.c. power circuit Positive (of negative earthed) circuit Negative (of negative earthed) circuit Positive (of positive earthed) circuit Negative (of positive earthed) circuit	 Brown Blue Blue Grey
Three-wire d.c. power circuit Outer positive of two-wire circuit derived from three-wire system Outer negative of two-wire circuit derived from three-wire system Positive of three-wire circuit Mid-wire of three-wire circuit[2] Negative of three-wire circuit	 Brown Grey Brown Blue Grey
Control circuits, ELV and other applications Phase conductor	Brown, black, red, orange, yellow, violet, grey, white, pink or turquoise
Neutral or mid-wire[3]	Blue

NOTES
(1) Power circuits include lighting circuits.
(2) Only the middle wire of three-wire circuits may be earthed.
(3) An earthed PELV conductor is blue.

Table 4.03 Table 51 of BS 7671

Before 2004, all UK installations used the insulation colours shown in Table 4.04. You will come across conductors with these colours of insulation for many years to come, but only in existing installations.

Conductor	Old colour
Phase	Red
Neutral	Black
Protective conductor	Green and yellow
Phase one	Red
Phase two	Yellow
Phase three	Blue
Neutral	Black
Protective conductor	Green and yellow

Table 4.04 Identification of conductors

Environmental factors

Many factors affect cable selection, some of which have already been mentioned:

- the risk of excessive ambient temperature
- the effect of any surrounding moisture
- the risk of electrolytic action
- proximity to corrosive substances
- the risk of damage by animals
- exposure to the elements
- the risk of mechanical stress or of mechanical damage
- aesthetic considerations.

Ambient temperature

Current-carrying cables produce heat; the rate at which that heat can be dissipated will be affected by the temperature surrounding the cable. If the cable is in a cold location, the temperature difference between cable and environment is high so substantial heat can be dissipated. However, if the cable is in a hot location, the temperature difference between the cable and its surrounding environment will be small, and little, if any, of the heat in the cable will be dissipated. Typical problem areas are boiler houses and plant rooms, thermally insulated walls and roof spaces.

Low temperatures can damage PVC cables. PVC cables stored in areas where the temperature has dropped to 0°C should be warmed slowly before being installed. However, if cables have been left out in the open and the temperature has been below 0°C (say, a heavy frost has attacked the cables), you must report this to the person in charge of the installation.

Moisture

Water and electricity do not mix, and you should take care at all times to avoid the movement of moisture into any part of an electrical installation, using watertight enclosures where appropriate.

Any cable with an outer PVC sheath will resist the penetration of moisture and will not be affected by rot. However, you should use suitable glands for termination of these cables. Mineral-insulated cable can be affected by moisture even when indoors; if it is not possible to terminate the cable, seal the end of the cable to stop any moisture getting in.

Electrolytic action

Having two different metals together in the presence of moisture can cause an electrolytic action, resulting in the deterioration of the metal. Take care to prevent this. An example is where brass glands are used with galvanised steel boxes in the presence of moisture. Metal-sheathed cables can also suffer when run across galvanised sheet-steel structures or tray. If aluminium cables are to be terminated onto copper busbars, the busbars should be tinned.

Corrosive substances

Metal cable sheaths, cable armouring, glands and fixings of cables can also suffer from corrosion when exposed to certain substances. Examples include:

- magnesium chloride, used in the construction of floors
- plaster undercoats containing corrosive salts/lime
- unpainted walls of lime or cement
- oak and other types of acidic wood.

Metalwork should be plated or given a protective covering. In any environment where a corrosive atmosphere exists, you may need to use special materials such as PVC-coated tray or PVC-sheathed cables and accessories.

Damage by animals

Cables installed in situations where rodents are prevalent should be given additional protection or installed in conduit or trunking. These animals often gnaw cables, chewing completely or partially through them and leaving them in a dangerous condition.

Installations in farm buildings should receive similar consideration; if possible, keep any installation well out of the reach of animals to prevent the effects of rubbing, gnawing and urine.

Exposure to the elements

Cables sheathed in PVC should not be installed in positions where they are exposed to direct sunlight as this causes them to harden and crack: the ultraviolet rays leaches out the plasticiser in the PVC, making it hard and brittle.

Similarly, PVC cables should not be installed in circumstances where they will be operating for long periods at temperatures below 0°C, such as when exposed to snow or frost.

Mechanical stress and mechanical damage

If cables are used in situations such as to provide an overhead supply from one building to another, they will be subjected to mechanical stress unless a **catenary wire** is used to support them along the way. Cables installed in this way should have 'drip loops' at either end, to allow for a degree of movement should the system be hit. The drip loop is also designed to allow water to run off the cable, if the cable is installed at a slight angle of degree. The loop allows water to drop away from walls.

> **Key term**
>
> **Catenary wire** – a supporting wire fixed between two buildings from which cable is 'hung' with suitable supports/clips

Flexible cables are often used to suspend luminaires. To avoid stress, the maximum weight that can be accommodated is given in Appendix 4 of BS 7671 in Table 4F3A, as shown in Table 4.05 (see page 254). Bear in mind that cables can also suffer from stress when subjected to excessive vibration, which can cause breakdown of the insulation.

Selecting the correct cable for protection from mechanical damage depends on the type of installation involved and the level of damage that is anticipated.

For example, in a domestic installation where the cables are to be concealed in walls and ceilings at a depth no less than 50 mm from any surface, the main function of the cable sheath is to protect the cable from light mechanical damage. Here you would probably use PVC/PVC and cpc (circuit protective conductor) cable.

Conductor (cross-sectional area mm²)	Current carrying capacity		Maximum mass
	1- phase a.c.	3- phase a.c.	
0.5	3 A	3 A	2 kg
0.75	6 A	6 A	3 kg
1	10 A	10 A	5 kg
1.25	13 A	10 A	5 kg
1.5	16 A	16 A	5 kg
2.5	25 A	20 A	5 kg
4	32 A	25 A	5 kg

Table 4.05 Flexible cords weight support (Table 4F3A, Appendix 4 BS 7671)

Where cables are to be installed on the surface of a building fabric or in an underground trench between buildings, they should have a metal sheath (for example, MICC) or armouring (for example, SWA) that will be resistant to any likely mechanical damage.

Aesthetic considerations

Although not necessarily electrically relevant, if it is possible to choose between more than one electrically acceptable system, it would seem sensible to choose the one that is most pleasing to the eye and sympathetic to the building that it is being installed in.

As an example, you wouldn't choose to install surface-mounted, grey-enamelled metal-clad switches on the wall of your living room when all of the wiring is concealed in the walls. White, flush-mounted plate switches would be much neater.

Cable types (aka wiring systems)

Here you will look at different types of cables and cords including:

- single-core cable
- PVC/PVC flat profile cable
- flexible cords
- mineral-insulated cable
- SWA and SWB cable
- data cables
- fibre-optic cable
- fire-resistant cables.

Single-core thermoplastic insulated unsheathed cable (6491X)

This is designed for installing into conduit, trunking and, when protected, within lighting fittings and control panels.

Its construction is solid or stranded copper conductor with insulation of PVC (-15° to +70°C), making it cheap and easy to install.

Figure 4.02 Single-core thermoplastic insulated unsheathed cable (6491X)

It is available in sizes from 1 mm² to 630 mm². Insulation colours include grey, black, blue, yellow, brown, white, orange, violet and green/yellow stripes when intended for use as a circuit protective conductor.

Single-core thermosetting insulated unsheathed cable (6491B)

This is designed for installing into conduit and trunking and, when protected, within lighting fittings and control panels, where smoke and acid gas emission would pose a major hazard in the event of a fire.

Its construction is solid or stranded copper conductor with thermosetting insulation (up to 90°C).

Figure 4.03 Single-core thermosetting insulated unsheathed cable (6491B)

It is available in sizes from 1.5 mm² to 630 mm². Insulation colours include grey, black, blue, brown, white, orange, violet and green/yellow stripes when intended for use as a circuit protective conductor.

Single-core insulated and sheathed cable (6181Y)

This is designed for surface wiring where there is little risk of mechanical damage. This cable is normally used as 'meter tails', for connecting the consumer unit/distribution board to the DNO (distribution network operator) supply equipment.

The construction of this cable is solid or stranded copper conductor with PVC insulation and a PVC sheath (-15° to +70°C).

Figure 4.04 Single-core insulated and sheathed cable (6181Y)

It is available in sizes from 1.0 mm² to 35 mm². Insulation colours are only brown and blue, with sheath normally only available in grey.

Figure 4.05 PVC-insulated PVC-sheathed flat profile cable with integral CPC (6243Y)

PVC-insulated PVC-sheathed flat-profile cable with integral cpc (6241Y, 6242Y, 6243Y)

This is designed for domestic and general wiring where a circuit protective conductor (cpc) is required for the circuit. Its two-core version is commonly referred to as 'twin and earth'. Normally concealed in walls and ceilings, it may be clipped direct to a surface where there is no risk of mechanical damage.

The construction of the cable is solid or stranded copper conductors with PVC insulation laid parallel with a plain and uninsulated copper cpc and then overall PVC sheath (-15° to +70°C).

It is available as single core (6241Y), two core (6242Y) and three core (6243Y) in sizes from 1.0 mm^2 to 16 mm^2. Insulation colours are only brown, blue, black and grey, with sheath normally only available in grey.

Circular flexible cords (3182Y, 3183Y, 3184Y, 3185Y)

This is designed for general-purpose use indoors or outdoors, in dry or damp situations. It is suitable for portable tools, immersion heaters, washing machines, vacuum cleaners, lawn mowers and refrigerators. However, it should not be used with heating appliances or where the sheath can come into contact with hot surfaces.

The construction of the cable is stranded copper conductors, PVC-insulated, 2, 3, 4 or 5 cores laid up and PVC-sheathed (0°C–70°C).

Available with conductor sizes of 0.75 mm^2 to 4.0 mm^2, the core colours are:

Figure 4.06 Circular flexible cords

- 3182Y two core: brown and blue
- 3183Y three core: brown, blue and green/yellow
- 3184Y four core: black, grey, brown and green/yellow
- 3185Y five core: black, grey, brown, blue and green/yellow.

Circular flexible cords (3092Y, 3093Y, 3094Y)

This is designed for general-purpose use indoors or outdoors, in dry or damp situations. It is suitable for portable tools, washing machines, vacuum cleaners, lawn mowers and refrigerators, especially in higher-temperature zones.

The construction of the cable is stranded copper conductors, heat-resistant PVC-insulated, 2, 3 or 4 cores laid up and heat-resistant PVC-sheathed (0°C–90°C).

Available with conductor sizes of 0.5mm^2 to 2.5mm^2, the core colours are:

- 3092Y two core: brown and blue
- 3093Y three core: brown, blue and green/yellow
- 3094Y four core: black, grey, brown and green/yellow.

Mineral-insulated cable

This is designed primarily for use where **circuit integrity** is required in arduous conditions or during a fire. MI cable is ideal for use in critical fire-protection applications such as alarm circuits, fire pumps and smoke control systems, and in process industries handling flammable fluids. The cable is also highly resistant to ionising radiation, and so finds applications in instrumentation for nuclear reactors and nuclear physics apparatus.

Because of its construction, mineral-insulated cable affords excellent mechanical protection, which again supports its use when buildings disintegrate during a fire.

The cable is rarely referred to by its name, but rather as MI cable or MICC cable (to denote a bare sheath) or MICV (to denote a PVC outer sheath has been applied). On site, it is probably more often referred to as 'pyro', after the original UK manufacturer, Pyrotenax.

The construction of the cable is solid copper conductors surrounded by compressed magnesium oxide powder insulation inside a solid copper sheath. In light-duty (500V) format, it is available with conductor sizes of 1.0 mm^2 to 4.0 mm^2, it can be provided with 2 to 7 conductors; in heavy-duty (750 V) format, it can be provided as a single core (1.5 mm^2 to 240 mm^2) and up to 19 core (1.5 mm^2). In both formats, it has an operating temperature range of -10°C to +250°C.

PVC/Steel Wire Armour/PVC cable (6942X, 3X, etc.)

This is designed for industrial wiring applications and mains distribution. This cable can be laid direct in the ground, in ducts, clipped to surfaces or mounted on tray. It is commonly referred to as PVC/SWA/PVC.

Figure 4.07 Mineral-insulated cable

Key term

Circuit integrity – the ability of a circuit to keep operating

Did you know?

Sometimes you will see documents that use a generic term of MIMS cable, meaning mineral-insulated metal sheath. However, MIMS also refers to a similar product where the sheath is made from metals other than copper.

Figure 4.08 PVC/Steel Wire Armour/PVC cable

The construction of the cable is stranded plain copper conductors, PVC-insulated, 2 to 5 cores laid up, extruded PVC bedding, galvanised steel wire armoured and PVC-sheathed (0°C–90°C). Available with conductor sizes of 1.5 mm^2 to 400 mm, its available colours are:

- two core: brown and blue
- three core: brown, black and grey
- four core: brown, black, grey and blue
- five core: brown, black, grey, green/yellow and blue.

XLPE can also be used as the conductor insulation for this type of cable.

Single-core PVC/Aluminium Wire Armour/PVC cable

Although in every other respect identical to its multi-core counterpart, a single-core armoured cable will always have a layer of aluminium wire armour (AWA) instead of steel wire armour (SWA). This is because steel in SWA has a much lower conductivity, giving it higher resistance than aluminium. If it were used in a single-core cable, the magnetic field generated by SWA would induce an electric current in the armour (eddy current), which would have a heating effect. AWA is non-magnetic and has a much better conductivity (lower resistance), so it can conduct these induced currents to earth more efficiently than steel.

Although now not readily available, a version of this cable was made using flat aluminium strip armouring.

XLPE can also be used as the conductor insulation for this type of cable.

XLPE/Steel Wire Armour/PVC cable (6945XL7W TO 69448XL7W)

Designed for industrial wiring applications such as control panel circuits, this multi-core cable can be laid direct in the ground, in ducts, clipped to surfaces or mounted on tray.

The construction of the cable is stranded plain copper conductors, XLPE-insulated, 5 to 48 cores laid up, extruded PVC bedding, galvanised steel wire armoured and PVC-sheathed (0°C–90°C).

Available with conductor sizes of 1.5 mm^2 to 4.0 mm^2, the insulation is white with black numbering.

Figure 4.09 XLPE/Steel Wire Armour/PVC cable 6945XL7W

Figure 4.10 XLPE/Steel Wire Armour/PVC cable 6947XL7W

Paper-insulated/lead-covered/ SWA/PVC cable

This is designed for use as a mains distribution cable and is still manufactured today for use on 3.3 kV and above. You will find that the intake position in many properties will be supplied by this type of cable.

Termination of these cables traditionally involved 'wiping' the lead sheath onto a brass cone within the termination – a process that required special training. However, there are now termination kits that remove the need for such specialised craftspeople.

PVC/Galvanised Steel Wire Braid/PVC cable

Galvanised Steel Wire Braided (GSWB) comes in a variety of forms, some with PVC insulation and some with XLPE. Perhaps the most common variation, as shown in Figure 4.11, is known as SY cable.

This is designed for use as an interconnecting cable for measuring, control and regulation in control equipment for assembly and production lines and conveyors. It can only be used outdoors if protected, and in dry conditions indoors. It is also suitable for fixed installations or for flexible use where there may be light mechanical stress.

The construction of the cable is stranded copper conductors with PVC insulation, 2 to 42 cores laid up as twisted cores, PVC bedding, GSWB, clear PVC sheath (0°C–70°C).

The cable is available with conductor sizes of 0.75 mm^2 to 1.5 mm^2.

Figure 4.11 PVC/Galvanised Steel Wire Braid/PVC cable

Data cables

There are three basic types of cabling used in data systems: coaxial, fibre-optic and twisted pair. Coaxial is widely installed in older networks but is not recommended for new network installations. Fibre-optic is used for high-speed networks and to connect networking devices separated by large distances. Twisted pair (such as Cat 5e) is currently the most common and recommended cabling type.

Used extensively for data transfer in computer networks and telephone systems, it has four pairs of wires that transmit and receive data along them at very high frequencies. Special termination ends are required for these cables.

Coaxial cable

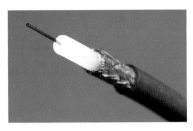

Figure 4.12 Coaxial cable

There are three main types of coaxial cable construction. One is semi-rigid, which has a solid, tubular conductor to contain the RF energy efficiently. Ribbon coaxial is actually many smaller coaxial cables placed inside one another. Flexible coaxial, the most common, has an outer braided conductor and is used when shielding is needed. How effective the shielding is depends on how it is braided.

Fibre-optic cable

Figure 4.13 Fibre-optic cable

As a child, you may have made a telephone out of two tin cans and some string. If you did, sound waves were created as the air vibrated in response to you speaking into the tin.

When the string was pulled tight and someone spoke into one of the cans, its bottom acted as a diaphragm, converting the sound waves into longitudinal mechanical vibrations, which vary the tension of the string. These variations in tension set up waves in the string that travel to the other can, causing its bottom to vibrate in a similar manner as the first can, thus recreating the sound.

A fibre-optic system works in a similar way. At one end of the system is a transmitter, the place of origin for information coming onto the fibre-optic line. The transmitter takes electronic pulse information coming from copper wire and then processes and translates that information into equivalently coded light pulses. A light-emitting diode (LED) or an injection-laser diode (ILD) generates the light pulses and then, using a lens, the light pulses are funnelled into the fibre, where they travel down the cable. A light-sensitive receiver on the other end of the cable then converts the pulses back into the digital ones and zeros of the original electronic signal.

Think of a fibre-optic cable in terms of being a long tube that is coated with a mirror on the inside. If you shine a flashlight in one end you can see light come out at the far end – even if it's been bent around a corner. Light pulses move easily down the fibre-optic line because of a principle known as total internal reflection. This states that, when the angle of incidence exceeds a critical value, light cannot get out of the glass; instead, the light bounces back in.

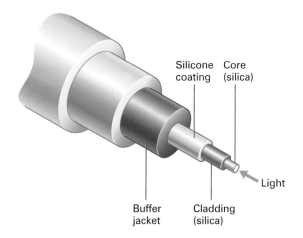

Figure 4.14 Cross section through a fibre-optic cable

The core must be a very clear and pure material for the light. The core can be plastic (only used for very short distances) but most are made from glass. Glass optical fibres are almost always made from pure silica, but some other materials, such as fluorozirconate, fluoroaluminate and chalcogenide glasses are used for longer-wavelength infrared applications.

Two types of fibre-optic cable are commonly used: single-mode and multi-mode.

The simplest type of optical fibre is called single-mode. It has a very thin core of about 5–10 microns (millionths of a metre) in diameter. In a single-mode fibre, all signals travel straight down the middle without bouncing off the edges. Cable TV, Internet and telephone signals are generally carried by single-mode fibres, wrapped together into a huge bundle. Cables like this can send information over 100 km (60 miles).

Another type of fibre-optic cable is called multi-mode. Each optical fibre in a multi-mode cable is about 10 times bigger than those in a single-mode cable. This means light beams can travel through the core by following a variety of different paths, in other words, in multiple different modes. Multi-mode cables can send information only over relatively short distances and are used (among other things) to link computer networks together.

Tight radius bends in fibre-optic cable should be avoided, as should 'kinks'. Jointing and termination requires specialist tools and equipment. Never look into the ends of the cable as the laser light could damage your eyes.

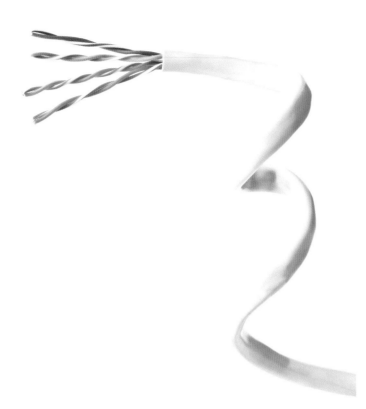

Figure 4.15 Cat (Category) 5e, 6 and 7 cable

Cat (Category) 5e, 6, 7 and 8 cable

In the past, there was very little choice in the type of cabling used for the design and installation of a multimedia networked home. For audio, there was speaker cable; for video and RF (radio frequency) signals there was coaxial cable; and for computer data there was Cat 5.

Although Cat 5e is still installed today, and indeed continues to meet most needs, Cat 8 cable represents the latest development of twisted-pair copper data communications transmission technology. Indeed all of these cables use four twisted pairs in their construction.

Of the older cable grades, Cat 5e can take a maximum bandwidth of 100M Hz, Cat 6 is capable of 400 MHz and Cat 7 625 MHz. Cat 8 is capable of carrying a bandwidth of up to 1400 MHz, so it easily meets and exceeds the IEC 61156-7 standard, which sets the benchmark for multimedia cables at 1200 MHz. Consequently, Cat 8 is normally referred to as 1.2 GHz cable, and is capable of taking a standard European TV signal with a bandwidth of 862 MHz without any form of signal manipulation.

Cat 8 cable (below) achieves its signal capacity by virtue of its precision manufacture. The cable is constructed from four colour-coded copper and polyethylene-insulated pairs.

These are twisted together to very tight tolerances and then each pair is wrapped in its own aluminium-bonded polyester tape shield. The four foil-screened pairs are then twisted together to provide a single-core element that is covered with a tinned copper wire braid for additional shielding. This cable has an off-white outer sheath made of a zero-halogen, low-smoke and flame-retardant compound that is fully verified for compliance with all relevant international standards.

Other types of cable

Low Smoke and Fume (LSF) cable

LSF is a generic term for cables that are made with insulation that produces low smoke and fumes under fire conditions. The insulation can be used on different sorts of cable and is widely used in public building installations.

Some local authorities may require the installation of this type of cable in special installations: black smoke from PVC insulation can obscure exit routes in the event of a fire and hydrogen chloride gas can be deadly to both people and sensitive equipment.

At one time, LSF was perceived as a 'type' of cable and data, signal and control cables were made from a modified version of PVC that could still give off black smoke and hydrogen chloride gas when burned.

We would now say that a low-smoke cable has a LSZH sheath, where LSZH stands for Low Smoke Zero Halogen and refers to the compound making up the sheath of a cable. In the event of a fire, this type of sheath will emit very low levels of smoke, and non-toxic levels of poisonous halogen gases (typically under 0.5 per cent HCl emission). This type of sheathing is mainly recommended for use in highly populated enclosed public areas.

Other equivalent names in use for this type of sheathing are LSHF (Low Smoke Halogen Free), LS0H (Low Smoke Zero Halogen) and 0HLS (Zero Halogen Low Smoke).

Fire-resistant cables

Although mineral-insulated cable is seen as the best performer in this category, there are several other types of cable that fulfil the criteria of fire-resistant cables.

FP.100

Systems based on standard PVC cables, even when protected by metal conduit, have been shown to fail within minutes in a fire. They are no longer permitted by BS5839-1 or BS5266-1 for fire alarm or emergency lighting systems, as they do not ensure prolonged operation in the event of a fire.

FP.100 was designed for use in metal conduits or trunking to achieve a highly robust fire-resistant wiring system. It has a maximum operating temperature of 90°C and, when burned, produces low levels of smoke and virtually no acidic gases.

Did you know?

In 1987, a fire broke out under an escalator in King's Cross Station, killing 31 people and injuring 60 more. After this tragedy, it became mandatory to use LSZH sheathing on all London Underground cables, as the majority of fatalities occurred through gas and smoke inhalation, rather than directly from the fire itself.

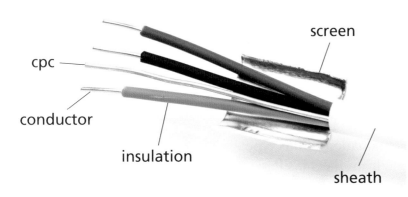

Figure 4.16 FP.200 Gold fire-resistant cable

FP.200 Gold

This cable is suitable for fire detection and fire alarm systems for buildings, and indoor and outdoor installation in suitably protected environments. It is appropriate for surface wiring, direct burial in plaster, clipping to tray or other installations requiring a dressable product. It has copper conductors with a composite insulation called Insudite.

FP.400 (Formerly CWZ cable)

FP.400 is a fire-resistant armoured cable that is installed much like ordinary SWA cable. It can maintain circuit integrity during a fire and produces low levels of smoke and virtually no acidic gases.

No special tools or accessories are needed for installation or termination. The cable has an operating range of 0°C to 90°C, and is suitable for indoor or outdoor installation, direct burial, trough, fixed direct, clipped to tray or ladder.

PVC-insulated and -sheathed flat twin flexible cord

This is only intended for light duty indoors, for table lamps, radios and TV sets where the cable may lie on the floor. It should not be used with heating appliances.

The construction is that of plain copper flexible conductor, PVC-insulated two cores laid parallel and sheathed overall with PVC. Core colours are brown and blue.

Figure 4.17 PVC-insulated and -sheathed flat twin flexible cord

PVC-insulated bell wire

As the name suggests, this cable is used for wiring bells, alarms and other indicators that operate at extra low voltage. The construction is one single-core plain soft copper conductor insulated with PVC. Twin-core wire is produced from two single-core wires laid parallel and insulated overall with PVC compound to form a figure-8 section.

The standard colour of this wire is white; core identification is by a coloured stripe on the insulation of one core. It is not suitable for mains voltage circuits.

Figure 4.18 PVC-insulated bell wire

Containment types (aka wiring enclosures)

In this section you will look at different types of cable containment systems including:

- conduit
- trunking
- cable tray
- cable basket
- ladder systems
- ducting
- modular wiring systems
- busbar systems (including Powertrack).

Conduit

Conduit is a system of pipework into which cables are pulled through between various pieces of electrical equipment. There are two common types: steel and PVC. Electrical conduit provides very good protection to the enclosed conductors from impact (steel obviously more so) and moisture, and can be concealed, surface-mounted or installed in a floor screed.

Figure 4.19 Galvanised steel conduit

With steel conduit, there are two types: black enamelled and galvanised. Both see use in commercial and industrial applications, with black enamelled being used indoors where there is no likelihood of dampness, and galvanised used in damp situations or outdoors.

Steel conduit is traditionally available with tube diameters of 16 mm to 32 mm, and PVC conduit is available in 20 mm through to 50 mm.

Both steel and PVC have a wide range of accessories such as boxes, bends, and saddles that remove some of the fabrication problems. PVC is the lighter and cheaper of the two systems, but steel can afford greater mechanical protection.

Once the conduit system is erected, PVC-insulated (non-sheathed) cables are normally run inside the steel tubing.

The appeal of both types is that, once installed, wiring can be removed from the conduit and replaced: for example, because of age or different circuit arrangements. However, sometimes the initial cost versus the life expectancy of the installation can prohibit its use.

Figure 4.20 Box, bend and saddle accessories (metal and plastic)

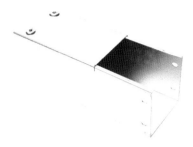

Figure 4.21 Steel trunking

Trunking

Trunking is a normally square/rectangular casing system with a removable lid, into which cables are then pulled through between various pieces of electrical equipment.

As with conduit there are two common types: galvanised steel and PVC. Both can be used indoors in commercial and industrial applications where there is no likelihood of dampness, and galvanised is used in damp situations or outdoors in its weatherproof version.

Electrical trunking provides good mechanical protection and is traditionally installed in conjunction with its relevant PVC/steel conduit counterpart; the trunking is used to carry many cables down a common route and conduit then branches off to supply individual circuits or items of equipment.

Steel trunking is traditionally available with dimensions of 50 mm × 50 mm through to 300 mm × 300 mm. PVC trunking can be more aesthetically pleasing so is manufactured in a wide range of styles and sizes, from 'standard' shaped trunking (maximum 100 mm × 100 mm) .

Unlike conduit, which is traditionally used as a complete system with unsheathed cables, PVC trunking is often used as surface-mounted protection for short drops to accessories from ceilings for other cables, such as data cabling or PVC/PVC cables. This is especially useful when circuits are added to an existing installation and cannot be concealed without damaging the building fabric.

Whatever type or style is used, both steel and PVC have a wide range of accessories such as boxes, bends, and saddles, which remove some of the fabrication problems. PVC still remains the cheaper of the two and because of the nature of its fittings can be easier to install, with some of the smaller types being self-adhesive.

Figure 4.22 PVC Trunking

Once the system is erected, PVC-insulated (non-sheathed) cables are normally run inside. As with conduit, the appeal of both types is that, once installed, wiring can be removed from the trunking and replaced: for example, because of age or different circuit arrangements.

Trunking design

There are many variations of trunking design. Some of the most common are:

- **Lighting trunking** – this differs from standard trunking as it uses a single flange in its design. This helps to accommodate trunking support brackets, which are normally held in turn via threaded rod supports from the structure of the building. The return flange also accommodates a clip-in trunking lid rather than the 'screw-in-place' version used on the standard design. Clamp or spring style brackets are also used to support lighting fittings directly to the trunking or chain support the fitting to a required height. Figure 4.28 shows lighting trunking.

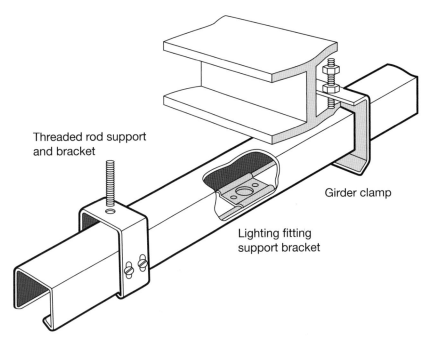

Threaded rod support and bracket

Girder clamp

Lighting fitting support bracket

Figure 4.23 Lighting trunking

- **Multi-compartment trunking** – to comply with BS 7671, in terms of segregation of circuit voltage bends, multi-compartment trunking is available in metal and plastic, ensuring segregation throughout. It is often used as a surface mounted dado trunking within offices, it typically contains power and data cables that are then terminated into data points or socket outlets as required to suit the office layout

- **Floor trunking** – this is similar in appearance to powertrack, with the exception that all circuit cabling must be installed by the contractor

Cable tray

There are two main designs of cable tray: single-flange and return-flange.

Single-flange tray

Single-flange tray is sometimes referred to as standard vertical-edge tray. It is only suitable for light-duty installation work or for use in food manufacturing environments, where its simple structure reduces build up of dust, etc. Made from galvanised steel and available in widths from 50 mm to 900 mm, the tray can be mounted either horizontally or vertically with cables then laid on the tray and secured using clips or ties through the numerous perforations in the tray.

Figure 4.24 Single-flange cable tray

The vertical flange height (the upstand on each side) will vary with each manufacturer, but can be expected to be between 13 mm for narrow widths and 20 mm on wider tray.

Return-flange tray

Return-flange cable tray differs from standard in that the vertical edge has a turned edge at the top, making the tray stronger and less likely to warp. There are two types of return flange tray: medium- and heavy duty.

Made from galvanised steel and available in widths from 50 mm to 900 mm, the tray can be mounted either horizontally or vertically to accommodate medium to large industrial-sized cables, which are then laid on the tray and secured using cleats or straps that are fixed through the numerous perforations in the tray.

The vertical edge height (the upstand on each side) will vary with each manufacturer, but can be expected to be about 25 mm for medium-duty tray and 50 mm for heavy-duty.

Figure 4.25 Return-flange cable tray

Cable ladder

Cable ladders are an effective and convenient method of transporting cables across long, unsupported spans or where the number of supports needs to be reduced. They may be used in the most adverse site conditions and can withstand high winds, heavy snow, sand or dust settlement, or high humidity.

Cable ladders comprise a series of prefabricated 'ladder' sections, usually available in different widths, together with a comprehensive range of ancillary components and accessories which minimise site installation time and costs. The result is a racking network of considerable strength and flexibility.

Figure 4.26 Cable ladder

Cable ladders are manufactured throughout in 2 mm mild steel with coupling plates made from 3 mm mild steel. The ladder side channels are strengthened by reinforcing inserts, to increase the lengthways **torsional** rigidity. Rungs are slotted to take most available types of cable cleat, cable tie, pipe or conduit clamp. Cable-ladder design allows the maximum airflow around the cables and so prevents possible de-rating of power cables. Cable ladders may be mounted in virtually any direction, as required.

> **Key term**
>
> **Torsional** – to do with twisting, when 'torque' or turning force is applied

Cable basket

Cable basket is not unlike cable ladder in terms of its applications. Made from a steel-wire basket, it requires similar installation methods and techniques.

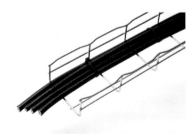

Figure 4.27 Cable basket

To cut the basket to form bends or tees, you would usually use bolt cutters. Any cuts then need to be made smooth, as with tray or ladder systems.

Cable ducting

BS 7671 defines cable ducting as being 'an enclosure, other than conduit or trunking, intended for the protection of cables that will be drawn in after erection of the ducting'. From that definition, it could be argued that ducting sounds like conduit or trunking – and certainly, when you look at the ducting products available from most manufacturers, it looks like conduit and trunking.

Cable ducting is usually earthenware or concrete buried in slab or ground, utilised to allow cables with multiple distribution circuits to be run together. Ducts are usually easier to access and can host many cables in common runs due to their capacity.

Although underfloor steel cable trunking is often referred to as electrical ductwork, circular PVC cable ducting is used as the means of installing cables in underground situations, predominantly by the supply industry and telecommunications and digital information providers.

Runs of ductwork then tend to run between sunken inspection chambers, which allows inspection and aids the installation process.

When it comes to the installation of network installations, there has also been the development of what are termed 'blown fibre' installations, where the fibre-optic cables are not 'drawn in' to the ducting system, but are instead air propelled along the length of a duct.

Busbar systems

A busbar system refers to conductors that take the form of a solid bar or bars of aluminium or copper conductor. The bars may be exposed or enclosed. The system may have one or more joints to assure proper length and configuration, and one or more take-off points that are connected to the relevant equipment. The exposed busbar system tends to be used for overhead equipment such as cranes, loading bridges, container-handling equipment or electric hoists.

The exposed bars can also be used as a system within large switchboard panels or within busbar chambers as a means of connecting together items of equipment in purpose-made arrangements on site. This could be seen

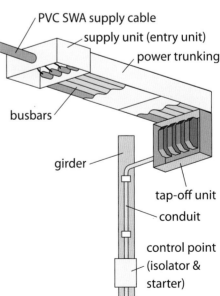

Figure 4.28 Busbar trunking showing tap-off unit

as enclosing them as they are now within a panel or enclosure, but enclosed busbar systems are more commonly referred to as busbar trunking.

Busbar trunking is a popular means of three-phase power distribution in machine shops, laboratories and many industrial situations. It consists of a broad, flat trunking in which three or four busbars are rigidly fixed onto moulded block insulators.

The conductors used in busbar trunking are generally made of copper or sometimes aluminium, and can be sleeved with insulating material or left bare. The size of each copper bar will vary according to the current-carrying capacity required; their shape can be either round, oval or rectangular.

A complete range of fittings is available, such as feed units, right-angled bends, tee pieces and crossovers. These fittings, which allow the trunking to change route, are self-contained assemblies complete with busbars and couplings.

This system has several advantages over distribution cables.

- The contractor can achieve savings in material costs (cable tray and multiple fixings) and in labour costs associated with multiple runs of cable.
- Installation takes less time as busbar trunking requires fewer fixings per metre run than cable.
- Repositioning of distribution outlets is simpler.
- The system is easy to extend.
- It is more aesthetically pleasing in areas of high visibility.

The supply that feeds the busbar trunking enters via a feed unit, which has a gland plate capable of accepting an SWA cable. Busbar trunking can also be centre-fed, the centre feed being more popular as this method makes full use of the busbar capacity.

One major advantage of busbar trunking is the ability to 'tap off' and provide a supply to a piece of equipment – something that would normally entail installing a new cable. Tap-off units will contain the device providing protection (e.g. HRC fuse or MCB) to the outgoing circuit terminated at the unit that distributes the power to the required load which also offers the flexibility to accommodate changes in requirements after the initial installation. The tap-off units can be either fixed or be of the plug-in type, and can be side-mounted or under-slung. Sets of drillings are provided at regular intervals along the busbar trunking, and blanking-off plates are supplied to cover off the apertures not initially used.

Another use of busbar trunking is to provide what is called a 'rising mains' within multi-storey developments for hospitals, offices, flats, and so on. In this system, purpose-made busbar trunking is run vertically through the walls of the building (see Figure 4.29). It is fed and controlled usually from the bottom at the service entry, and has a fixed fuse box mounted at each floor.

The essential difference between horizontal type busbar trunking and vertical rising main type busbar trunking is the substantial insulated support rack at the base of each riser, which is designed to carry the full weight of the copper conductors, which are then free to expand upwards.

Powertrack

While busbar trunking systems can be designed to handle heavy loads, lighter-duty distribution can be handled by Powertrack systems, which use solid busbars throughout their length. There are three common variations: suspended lighting, wall mounted and underfloor.

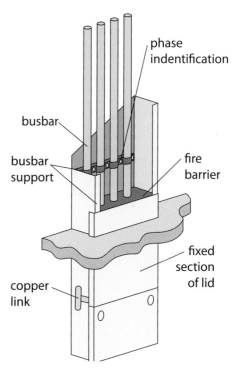

Figure 4.29 Rising mains busbar trunking

Wall-mounted power tracks tend to be used in offices and laboratories. They have the advantage that appliances and equipment can be plugged in anywhere along the length of the track. The socket outlet is simply inserted into the slot and then twisted to lock it in place and make it part of the circuit.

Underfloor track systems tend to be used in large, open-plan offices where a common run of track is fed from the distribution board via track feed boxes that accept MI cable, armoured cables or single-core in conduit.

The first length of track connects to the track feed by snap-fitting the integral track connector into the track feed outlet socket. Lengths of track fit together by simply snap-fitting the track connector of one end into the shuttered end of the previous track length.

Access to power is provided along the Powertrack by plugging tap-off units into shuttered socket outlets. These tap-off units feed all types of conventional floor service outlet boxes or feed directly through the floor to workstations, via 4 mm^2 insulated conductors contained in 2.8 metres of flexible metal or **VO-rated** nylon conduit.

Although this system has the advantage of saving installation time, as it is under the floor and therefore less accessible, it is slightly less flexible than the wall-mounted version.

Key term

VO-rated – a flame resistant plastic made from materials with flame protection above industry standards

Lighting track is similar to underfloor, in that is normally seen in the form of a steel trunking with tap-off outlets along its length. The difference is that it is suspended and its connections accommodate the required luminaires.

However, with advancing technology, lighting track systems can now be provided as **addressable**. This technology enables fully addressable lighting control modules to combine with presence and absence detectors, scene set switching and timing scheduling, all of which can help with energy conservation and therefore Building Regulations compliance.

> **Key terms**
>
> **Addressable** – digitally programmable
>
> **Modules** – sections that link together to form a whole.

Modular wiring

The need for modular wiring

Modular buildings are becoming increasingly popular. In part, this is due to the UK government encouraging what is now referred to as Modern Methods of Construction (MMC). MMC reflects technical improvements in prefabrication, encompassing a range of on- and off-site construction methods, including 'plug and play' wiring systems. Primarily it involves the manufacture of buildings in factories, with potential benefits such as faster construction, fewer housing defects, and reductions in energy use and waste.

Today, modular building techniques mean that building services such as a complete boiler room or even the control tower at Heathrow Airport can be built off site, delivered and then craned into position. A whole structural part of a house can be delivered to site as a single unit, complete with water, heating, lighting and other equipment ready to connect to the main water, lighting and power systems.

Depending on the design and materials used, with these methods a house can be assembled in less than a month on site. Time savings can be high, but initial costs can also be high. However, if co-ordinated and designed correctly, savings in the region of 50 per cent can be expected. At the same time there are reductions in emissions, thus helping with climate control.

However, such flexibility needs a similarly flexible approach to wiring. Modular wiring systems have also developed, whether for use in modular buildings or traditionally constructed projects.

How modular wiring works

Think about the lighting installation for the university warehouse that you looked at in Unit ELTK03, which needed to be wired with PVC singles in galvanised conduit.

Now imagine that every run of wiring – whether to a switch, socket or luminaire – was provided to you as a pre-measured armour-protected cable, fully marked up and with a plug on the end of it.

Imagine that all equipment, including the distribution board, was already connected up as the consultant intended and was provided with plugs on for you to connect to the previously provided cable runs. Welcome to the world of modular wiring!

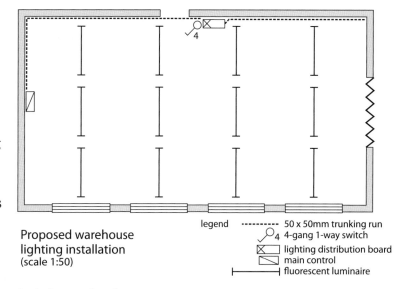

Proposed warehouse lighting installation (scale 1:50)

legend	
- - - - -	50 x 50mm trunking run
4-gang 1-way switch	
lighting distribution board	
main control	
fluorescent luminaire	

Scale layout drawing

The application of systems, circuits and equipment

Lighting systems

There are numerous combinations of wiring system, wiring enclosure and equipment that can constitute a lighting circuit. In this section you will look at the most common types of each.

Technically, any of these elements could be used in a variety of situations. However, apart from modular systems, you will most commonly find:

- PVC/PVC twin and cpc used in domestic installations
- conduit/trunking with PVC singles used in commercial and industrial applications
- MI cable and SWA cables being used in specialist or outdoor circumstances.

As for the luminaire, you can expect to see:

- either pendant or recessed LED or halogen lighting in domestic indoor situations
- surface-mounted batten-type or recessed modular-type fluorescent fittings in office/commercial environments
- suspended/recessed halogen spotlighting in shops
- halogen strip lighting for outdoor security and construction site lighting
- specialist high-pressure sodium or metal halide fittings used for stadium and street lighting applications.

Progress check

1 When selecting a cable for an electrical installation, what are the main criteria to consider?

2 Identify the main factors affecting the choice of size of conductor to be selected.

3 What are the significant advantages of PVC/SWA over PVC/PVC cable?

4 Identify suitable fire-resistant cables for fire alarm installations.

5 What are the benefits, of utilising busbar distribution systems in a factory for machinery?

Switching arrangements

It would not be practical to look at all the possible switching arrangements, as there are simply far too many. Instead you will look at the most common arrangements:

- how each one operates
- how each one would be wired when using either PVC singles or PVC/PVC multi-core cables
- how the wiring colours relate to the old and current systems.

With switching, the supply cable that comes from the distribution board to the switch is called the 'switch feed'; the control cable that leaves the switch and goes to the light is the 'switch wire'.

One-way switching

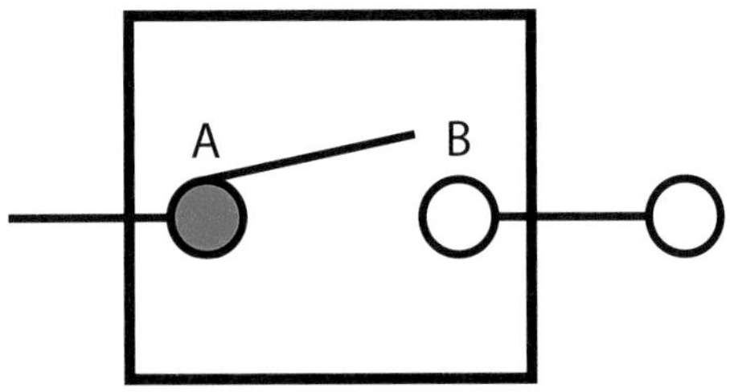

The most basic circuit possible is where one switch breaks the supply to one light. A one-way switch is the most basic type of switch to achieve this.

As Figure 4.30 shows, one terminal of the one-way switch (terminal A) receives the switch feed, and the switch wire leaves from the other terminal (terminal B) and goes directly to the luminaire. In Figure 4.30, the luminaire is not lit as the switch contact is in the 'open' position.

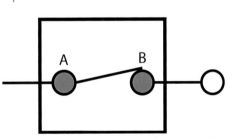

Figure 4.30 One-way switching: switch 'open'

Figure 4.31 One-way switching: switch 'closed'

Figure 4.31 shows how, when you operate the switch, the switch contact 'closes' and is held in place mechanically allowing electricity to flow to the light. The luminaire will now be lit.

So:

- supply the switch feed terminal, (A)
- operate the switch and it comes out at the switch wire terminal, (B).

Figure 4.32 shows one light controlled by a one-way switch. Figure 4.33 shows how you can connect a second light in parallel

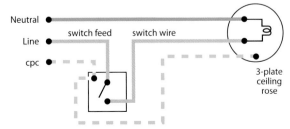

Figure 4.32 One-way switching for wiring with single-core cables (new cable colours)

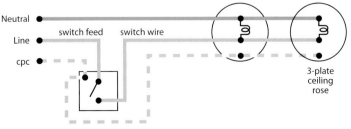

Figure 4.33 Extra lighting fed from the same switch, wired in parallel (new cable colours)

to the first, so that the one-way switch is controlling two lights. Both show the current cable colours.

However, Figure 4.34 shows what the circuit would look like using the old cable colours. Be aware that you may come across these colours in existing installations for many years to come.

In summary, the one-way switch lets us switch lights on or off, but from one location.

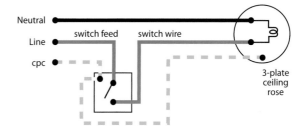

Figure 4.34 One-way switching for wiring with single-core cables (old cable colours)

Two-way switching

Sometimes you will need to switch a light on or off from more than one location: for example, at opposite ends of a long corridor. Here you must use a different switching arrangement, usually a two-way switch circuit. In this type of circuit, the switch feed is taken to one terminal of a two-way switch, and the actual switch wire goes from a second two-way switch to the luminaire(s). Each switch has three terminals. Two wires, known as 'strappers', then link both two-way switches together.

This is how each two-way switch operates.

You supply the switch feed terminal, point (A).

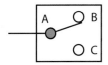

Depending on the switch contact position, the electricity will come out on either terminal B or terminal C. In the following diagram, it is shown energising terminal B.

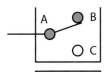

If you now operated the switch, the contact would move across to energise terminal C.

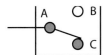

Take a look at the next diagram, which shows actual terminal marking. You can see that, by connecting both two two-way switches together, at each switch you can now either energise or de-energise the switch wire going to the light. This is why this

> **Remember**
>
> Note that actual switch terminals are not marked in this way. The letters used in the diagram are for ease of explanation only.

system is ideal for controlling lighting on corridors or staircases. In this diagram, the luminaire is off.

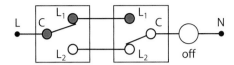

You can now bring the light on by operating the first switch:

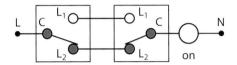

Alternatively, you can do so by operating the second switch:

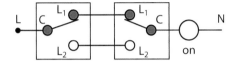

Figures 4.35 and 4.36 show the full circuit wired with single-core cable. Figure 4.35 uses the current cable colours; Figure 4.36 uses the old colours. In both, the light will be on.

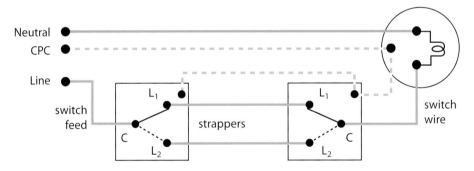

Figure 4.35 Full circuit wired with single core cable (current colours)

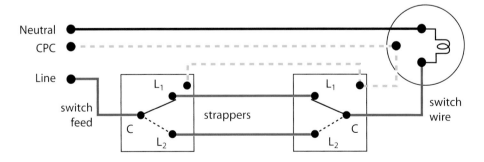

Figure 4.36 Circuit wired with single core cable (old colours)

You can use a two-way switch for other purposes: for example, to switch between two separate functions. As an example, the next diagram shows how a two-way switch on a desk can be used to indicate whether a doctor's treatment room is available.

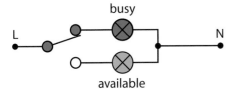

Figure 4.37 Two way switch on a desk to show doctor availability – the switch activates the lights

Intermediate switching

Two-way switching allows us to control luminaires from two locations, such as either end of a corridor. If more than two switch locations are required (for example, the corridor has another corridor coming off it), then you need a third location from which you can control the luminaires. To do this you use intermediate switches.

Intermediate switches have four terminals and are wired into the strappers between the two-way switches. The intermediate switch cross-connects the strapping wires, enabling you to route a supply to any terminal, irrespective of the two-way switch contact positions.

Each intermediate switch has two contacts that move between two positions. In position one, both contacts lie in the direction of the strapper wires. If terminal A is live, so is terminal B.

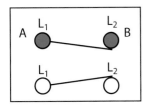

Figure 4.38 Position one

In position two, the contacts cross over to connect the strapper wires. Now if terminal A is live, terminal C has become live.

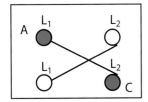

Figure 4.39 Position two

Here is how you might apply this to a circuit.

Imagine that you have two corridors joining each other as shown below. You are walking in the direction of the arrow and can see that the main corridor lights are turned off. The main corridor is 100 m long, with a two-way switch at each end. There is also an intermediate switch where the two corridors meet.

This diagram on the right shows a circuit that represents the wiring problem.

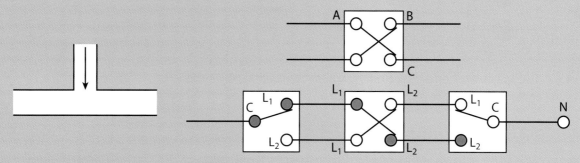

As you can see, the contacts in the intermediate switch are crossing, which is routing the supply to terminal L2 in the second two-way switch, so the light is off.

However, if you operate the intermediate switch as shown in the next diagram, the contacts revert to their other position and the supply is routed to terminal L2 in the second two-way switch, so light will come on.

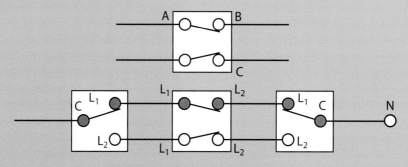

Figure 4.40 Intermediate switching at each junction of a corridor

When wired in single-core cable (current colours) the circuit will look like this:

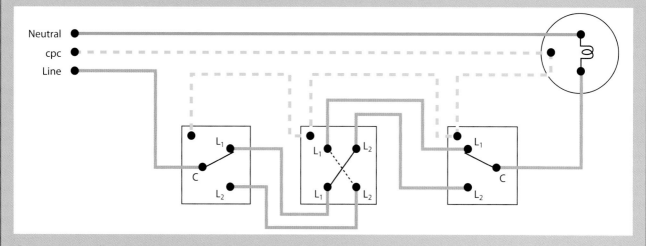

Figure 4.41 Full circuit in new cable colours

Wiring with multi-core cable

In this section, **multi-core** refers to PVC/PVC flat cable with cpc:

* with two cores and cpc (ref. 6242Y, often called 'twin and earth'), and

* with three cores and cpc (ref. 6243Y).

In the old colours, it would have been correct to use cable that contained two red conductors (switch feed and switch wire) from any switch location up to the luminaire. However, for ease, and as shown in Figure 4.42, it was common practice for many to use a cable containing red and black conductors, as this would be the most available combination on site.

Key term

Multi-core – having more than one core

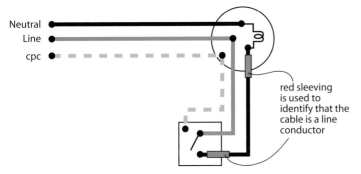

red sleeving is used to identify that the cable is a line conductor

Figure 4.42 Wiring with multi-core cables: old cable colours

The black conductor should be sleeved at both ends to indicate that it is a 'live' conductor. Be aware that some electricians never sleeved these conductors, so when removing a luminaire on an existing older installation you may find luminaires that are connected across two black conductors. One of these conductors will be a switch wire and the other the neutral.

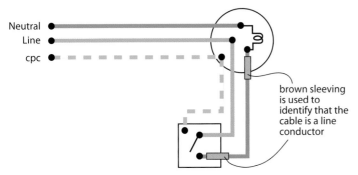

brown sleeving is used to identify that the cable is a line conductor

Figure 4.43 Wiring with multi-core cables: new cable colours

Although 'twin brown' cable is available, the same problems can exist on installations that use the current cable colours. On any multi-core cable installation where any conductors other than brown are used as a phase conductor, these should be fitted with a brown sleeve at their terminations.

Using this type of cable instead of singles is slightly more complicated because you are restricted as to where you plan your runs. You can use a **'loop in'** method with this type of installation as there are no joint boxes installed and all terminations are readily accessible at the switches and ceiling roses.

In the joint box system, once popular in local council installations, normally only one cable is run from the joint box to each wiring outlet. Where such joint boxes are installed beneath floors they should be made accessible via a screwed trap in the floorboard directly above the joint box. All conductors should be correctly colour identified.

All conductors must be contained within a non-combustible enclosure at wiring outlets (the sheathing of the cable must be taken into the wiring accessory). Throughout the lighting installation a cpc must be installed and terminated at a suitable earthing terminal in the accessory or box.

Where an earthing terminal hasn't been provided in a PVC switch **pattress**, the cpc may be terminated in a connector. Where the sheathing is removed from a multi-core cable, the cpc must be fitted with an insulating sleeve (green and yellow); this provides equivalent insulation to that provided by the insulation of a single core non-sheathed cable of appropriate size complying with BS 6004 or BS 7211.

Two-way switching and the conversion circuit

Installing two-way switching using PVC/PVC and cpc cable requires the installation of a three-core and cpc between the two switches.

You can also use this method when converting an existing one-way switching arrangement into a two-way switching arrangement.

In the old cable colours, the three coloured conductors were red, yellow and blue. In the current cable colours, the three conductors are coloured brown (L1), black (L2) and grey (L3).

To show both the two-way and conversion process, we will use the example of a one-way circuit that has been installed using only brown and blue cable, which now needs to be a two-way circuit.

Conversion of one-way switching to two-way switching

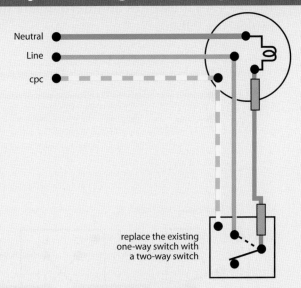

replace the existing one-way switch with a two-way switch

Step one: First replace the existing one-way switch with a two-way switch. You should then connect the new two-way switch as shown above, noting that the supply has been taken to terminal L1, not the common terminal as usual.

Note that in the diagram the light is off.

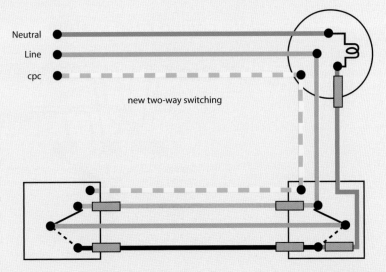

new two-way switching

Step two: Now install a three-core and cpc PVC/PVC flat cable between the existing switch location and the new two-way switch location, then complete the circuit connections as shown above.

Again, the diagram shows that, with the switches in these positions, the light will be off.

Intermediate switching

Figure 4.44 illustrates an intermediate lighting circuit using multi-core cables.

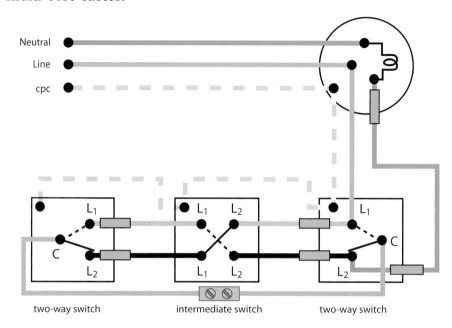

Figure 4.44 Intermediate switching

Three-plate ceiling rose system

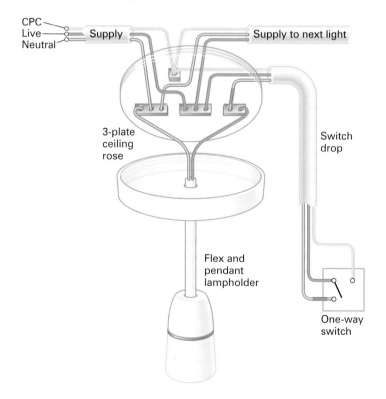

Figure 4.45 Ceiling rose system

Probably the most common domestic wiring system, the three-plate or 'loop in' system works by taking the PVC/PVC and cpc cable from the supply to the first lighting point and then looping the cable between this and every other lighting point on the circuit.

The cables are terminated inside a three-plate (four terminal) ceiling rose at each lighting point as shown. The four terminals are:

- line in/out
- neutral in/out
- switch wire
- cpc.

A two-core PVC/PVC flexible cable is then connected from a pendant lampholder across the switch wire and neutral terminals within the ceiling rose as shown.

Power (final) circuits

BS 7671 defines a final circuit as being 'a circuit connected directly to current-using equipment or to socket outlets/other outlet points for the connection of such equipment'. BS 7671 then goes on to define current-using equipment as 'something which converts electrical energy into another form such as light, heat or motive power'.

The list of circumstances that meet these criteria goes on forever. This section will concentrate on the use of socket outlets and fused-connection units and the installation of rings, radials and spurs.

The ring final circuit

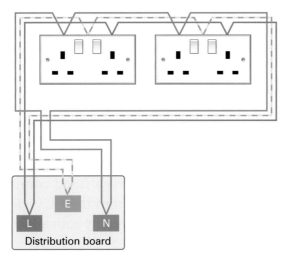

Figure 4.46 Ring final circuit

The ring main was introduced just after the Second World War. The need for electricity had to be matched against the low amount of copper available, so the circuit design allowed a ring to be created by simply linking two existing radial circuits together.

As shown in Figure 4.46, in the ring final circuit, the line, neutral and cpc start from their respective terminals at the distribution board and pass through the corresponding terminals of each socket outlet before returning to the same terminals at the distribution board, thus forming a ring where they are protected by a 30 A or 32 A protective device.

Appendix 15 of BS 7671 sets out the options for the design of both ring and radial circuits for domestic premises in accordance with Regulation 433.1: using socket outlets and fused connection units.

Although an unlimited number of socket outlets can be installed, the load current in any part of the circuit should be unlikely to exceed, for long periods, the current carrying capacity of the cable (Regulation 433.1.5).

This can generally be achieved by:

- not supplying immersion heaters, electric space heaters or similar from the ring
- positioning the sockets so that the load is reasonably shared around the ring
- connecting cookers, ovens and hobs with a rated power greater than 2 kW on their own, dedicated radial circuit
- taking account of the floor area being served by the ring – as a rule of thumb, a limit of 100 m² has been adopted.

Here are the terms used in Figure 4.47:

- A – a BS 1363 single or twin 13 A socket outlet. Each outlet of twin or multiple sockets is regarded as one socket outlet.
- B – a BS 1363 socket fed via an unfused spur. An unfused spur should only feed one single or twin socket outlet. The unfused spur may also be connected to the origin of the circuit in the distribution board.
- C – a fused connection unit (FCU) supplying fixed equipment. The fuse in the FCU must not exceed 13 A.
- D – A BS 1363 socket outlet fed via a fused connection unit. The number of sockets supplied from the FCU depends on the load characteristics, taking diversity into account.

- E – A spur made using a junction box and FCU that is not directly connected to the ring.

The FCU is connected to the ring via a junction box that must have screw terminals and be accessible for inspection. The number of sockets that may be installed depends on the load, taking diversity into account.

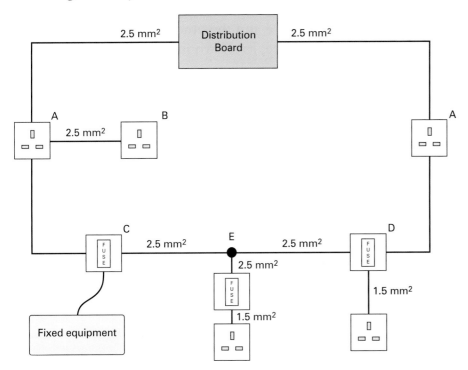

Figure 4.47 Ring final circuit

The conductor sizes shown in Figure 4.47 relate to the live conductors only within flat 'twin and earth' cable. A reduced circuit protective conductor size is permitted (for example, 2.5 mm² live conductors with 1.5 mm² cpc).

Wherever possible, cables should be fixed in such a position as not to be covered by thermal insulation. Should a cable be partially or completely covered by thermal insulation then reference should be made to Regulation 523.7.

Where more than one ring main is installed in the same premises, it is good practice to reasonably share the load over the ring main circuits so that the assessed load is balanced. Care must also be taken to ensure that the requirements of Regulations 411.3.3 and 522.6.6 to 522.6.8 have been met in terms of additional protection by an RCD.

Radial circuits

In this type of circuit, the line, neutral and cpc start at the distribution board connected via an appropriate overcurrent protective device, then pass through the respective terminals of each socket outlet in the circuit and finish at the last outlet.

There are two types of radial circuit: one protected by a 20 A overcurrent protective device and the other by a 30 A/32 A overcurrent protective device, as shown in Figure 4.48.

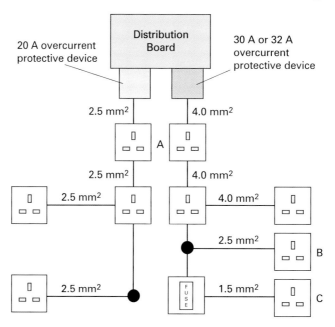

Figure 4.48 Radial circuits with overcurrent protective device

Here are the terms used in Figure 4.48.

- A – a BS 1363 single or twin 13 A socket outlet. Each outlet of twin or multiple sockets is regarded as one socket outlet.
- B – a BS 1363 socket fed via an unfused spur. An unfused spur should only feed one single or twin socket outlet and only 2.5 mm² cable should be used. The unfused spur may also be connected to the origin of the circuit in the distribution board
- C – a BS 1363 socket outlet fed via a fused connection unit. The number of sockets supplied from the FCU depends on the load, taking diversity into account.
- ● – a junction box. Junction boxes with screw terminals must be accessible for inspection, testing and maintenance. Alternatively, use maintenance-free terminals/connections (Regulation 526.3).

The conductor sizes shown relate to the live conductors only within flat 'twin and earth' cable. However, reduced circuit protective conductor size is permitted (for example, 4.0 mm² live with 1.5 mm² cpc). Where the radial circuit is protected by a 20 A overcurrent protective device, the floor area being served is generally limited to 50 m². Where the radial circuit is protected by a 30/32 A overcurrent protective device, the floor area being served by the ring is generally limited to 75 m².

You must take care to make sure that the requirements of Regulations 411.3.3 and 522.6.6 to 522.6.8 have been met in terms of additional protection by an RCD.

Wherever possible, cables should be fixed in such a position as not to be covered by thermal insulation. If a cable is partially or completely covered by thermal insulation, reference should be made to Regulation 523.7.

Distribution systems (sub-mains)

Different people define 'sub-mains' in different ways. Here is one example, which we will use for this book.

Working life

Imagine you are working on a large house with a separate outbuilding that is intended for use as a garage and small workshop. The Distribution Network Operator (DNO) generally provides the supply to a building, so you can assume that they have provided the supply to this house. You can call this the main supply intake position, and you connect the main eight-way distribution board to it.

As one of the electricians on site, you are required to install a PVC/SWA/PVC cable from this distribution board to the garage/workshop, where you must terminate the cable in a smaller four-way distribution board. This cable and smaller distribution board could be termed a sub-main.

This example relates to a house, but the principle holds true for any size or type of building. If you accept that a sub-main is the distribution of electricity from a main intake position to a smaller sub-division of an installation, the range of installation possibilities is huge.

Generally, for commercial and industrial situations, you could expect to see PVC/SWA/PVC cable or XLPE/SWA/PVC cable, MI cable or busbar trunking being used. Depending on the installation, you could also see tray plate or basket systems being used to support such cabling.

Emergency management systems (standby power supplies)

A new section within BS 7671 – Chapter 56 – covers general requirements for the selection and erection of supply systems for safety services (such as emergency lighting, fire detection and alarm systems) and essential medical and industrial systems (such as air traffic control).

An electrical safety service supply is classed as either being:

- non-automatic – its operation is started by an operator
- automatic – its operation is automatic and independent of an operator.

Automatic systems are then classified according to their changeover time, such as:

- no break – automatic supply with continuous supply within specified conditions
- very short break – automatic supply available within 0.15 seconds
- short break – automatic supply available between 0.15 and 0.5 seconds
- medium break – automatic supply available between 5 and 15 seconds.

Figure 4.49 A diesel generator

Traditionally there have been two sources: batteries (normally for fire alarms and emergency lighting) and diesel generators for essential services). A third technology (rotary systems) provides an 'overlap' between them.

However, with the need for cleaner emissions, some manufacturers now make use of fuel cell technology. Whereas conventional generators use internal combustion engines to rotate an alternator, fuel cells generate power by producing electrons directly, with few moving parts. As a result, they have the potential to be very efficient and reliable. Moreover, fuel cells are comparatively quiet.

A fuel cell converts the chemicals hydrogen and oxygen into water and in the process produces electricity. This makes them ideal for indoor use (for example, in an IT facility, or where a diesel generator is difficult or impossible to install.

You should also be aware of some important terminology, such as auxiliary supplies, back-up supplies, standby supplies, emergency supplies and uninterruptible power supplies (UPS). Hospital facilities provide a good example of these.

If a mains failure occurs, hospital operations are typically protected by standby generators within the hospital grounds. These will provide power to allow essential services to continue as normal. During a mains failure, fire alarms and emergency lighting need to keep operating too, in both essential and non-essential areas, and this could be provided by a central battery system. This would be considered an auxiliary supply (or back-up, standby or emergency supply).

UPS

A UPS maintains a continuous supply to connected equipment by supplying power from a separate source (such as batteries). It is different from an auxiliary power supply or standby generator, as these don't provide instant protection from a momentary power interruption. Integrated systems that have both UPS and standby generators are often referred to as emergency power systems.

You may also see some official systems within the hospital having UPS protection, where there is an issue over quality of supply, or the need to ensure power while a generator starts. Here the UPS battery pack should be sized to allow the powering up of on-site generators and/or sufficient time to allow any medical procedure to be safely finished.

There are two main types of UPS: off line and on line.

- An off-line UPS remains idle until a power failure occurs, and then switches to its own power source, almost instantaneously.
- An on-line UPS (sometimes called double conversion) continuously powers the protected load from its reserves (usually lead-acid batteries), while at the same time refilling the reserves from the main supply. It also provides protection against power fluctuations and for this reason is used with IT equipment where a 'steady' supply is required.

Rotary systems

Bearing in mind the slight lag between loss of supply and the standby supply kicking in, some installations use a secondary rotary system in conjunction with the diesel generator.

The rotary system uses the inertia of a continuously spinning flywheel to provide short-term provision (for example, 120 kW for 20 seconds) in the event of power loss. The flywheel can also act as a buffer against power fluctuations, as these are not really capable of affecting the speed of the flywheel.

However, as a rotary system is only capable of providing reserve power for a few seconds, it is traditionally used in conjunction with standby diesel generators, providing back-up power only for the brief period of time the engine needs to start running.

Regulation 560.6 requires electrical sources for safety services 'to be installed as fixed equipment in such a manner that they cannot be affected by a failure of the normal supply'. They must be placed in a suitable, well-ventilated location to allow exhaust gas to escape, and must only be accessible to skilled or instructed persons.

With a central battery system, there is no upper limit of the supply capacity. However, Regulation 560.6.10 requires battery design to have a declared life of 10 years.

Regulation 560.7 also requires that circuits of safety services must:

- be independent of other circuits
- not pass through zones exposed to risk of explosion
- not pass through zones exposed to fire risk unless they are fire resistant (last resort)
- not be installed in shafts unless for a rescue service lift.

Regulation 560.7.10 also requires that a drawing showing the electrical safety service provision must be provided and displayed at the origin of the installation, giving the exact location of all connected equipment, circuit designation and switching arrangements.

Progress check

1 Describe briefly how a one-way lighting circuit could be converted to allow two switching positions.

2 How may additional socket outlets be added to a ring circuit?

3 To ensure the supply to essential equipment is maintained without loss what system should be employed?

Security systems (emergency lighting, fire alarm and security systems)

Emergency lighting

Emergency lighting is lighting required in an emergency because the main power supply has failed. Even if the failure is due to nothing more than a power cut, a lack of light could become dangerous for people in a building, either because of physical danger or through people panicking. Emergency lighting should operate fully automatically, affording a level of light that allows people to evacuate a building safely.

Emergency lighting should be planned, installed and maintained to the highest standards of reliability and integrity, so that it will operate satisfactorily when called on. It must be installed in accordance with the British Standard Specification BS 5266: Part 1: 1999 – Code of Practice for Emergency Lighting, which includes in its scope residential hotels, clubs, hospitals, nursing homes, schools and colleges, licensed premises, offices, museums, shops and multi-storey dwellings.

Emergency lighting is not required in private homes because the people living there know their surroundings. However, in public buildings, people are in unfamiliar surroundings and in an emergency they will need a well-illuminated and easily identified exit route.

The Workplace (Health, Safety and Welfare) Regulations also have a bearing on emergency lighting, stating: 'Automatic emergency lighting, powered by an independent source, should be provided where sudden loss of light would create a risk.'

> **Remember**
>
> Text-only signs are not acceptable and should have been replaced with 'running man' signs from December 1998.

Emergency lighting terminology

For the purposes of the European Standard EN 1838, emergency lighting is regarded as a general term. There are actually several types.

- **Emergency escape lighting**: provided to enable safe exit in the event of failure of the normal supply.
- **Standby lighting**: provided to enable normal activities to continue in the event of failure of the normal mains supply.
- **Escape route lighting**: provided to enable safe exit for occupants by providing appropriate visual conditions and direction-finding on escape routes and in special areas/ locations, and to ensure that firefighting and safety equipment can be readily located and used.

- **Open area (or anti-panic area) lighting**: provided to reduce the likelihood of panic and to enable safe movement of occupants towards escape routes by providing appropriate visual conditions and direction finding.
- **High-risk task area lighting**: provided to ensure the safety of people involved in a potentially dangerous process or situation and to enable proper shut-down procedures to be carried out for the safety of other occupants of the premises.

Lighting types

Emergency lighting comes in two main formats: individual, self-contained systems with their own emergency battery power source, and centralised battery back-up systems. Within these formats, there are then three types available: maintained, non-maintained and sustained.

Maintained

The same lamp is used by both the mains and the emergency back-up system, so it operates continuously. The lamp is supplied by an alternative supply when the mains supply fails.

The advantage of this system is that the lamp is continuously lit, so you can see whether a lamp needs replacing. The disadvantage is that, although the lamp is lit, you do not know whether it is being powered by the mains supply or by the batteries. It is common to find a buzzer and indicator lamp that show which supply is being used.

Emergency lighting should be of the maintained type in areas in which the normal lighting can be dimmed, such as theatres or cinemas, or where alcohol is served.

Non-maintained

The emergency lighting lamp only operates when the normal mains lighting fails. Failure of the mains supply connects the emergency lamps to the battery supply.

The disadvantage of this system is that a broken lamp will not be detected until it needs to operate. It is common to find an emergency-lighting test switch available that disconnects the mains supply for test purposes.

Sustained

An additional lamp housed in the mains luminaire is used only when the mains fails.

The duration of the emergency lighting is normally three hours in places of entertainment and for sleeping risk, or where evacuation is not immediate. However one hour's duration may be acceptable in some premises if evacuation is immediate and reoccupation is delayed until the system has recharged.

You may also come across the types of system and emergency duration abbreviated as:

- M3 (a maintained system of emergency, duration three hours)
- NM2 (a non-maintained system of emergency, duration two hours)
- S1 (a sustained system of emergency, duration one hour).

Siting of luminaires

BS 5266 and IS 3217 provide detailed guidance on where luminaires should be installed and what minimum levels of illuminance should be achieved on escape routes and in open areas. It also specifies what minimum period of duration should be achieved after failure of the normal mains lighting.

Local and national statutory authorities, using legislative powers, usually require escape lighting. Escape-lighting schemes should be planned so that identifiable features and obstructions are visible in the lower levels of illumination that will occur during an emergency.

Current UK regulations require the provision of a horizontal illuminance at floor level on the centre line of a defined escape route, of not less than 0.2 lux (similar to the brightness of a full moon). In addition, for escape routes of up to 2 m wide, 50 per cent of the route width should be lit to a minimum of 0.1 lux. Wider escape routes can be treated as a number of 2 m-wide bands.

Emergency escape lighting should:

- indicate the escape routes clearly, allowing for changes of direction or level
- provide illumination along escape routes to allow safe movement towards the final exits
- ensure that fire alarm call points and firefighting equipment can be readily located.

Standby lighting is required in, for example, hospital operating theatres and in industry, where an operation or process, once started, must continue even if the mains lighting fails.

Additional emergency lighting should also be provided in:

- lift cars – potential for the public to be trapped
- toilet facilities – particularly disabled toilets – and open tiled areas over 8 m²
- escalators – to enable users to get off them safely
- motor, generator, control or plant rooms – these require battery-supplied emergency lighting to help any maintenance or operating personnel
- covered car parks along the normal pedestrian routes.

BS 5266 requires 1 lux average over the floor area. The European standard EN 1838 requires 0.5 lux minimum anywhere on the floor level, excluding the shadowing effects of contents. The core area excludes the 0.5 m to the perimeter of the area.

In accordance with The Building Regulations, emergency lighting is required for areas larger than 60 m² or open areas with an escape route passing through them.

High-risk task area lighting

BS 5266 requires that higher levels of emergency lighting are provided in areas of particular risk, although no values are defined.

The European standard EN 1838 says that the average horizontal illuminance on the reference plane (not necessarily the floor) should be as high as the task demands in areas of high risk. It should not be less than 10 per cent of the normal illuminance, or 15 lux, whichever is the greater. It should be provided within 0.5 seconds and continue for as long as the hazard exists.

This can normally only be achieved by a tungsten or permanently illuminated maintained fluorescent lamp source. The required illuminance can often be achieved by careful location of emergency luminaires at the hazard, and may not require additional fittings.

Maintenance and testing

Essential servicing should be defined to ensure that the system remains at full operational status. This would normally be performed as part of the ICEL 1008 testing routine (see Table 4.06), but for consumable items, such as replacement lamps, spares should be provided for immediate use.

Daily	• Visually check all maintained lamps are operating and that all system healthy indicators on Central Power Supply Systems are illuminated • Check any recorded system fault is given urgent attention and record all corrective actions in a logbook
Monthly	• Check all luminaries and other emergency lighting equipment are in a good condition, all lamps and light controllers are clean, undamaged and not blackened • Briefly test all emergency lighting equipment are in a good condition, by simulating a failure of the normal lighting supply. The test should not exceed a quarter of the equipment rated duration. Check that all equipment functions correctly • Once the mains supply is restored, check all supply healthy indicators are illuminated again
Six monthly	• Carry out inspection and testing as described in the monthly test schedule, but conduct a test of the equipment for one-third of its rated duration
Annually	• Conduct a full system test with a competent service engineer, including a full rated test of the system • Comply the installation and system with the requirements of BS 5266 and document this

Table 4.06 Maintenance and testing requirements

Fire alarm systems

A correctly installed fire alarm system installation is of paramount importance: life could be lost and property damaged as a result of carelessly or incorrectly connected fire detection and alarm equipment.

The subject is detailed, so this section just gives an overview of requirements.

BS 5839 Part 1 classifies fire alarm systems – better described as fire detection and alarm systems – into three general types:

Figure 4.50 Firefighters at work

- Type M: manual 'break-glass' contacts operated by occupants operate sounders for protection of life; no automatic detection

- Type L: automatic detection systems for the protection of life

- Type P: automatic detection systems for the protection of property.

It is essential that the installation of fire alarm systems is carried out in compliance with the requirements of BS 5839 Part 1, BS 7671 and manufacturers' instructions, but remember: local government can enforce even stricter requirements in the interests of public safety.

Remember

Fire alarm systems can be designed and installed for either property protection (Type P) or life protection (Types L and M).

BS 5839 and BS 7671 (528.1) state that fire alarm circuits must be segregated from other circuits and, in order to comply with BS 7671, a dedicated circuit must be installed to supply mains power to the fire alarm control panel.

Property protection

A satisfactory fire alarm system for the protection of property will:

- automatically detect a fire at an early stage
- indicate its location, and
- raise an effective alarm in time to summon firefighting forces (both the resident staff and the fire service).

The general attendance time of the fire service should be less than 10 minutes. Therefore an automatic direct link to the fire service is a normal part of such a system.

Protection for property has two classifications:

- P1 – all areas of the building must be covered with detectors with the exception of lavatories, water closets and voids less than 800 mm in height, such that spread of fire cannot take place in them prior to detection by detectors outside the void
- P2 – only defined areas of high risk are covered by detectors; a fire-resisting construction should separate unprotected areas.

Life protection

A satisfactory fire-alarm system for the protection of life can be relied upon to sound a fire alarm in sufficient time to enable the occupants to escape. Life protection is generally classed as follows:

- **M** – the most basic and minimum requirement for life protection. It relies on manual operation of call points and requires people to activate the system.

Such a system can be enhanced to provide greater cover by integrating any, or a combination, of the following:

- **L1** – same as P1 (see above)
- **L2** – only provides detection in specified areas where a fire could lead to a high risk to life, such as sleeping areas, kitchens and day accommodation, and places where the occupants are especially vulnerable owing to age or illness or are unfamiliar with the building. An L2 system always includes L3 coverage.

- **L3** – protection of escape routes, so the following areas should be included:
  - corridors, passageways and circulation areas
  - all rooms opening onto escape routes
  - stairwells
  - landing ceilings
  - the top of vertical risers, such as lift shafts
  - at each level within 1.5 m of access to lift shafts or other vertical risers.

Types of fire alarm system

All fire alarm systems operate on the same general principle: if a detector detects smoke or heat, or if a person operates a break-glass contact, the alarm will sound. You will look at the devices that may be incorporated into a fire alarm system later. Most fire alarm systems belong to one of three categories: conventional, addressable, radio addressable and analogue.

Conventional

In this type of system, a number of devices (break-glass contacts/detectors) are wired as a radial circuit from the control panel to form a zone (for example, one floor of a building). The control panel will have lamps on the front to indicate each zone; if a device operates, the relevant zone lamp will light up on the control panel. However, the actual device that has operated will not be indicated.

Identifying accurately where the fire has started will depend on having a number of zones and knowing where in the building each zone is. Conventional systems are therefore normally found in smaller buildings or where a cheap, simple system is required.

Addressable

The basic principle here is the same as for a conventional system, the difference being that, by using modern technology, the control panel can identify exactly which device initiated the alarm. These systems have their detection circuits wired as loops, with each device having an 'address' built in. This allows the fire services to get to the source of a fire more quickly.

Radio addressable

These systems are the same as addressable systems, but have the advantage of being wireless, which can reduce installation time.

Analogue

Sometimes known as intelligent systems, analogue systems incorporate more features than either conventional or addressable systems. The detectors may include their own mini-computer, which evaluates the environment around the detector and lets the control panel know whether there is a fire, a change in circumstance likely to lead to a fire, a fault, or even if the detector head needs cleaning. These systems are useful in preventing the occurrence of false alarms.

Fire prevention systems

One recent innovation has been the introduction of the fire reduction system. This type of system is still under development, but works by reducing levels of one of the main components in the fire triangle – oxygen – seeking to create a 'fire free' area. Although not without problems, these systems could be appropriate in critical areas such as historical archives or identified unmanned areas, such as chemical storage.

Zones

To ensure a fast and unambiguous identification of the source of fire, the protected area should be divided into zones. Although less essential in analogue addressable systems, here are some broad guidelines.

- If the floor area of each building is not greater than $300\,m^2$, the building only needs one zone, no matter how many floors it has. This covers most domestic installations.
- The total floor area for one zone should not exceed $2000\,m^2$.
- The search distance should not exceed $30\,m$. This means the distance that has to be travelled by a searcher inside a zone to determine visually the position of a fire should not be more than $30\,m$. The use of remote indicator lamps outside doors may reduce the number of zones required.
- Where stairwells or similar structures extend beyond one floor but are in one fire compartment, the stairwell should be a separate zone.
- If the zone covers more than one fire compartment, the zone boundaries should follow compartment boundaries.
- If the building is split into several occupancies, no zone should be split between two occupancies.

System devices

The control panel

This is the heart of any system, as it monitors the detection devices and their wiring for faults and operation. If a device operates, the panel operates the sounders as well as any other related equipment and gives an indication of the area in which the alarm originated.

Figure 4.51 Control panel

Break-glass contacts (manual call points)

The break-glass call point is a device to enable personnel to raise the alarm in the event of a fire, by simply breaking a fragile glass cover (housed in a thin plastic membrane to protect the operative from injury sustained by broken or splintered glass). A sturdy thumb pressure is all that is required to rupture the glass and activate the alarm. Break-glass contacts are now available in wireless format.

Here is some guidance on the correct siting and positioning of break-glass call points.

- They should be located on exit routes and, in particular, on the floor landings of staircases and at all exits to the open air.

- They should be located so that no person need travel more than 30 m from any position within the premises to raise the alarm.

Figure 4.52 Break-glass fire point

- Generally, call points should be fixed at a height of 1.4 m above the floor, at easily accessible, well-illuminated and conspicuous positions, free from obstruction.

- The method of operation of all manual call points in an installation should be identical, unless there is a special reason for differentiation.

- Manual and automatic devices may be installed on the same system, although it may be advisable to install the manual call points on separate zones for speed of identification.

Automatic detectors

When choosing the type of detector to be used in a particular area, it is important to remember that the detector has to discriminate between fire and the normal environment existing within the building: for example, fumes from fork-lift trucks in warehouses or steam from kitchens and bathrooms.

There are several automatic detectors available.

Heat detectors (fixed-temperature type)

The fixed temperature heat detector is a simple device designed to activate the alarm circuit once a predetermined temperature is reached. Usually a choice of two operational temperatures is available: either 60°C or 90°C. This type of detector is suitable for monitoring boiler rooms or kitchens where fluctuations in ambient temperature are commonplace.

Heat detector (rate of rise type)

This type of detector responds to rapid rises of temperature. It samples the temperature difference between two heat-sensitive thermocouples or **thermistors** mounted in a single housing.

Smoke detectors

Smoke detectors may be either of the ionisation or the optical type.

The ionisation detector is very sensitive to smoke with fine particles, such as that from burning paper or spirit, whereas the optical detector is sensitive to optically dense smoke with large particles, such as that from burning plastics.

The optical smoke detector, sometimes known as the photoelectric smoke detector, operates by means of the 'light scattering' principle. A pulsed infrared light is targeted at a photo receiver, but separated by an angled non-reflective baffle positioned across the inner chamber. When smoke and combustion particles enter the chamber, light is scattered and reflected onto the sensitive photo receiver, triggering the alarm. Optical smoke detectors are ideal for detecting visually smoky fires, which are likely to be slow, smouldering fires.

Detector heads for fire alarm systems should only be fitted after all trades have completed work, as their work could create dust, which impairs the detector operation. Strict rules exist about the location of smoke detectors. Smoke detectors are not normally installed in kitchens, as burning foods like toast could activate the alarm.

Manufacturers are also now starting to produce 'all in one' sensors that are capable of detecting the products of combustion, such as heat, smoke and carbon monoxide.

Alarm sounders

These are normally either a bell or an electronic sounder, which must be audible throughout the building in order to alert and evacuate the occupants of the building.

Key term

Thermistor – an electronic device whose resistance changes quickly in response to changes in temperature

Figure 4.53 Smoke detector

Here is some guidance for the correct use of alarm sounders.

- A minimum level of either 65 dBA, or 5 dBA above any background noise likely to persist for a period longer that 30 seconds, should be produced by the sounders at any occupiable point in the building.

- If the alarm system is to be used in premises such as hotels or boarding houses where it must wake sleeping people, the sound level should be a minimum of 75 dBA at the bedhead.

- All audible warning devices used in the same system should have a similar sound.

- It may be preferable to have a large number of quieter sounders rather than a few very loud sounders. At least one sounder will be needed for each fire compartment.

- The level of sound should not be so high as to cause permanent damage to hearing.

Figure 4.54 Alarm bell

Wiring systems for fire alarms

BS 5839 Part 1 recommends eleven types of cable that may be used where prolonged operation of the system in a fire is not required. However, only two types of cable may be used where prolonged operation in a fire is required.

It is obvious that the cabling for sounders (and any other device intended to operate once a fire has been detected) must be fireproof. However, detection wiring can be treated differently, as such wiring is only necessary to detect the fire and subsequently sound the alarm.

In reality, fire-resistant cabling tends to be used throughout a fire alarm installation for both detection and alarm wiring. As an example, MICC cable used throughout the system is considered by many as the most appropriate form of wiring, but there are alternatives, such as Firetuf or FP 400.

Whatever the cable type and the circuit arrangements of the system, all wiring must be installed in accordance with BS 7671. Where possible, cables should be routed through areas of low fire risk; where there is risk of mechanical damage, they should be protected accordingly.

Because of the importance of the fire alarm system, it is wise to leave the wiring of the system until most of the constructional work has been completed. This will help prevent accidental damage occurring to the cables. Similarly, keep the control panel and activation devices in their packing cartons, and only remove them when building work has been completed in the area where they are to be mounted.

Standby back-up for fire alarm systems

The standby supply, which is usually a battery, must be capable of powering the system in full normal operation for at least 24 hours; at the end of that time, they must still have sufficient capacity to sound the alarm sounders in all zones for a further 30 minutes.

Table 4.07 shows the typical maintenance checks for a fire alarm system recommended by BS 5839 Part 1.

Daily inspection	Annual test
• Check that the control panel indicates normal operation • Report any fault indicators or sounders not operating to the designated responsible person	• Repeat the quarterly test • Check all call points and detectors for correct operation • Enter details of test in logbook
Weekly test	**Every two to three years**
• Check panel key operation and reset button • Test fire alarm from a call point (different one each week) and check sounders • Reset fire-alarm panel • Check all call points and detectors for obstruction • Enter details of test in logbook	• Clean smoke detectors using specialist equipment • Enter details of maintenance in logbook
Quarterly test	**Every five years**
• Check all logbook entries and make sure any remedial actions have been carried out • Examine battery and battery connections • Operate a call point and detector in each zone • Check that all sounders are operating • Check that all functions of the control panel are operating by simulating a fault • Check sounders operate on battery only • Enter details of test in logbook	• Replace battery (see manufacturer's information)

Table 4.07 Typical maintenance checks for a fire alarm system

Progress check

1 Explain the difference between:
   a) emergency escape lighting
   b) escape route lighting.

2 Identify the following classifications of emergency lighting systems:
   • M3
   • NM2
   • S1

3 Identify the following BS 5839 Part 1 fire alarm systems:
   • Type M
   • Type P

4 Explain the difference between ionisation and optical type smoke detectors.

5 Identify the minimum sounder levels for fire alarm systems set in BS 5839 Part I.

Closed circuit television (CCTV) and camera systems

There are many different types of CCTV systems in use today, ranging from those suitable for domestic properties through to sophisticated multi-camera/multi-screen monitoring for large commercial and industrial premises. Here you will look at the component parts that make up a typical system.

Wireless CCTV

These systems do not require cabling back to a monitor or video recorder; they have an in-built transmitter that transmits the image seen back to these pieces of equipment. Typically they can transmit 100 m outdoors and 30 m indoors. They do, however, still require a power supply (usually 9–12 V d.c.), which is usually obtained via a small power supply connected to the mains. These systems are useful where it is difficult to install video cable back to the monitor or video recorder, but they can suffer from interference problems.

Wired CCTV

These systems do require cabling back to the monitor or video recorder, but can be positioned many hundreds of metres away from them. Usually the same cable will provide power and the video signal back to the recording device, so all the power supplies for many cameras can be located at one central control point.

Cameras

There are many different types of camera available, ranging from very cheap (less than £100) to those costing many thousands of pounds. There are two common types: CMOS and CCD. The CMOS type is the cheapest, but the images produced are not very clear or sharp. The CCD camera, on the other hand, produces very clear and sharp images from which people are easily identifiable.

Most cameras are installed outside, so virtually all cameras available are weatherproof; if they are not, they will need to be fitted into a weatherproof housing. Nearly all cameras have the lens integrated into the camera and are sealed to prevent moisture getting in, so they do not need a heater built in to keep the lens dry. As the lens is sealed into the camera, it cannot be adjusted, so only one field of view is possible.

Colour and monochrome types of camera are both available. Colour cameras can only transmit colour if the light level is high, so usually they will not transmit colour images at night. Monochrome cameras, on the other hand, can incorporate infrared (IR) sensitivity, allowing for clearer images where discreet IR illumination is available.

Figure 4.55 Indoor monochrome camera

Light levels

Light levels available where the camera is to be used are an important consideration. When choosing a suitable camera for a particular environment, it is best to select one that is specified at approximately 10 times the minimum light level for the environment. One that is specified at the same level of light will not produce the clear images needed, because the camera will not have enough light to 'see'.

Monitoring and recording

Most CCTV systems use several cameras, each relaying images back to a central control where they are either viewed or recorded. There are three methods for recording or viewing these images: a video switcher, a quad processor or a multiplexer.

Switcher

A CCTV switcher is a device that allows the operator to switch between cameras one at a time. The image can be either viewed or recorded onto a video recorder, but only one image can be accessed at a time.

Figure 4.56 CCTV switcher

Quad processor

This device enables four camera images to be viewed on one screen at the same time, or one image or all four to be recorded at the same time. However, the quality of the image when recording all four is not as good.

Multiplexer

This device allows simultaneous recording of multiple full-sized images onto one VCR, or can allow more than one camera image to be displayed at the same time without losing picture quality. For recording purposes, a slower-moving tape can record the images for long periods of time. The time lapse can be set for either 24, 240 or 960 hours of recording on standard tapes.

Increasingly, images are recorded digitally onto DVD on a multiplexer (above), which allows faster retrieval of information as well as allowing higher-quality image reproduction without the degradation that is experienced with tape.

Other systems

PC-based systems

By adding a video-capture card and surveillance software to a PC, a powerful digital system can be created. Some of the advantages are:

- less manpower to monitor system
- management can review security remotely to review trends and risks
- security status of the property can be checked remotely, via a linked cellphone.

Motion detectors

The camera and recording facilities are only activated when movement is detected within the camera's range. Typically this is activated by the use of passive infrared sensors (PIRs) similar to those used on security lighting and alarm systems.

Intruder alarm systems

Alarm systems are increasingly seen as standard equipment in a house or office, as they act as a deterrent to some intruders. People feel more secure when they have an alarm installed, and in most cases it will reduce their insurance premiums.

Figure 4.57 Motion detector

There are two ways to protect a property: perimeter protection and space detection. Perimeter protection detects a potential intruder before they gain entry to the premises, whereas space detection only detects when the intruder is already on the premises. Sometimes both types are used together for extra security.

In this section, you will look at some of the component parts of an alarm system and some of the detection devices available, and what they do.

Proximity switch

A proximity switch is a two-part device: one part is a magnet and the other contains a reed switch. The two parts are fixed side by side (usually less than 6 mm apart) on a door or window. When the door or window is opened, the reed switch opens (because the magnet no longer holds it closed) and activates the alarm panel. The switch can be surface-mounted or can be recessed into the door or window frame. This device is generally used for perimeter protection as it is detecting an attempt to enter. However, many systems will use these devices on internal doors as well as external entry points.

Inertia switch

An inertia switch detects the vibration created when a door or window is forced open. This sends a signal to the alarm panel and activates the sounder. The sensitivity of these devices can be adjusted, and they are used for perimeter protection. These need a 12 V d.c. supply to operate.

Passive infrared

Passive infrared devices are used to protect large areas of space and are only activated when the intruder has already gained entry. The device monitors infrared, which detects the movement of body heat across its viewing range; this in turn sends a signal to the panel and activates the sounder. These can be adjusted for range and, by fitting different lenses, the angle of detection can also be adjusted. They need a 12 V d.c. supply to operate.

Ultrasonic devices

Ultrasonic devices send out sound waves. They receive back the same waves when no one is in the building; however, when an intruder enters the detection range, the sound waves change (because of deflection) and trigger the alarm panel.

These devices also require a 12 V d.c. supply for operation. They are used for space detection systems and commonly found in car alarm systems.

Microwave detectors

A microwave detector emits microwaves from a transmitter and detects any reflected microwaves or reduction in beam intensity using a receiver. The transmitter and receiver are usually combined inside a single housing for indoor applications, and separate housings for outdoor applications.

To reduce false alarms, this type of detector is usually combined with a passive infrared detector.

Photoelectric beams

Photoelectric beam systems detect the presence of an intruder by transmitting visible or infrared light beams across an area, where these beams may be obstructed. To improve the detection surface area, the beams are often employed in stacks of two or more.

The technology can be an effective long-range detection system, if installed in stacks of three or more where the transmitters and receivers are staggered to create a fence-like barrier. Systems are available for both internal and external applications typically to the sides of windows.

To prevent an intruder using a secondary light source to 'fool' the detector, most systems send and detect a modulated light source.

Fibre-optic detectors

Fibre-optic detectors work a bit like a pressure switch: pressure applied on the cable will cause it to distort, changing the route of the light, which is then detected and used to trigger the alarm.

Control panels

Control panels are the 'brains' of the system to which all the parts of the system are connected. They used to be key-operated but nowadays they virtually all use a digital keypad, either on the panel itself or mounted remotely elsewhere in the building, for switching the system on or off.

The panels can all be programmed so that entry and exit route zone delays can be adjusted, new codes selected for switching on/off, automatic telephone diallers set to ring any phone selected, and so on.

Control panels have a mains supply installed that is reduced (via a transformer) to 12 V d.c. for operation of all the component parts that need it, and a rechargeable battery back-up is provided in case of mains failure.

Audible and visual warning devices

When an alarm condition occurs, a means of attracting attention is needed, either audibly or visually or sometimes both. The most common audible sounder is the electronic horn, which will sound for 20 minutes (the maximum allowed by law) before being switched off by the panel automatically. The panel then re-arms itself and monitors the system again.

To help identify which alarm has sounded (especially when there are several in the same area), a visual warning is usually fitted to the sounder box (the red lens), which activates at the same time. This is a xenon light (strobe light), which can be obtained in a variety of colours. This light usually remains on after the alarm has automatically been reset to warn the occupant that an alarm condition has occurred. It is only reset when the occupant resets the control panel itself.

Water heating systems

It can be argued that there are only two main methods of heating water:

- storage, where you heat a large amount of water and store it in a tank ready for use
- instantaneous, where you only heat the water you need, when you need it.

With both of these types, it is important to ensure that the exposed and extraneous conductive parts are adequately bonded to earth: water and electricity do not mix together well! It is also important to ensure that the cables selected are of the correct size for full load current, since no diversity is allowed for water heaters.

For storage, large tanks of stored water (typically 137+ litres) are heated using an immersion heater and then controlled to be on or off via a timer switch or an on/off switch. The temperature of the water is regulated by a stem-type thermostat, incorporated within the housing of the heating element. This type of heater is normally found in domestic situations, although larger multiple immersion heaters can be used in commercial/industrial situations.

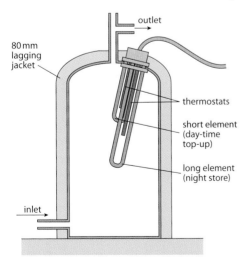

Figure 4.58 Dual-element immersion heater, hot water

The heater in a domestic situation must be fed from its own fuse/ MCB (Miniature circuit breaker) in the consumer unit and have a double-pole isolator fitted next to the storage tank. The final connection to the heating element must be made with heatproof flexible cable due to the high ambient temperatures where the water tank is normally located.

This type of system can sometimes have two elements: one element is controlled via a separate supply, which only operates at night time (Economy 7 or white-meter supply) when cheap electricity is available, heating a full tank of water ready for use the next day; the other is switched on as and when needed during the day to boost the amount of hot water available.

There are many variations of water system available. Here are some of the factors to consider:

- how much hot water is needed and where
- whether the water heating system needs to be part of the central heating system
- speed of hot water delivery
- flow rate of water (for a shower, for example)
- what space is available.

Each system can be powered from the boiler/combi-boiler, but electrically there are several options.

Cistern-type (storage)

Where larger volumes of hot water are needed (for example, in a large guest house), a cistern-type water heater (9 kW+) is used which is capable of supplying enough hot water to several outlets at the same time.

Non-pressure (storage)

Non-pressure water heaters, which are typically rated at less than 3 kW and contain less than 15 litres of water, heat and then store water ready for use and are usually situated directly over the sink. Typically they can be seen in a small shop or hairdresser's salon, but are sometimes used in toilet blocks in offices or factories.

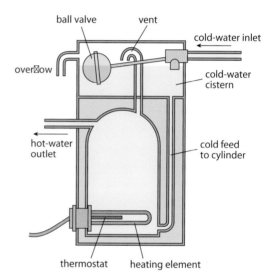

Figure 4.59 Cistern-type storage

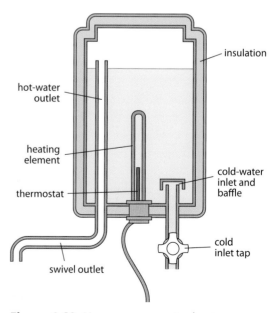

Figure 4.60 Non-pressure water heater

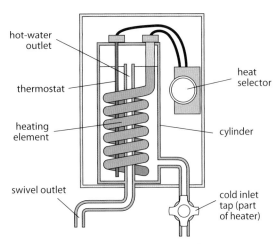

hot-water outlet

thermostat

heating element

swivel outlet

heat selector

cylinder

cold inlet tap (part of heater)

Figure 4.61 Instantaneous water heater

Instantaneous

Instantaneous water heaters heat only the water that is needed. This is done by controlling the rate of flow of water through a small internal water tank, which has heating elements inside it; the slower the flow of the water, the hotter the water will become. The temperature of the water can be continuously altered or stabilised locally at whatever temperature is selected.

This is exactly how an electric shower works, and showers in excess of 10 kW are currently available. The shower-type water heater must be supplied via its own fuse/MCB in the consumer unit and have a double-pole isolator located near the shower.

Electric heating

The type of electric heating available falls into two main categories: direct-acting heaters and thermal storage devices.

Direct-acting heaters

Direct acting heaters are usually just switched on and off when needed, although some of them can be thermostatically controlled. They fall into two basic categories: radiant and convection.

Radiant heaters

Radiant fires work on the simple principle of an element being heated to high temperature and heat then radiating away from the device. Radiant type heaters reflect heat and come in a variety of shapes, sizes and constructions.

The 'traditional' electric fire

This has a heating element supported on insulated blocks, with a highly polished reflective surface behind it; these range in size from about 750 W to 3 kW.

Infrared heater

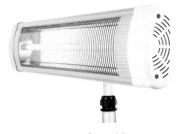

Figure 4.62 Infrared heater

This consists of an iconel™-sheathed element or a nickel-chrome spiral element housed in a glass silica tube, which is mounted in front of a highly polished surface. Sizes vary from about 500 W to 3 kW; the smaller versions are usually suitable for use in bathrooms and may be incorporated with a bulb to form a combined heating and lighting unit.

Oil-filled radiator

This consists of a pressed-steel casing in which heating elements are housed; the whole unit is filled with oil. Oil has a lower specific gravity than water, so heats up and cools down more quickly. Surface temperature reaches about 70°C, and power sizes range from about 500 W to 3 kW.

Tubular heater

This is a low-temperature unit designed to supplement the main heating in the building. It consists of a mild steel or aluminium tube of about 50 mm diameter, in which a heater element is mounted. The elements are rated at 200 W to 260 W per metre length, and can range in length from about 300 mm to 4.5 m. The surface temperature is approximately 88°C.

Underfloor heating

This consists of heating elements embedded under the floor that heat up the floor surface. The floor becomes a large, low-temperature radiant heater. A room thermostat controls the temperature within the room; the floor temperature does not normally exceed 24°C.

For laminate flooring, a carbon-heating film roll has heating elements rated at between 130 W/m^2 and 160 W/m^2. The roll is supplied in 1000 mm width and standard lengths between 2 and 6 metres. Each heating element is supplied with standard 5 m leads for connection to the control unit.

Convection heaters

When heated, air decreases in density. This makes the surrounding air cooler and denser. This in turn makes it heavier, so it falls beneath the hot air, forcing it upwards. This is convection.

In a convection heater, the heating element heats the air next to it. By convection, this sets up a constant current of hot air that leaves the appliance through vent holes and heats up the surrounding space.

Convection heaters consist of a heating element housed inside a metal cabinet that is insulated both thermally and electrically from the case, so that the

Figure 4.63 Oil-filled radiator

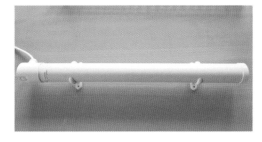

Figure 4.64 Tubular heater

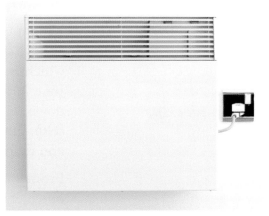

Figure 4.65 Convection heater

heat produced warms the surrounding air inside the cabinet. Cool air enters the bottom of the cabinet and warm air is passed out at the top of the unit, at a temperature of between 80 and 90°C. A thermostatic control is usually fitted to this type of heater.

Fan heaters

These operate in the same way as a convector heater, but use a fan to force the warm air into the room. Fan heaters usually have a two-speed fan incorporated into the casing and up to 3 kW of heating elements.

Thermal storage devices

Electric storage heaters consist of several heating elements mounted inside firebricks, which in turn are surrounded by thermal insulation such as fibreglass, all housed inside a metal cabinet. The firebricks are made from clay, olivine, chrome and magnesite, which have very good heat retaining properties.

Connected to a special tariff meter in the property, the bricks are heated up during 'off peak' hours (normally overnight, usually Economy 7 tariff) and the heat is stored within the bricks. The outlet vent is then opened the following day and allows the warm air to escape and hence heat up the room.

Economy 7

Because of the costs and process of electricity generation, large power stations need to run 24 hours a day and are therefore producing electricity overnight when most of us have little need for it.

To encourage overnight energy usage, particularly for storage heating, the industry came up with the term Economy (meaning cheap) and 7 (meaning available for 7 hours overnight) to offload the energy that the industry had produced.

Economy 7 electricity works for 7 hours at night between 12–7am and costs about 4p per unit compared with 'on peak' electricity which is more expensive at around 10 to 15p per unit as of June 2011, although it is possible these prices may go up or down. However, for the generating companies it is also better to sell it at this price than to have to keep constantly powering down and refiring their generating plant.

From a user perspective, it requires the property to have either one meter with two readings or two separate meters to record usage against the 'on peak' and 'off peak' tariffs.

Economy 7 can also be used to heat water, and run washing machines and dishwashers using timers to make the most of cheap rate electricity.

Environment and building management control

The previous pages looked at forms of electric heating that could be argued to be task specific, in that, with the exception of storage heaters, an electric heater is normally providing heat to an individual rather than trying to heat a whole room.

This section will look at systems intended for the building rather than the individual.

Central heating systems and controls

Before looking at the controls associated with them, perhaps it would be wise to have a very brief overview of what domestic central heating systems are available as installed by plumbing contractors.

Pumped systems

The water is heated by the boiler and sent round the pipework by means of a pump. The water is then directed to either the radiators or the hot water cylinder by means of a motorised valve.

Gravity systems

In older properties, gravity circulation is used to heat the hot water cylinder. This works on the convection theory mentioned earlier, in that water will expand when heated, will weigh less than cold water, will sink to the bottom of the system and therefore cause the hot water to be pushed up.

The gravity system therefore doesn't have a pump, but does need larger pipework.

This system usually has a water tank in the loft to keep the system topped up when water is lost by evaporation or leakage.

Sealed systems

This is a closed system and it common for the boiler to be a combination type rather than connected individual components. A pressure vessel handles expansion and contraction of the water and there is a safety valve to relieve excess pressure instead of the vent pipe found in gravity systems.

As these systems run at higher temperature than other systems, radiators can be a little smaller.

Hot water

When using a standard boiler, the hot water supply is heated by a pipe which runs from the boiler, coils through the hot water cylinder, and returns to the boiler. The water in the cylinder is completely separate from that which goes through the boiler.

The temperature of the hot water is controlled by a cylinder thermostat which switches the pump on and off or opens a motorised valve.

In the case of a combination boiler, mains water is only heated on demand when a tap is turned on, effectively acting like an instantaneous electric water heater.

Heating controls

It is probably fair to say that it will be the controls involved in the system that will make it easier to install, easy to use and generally make it is as efficient as possible.

It is probably also fair to say that control systems are many and varied, but that perhaps the most common controls are based around a concept introduced about 30 years ago by Honeywell and known as the 'Sundial' system. The two most common variants being the 'S' and 'Y' plan systems, although the 'C' Plan exists for gravity systems.

'S' Plan

The Sundial 'S' Plan uses two two-port valves (one for heating, one for hot water) to provide independent temperature control of heating and hot water circuits in fully pumped central heating installations. Time control must be provided by a programmer.

In terms of its operation, on demand for heat from either thermostat, the respective zone valve will be energised to open. Just before the valve reaches its fully open position, the auxiliary switch will be closed and switch on both pump and boiler.

When both thermostats are satisfied, the valves are closed and the pump and boiler switched off.

Figure 4.66 shows a wiring diagram for such a system.

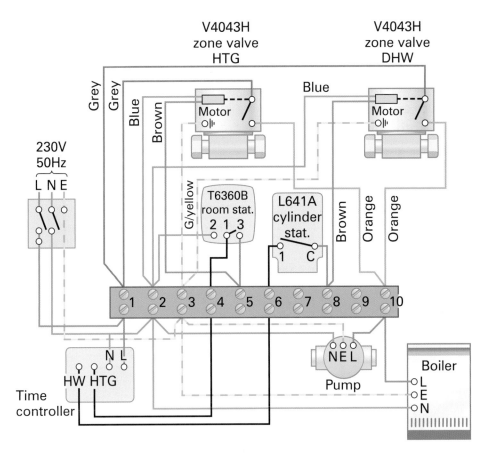

Figure 4.66 Wiring diagram for an 'S' Plan temperature control system

'Y' Plan

The Sundial 'Y' Plan uses one 3-port mid position valve (it handles both hot water and heating) to provide independent temperature control of both heating and hot water circuits in fully pumped central heating installations. Time control must be provided by a programmer.

Hot water only requirement

On demand for heat from the cylinder thermostat, the valve remains open to DHW only and the pump and boiler are switched on.

Heating only requirement

On demand for heat from the room thermostat, the valve motor is energised so that the CH port only is opened and the pump and boiler are switched on.

Heating and hot water requirement

When both thermostats demand heat, the valve plug is positioned to allow both ports to be open and the pump and boiler switched on. When neither thermostat is demanding heat, the pump and boiler are off. The valve remains in the last position of operation whilst the time control is in the "on" position.

Figure 4.67 shows a wiring diagram for such a system.

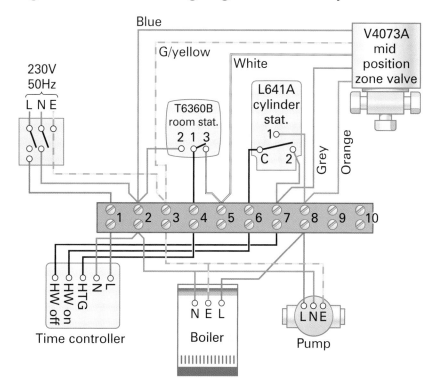

Figure 4.67 Wiring diagram for a 'Y' Plan temperature control system

Heating, ventilating and air conditioning

Although domestic central heating is installed by plumbers, at a commercial and industrial level this tends to become the work of the HVAC contractor.

Some of the technology is similar to that of the plumber, but on a larger scale. For example, in terms of heating, buildings may use boilers and radiators. However, whereas in domestic installations the boiler may be wall mounted as part of the kitchen units, a hospital or university may have a dedicated building to house the boiler.

One system particular to HVAC is the forced air system, which sends heated air through duct work. During warm weather the same duct work can be used for air conditioning and the air can also be filtered or put through air cleaners.

Intelligent (smart) buildings

With developments in technology you can now set up a building to better and more efficiently meet your needs and monitor and adjust its performance against those needs.

The concept of the intelligent building (or, as it is often referred to, the Building Management System) is to get the individual systems in a building – whether it be lighting, audio, heating or ventilation – to 'talk' to each other. For example, if no one is in a room, a PIR detects this, informs the system and the system then switches off the lighting and heating for that room.

Effectively this happens by linking all the systems to a controller, which is basically a purpose-built computer with input and output capabilities. These controllers come in a range of sizes and capabilities to control devices commonly found in buildings, and to control sub-networks of controllers.

The inputs allow the controller to read temperature, humidity, pressure, current flow, air flow, occupancy and other information. Once analysed, the outputs then allow the controller to send control signals out to the various parts of the system to make any adjustments necessary.

Controlling a building in this way can have an effect on climate change, especially when combined with good building design. For example, natural daylight can make people happier, healthier and more productive, as well as reducing the energy bill.

For offices, the number one complaint is that the workplace is too hot, with a close second being that it's too cold! Staff tend to compensate by adding fans, space heaters etc. A well thought-out building design and building management system that is talking to the HVAC system could easily prevent this.

Progress check

1   Describe briefly the difference between passive infrared and ultrasonic detectors used in security systems.

2   Explain briefly how an electric storage heater works.

3   What is the significant difference between a 'S' Plan and a 'Y' Plan central heating control system?

K5. Understand procedures for selecting and using tools, equipment and fixings for wiring systems, enclosures and equipment

At one time or another, you have probably started on something and then been frustrated at not having the right tool for the job: 'If I only had...'. Used properly and safely, good-quality tools let you work faster and more efficiently, so you need to look after them.

These days in the electrical trade you use materials that can last for more than 50 years – as long as the system is well designed, complies with appropriate safety regulations and is properly installed.

As part of an installation team, you could find yourself having to do jobs usually done by other trades, such as building and brickwork. To be a good electrician, you will need to get to grips with a number of different skills and learn how to use a wide range of tools.

It is not possible to list all the tools you might come across, so this section only describes the most common ones.

Hand tools, power tools and adhesives

There is a wide range of tools that you may use during your work as an electrician. Some of the most common are covered below.

Pliers

The main difference between electricians' pliers and any other sort of pliers is that, for obvious reasons, they have insulated handles. They have flat serrated jaws for gripping and bending, and oval serrated jaws for gripping pipes and cylindrical objects.

Long-nosed pliers also exist for working with smaller materials or in confined spaces and there are round-nosed pliers to form loops in conductors for fitting to post-type terminals.

Cutters and strippers

Cutters are designed to cut hard materials, such as wires and conductors.

Insulation strippers allow you to place a cable or flex into the jaws of the insulation stripper, using the calibrated markings to

Figure 4.68 Electricians' pliers

Figure 4.69 Side cutters

Figure 4.70 Insulation stripper

measure the amount of insulation to be removed. Simply close the handles with one hand while holding the cable in position with the other, and the insulation is removed.

Screwdrivers

The three main types of screwdriver tip are shown here, but there are other types for specialised jobs.

(a)

(b)

(c)

Figure 4.71 (a) Flared slotted for general use (b) parallel slotted (head same size as shaft) (c) cross-head (Phillips, Pozidriv®) gives better purchase

Hammers

Hammers are used to do three main jobs:

- to drive a fixing (such as a nail)
- to provide impact on another tool (such as a cold chisel)
- to alter the shape of a workpiece (for example, to bend a piece of metal).

It is important to use the right hammer for the job, and to make sure that the shaft (handle) is firmly fitted into the head.

Steel hammers are one-piece, so there is usually no problem. However, with wooden handles, which have wedges to tighten the shaft in the eye, always check that the wedges are not loose or missing and that the shaft is tight.

Here are the three types of hammer you're most likely to use.

Ball-pein

A ball-pein hammer has a rounded end or ball – used to shape metal and rivets. It can weigh up to 2 kg. Heavier ones are used with punches, cold chisels, and so on.

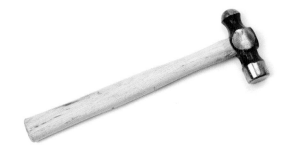

Figure 4.72 Ball-pein hammer

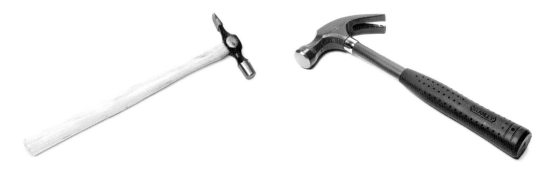

Figure 4.73 Cross-pein hammer   **Figure 4.74** Claw hammer

Cross-pein

Cross-pein hammers are often used by carpenters. The tapered end (pein) is used to start small nails held in the fingers. A cross-pein hammer can weigh up to ½ kg.

Claw hammer

A claw hammer is a general purpose hammer. The claw end is used to lever out nails.

Saws

Electricians are most likely to use hacksaws, tenon saws and flooring saws in their work.

Hacksaw

Figure 4.75 Hacksaw

A hacksaw is used for basic metal cutting, such as cutting tubes or sheets to length or size, or making thin cuts to help shape the metal. It consists of a frame, handle and blade. The blade is held in the handle, and tightened by a wing nut. In small, 'junior' hacksaws, the saw frame itself gives the tension and there is no wing nut.

Tenon saw

Figure 4.76 Tenon saw

A tenon saw is used mostly for cutting and making joints in timber. The metal strip along the top gives rigidity to the blade. To start a cut, angle the saw so that it cuts into an edge of the wood, then lower the angle for a straighter cut.

Flooring saw

The flooring saw is used for cutting the tongues of tongue and groove floorboards so they can be lifted. The saw has a curved blade that will slot into the gap between the boards.

Drills

A drill makes a hole when the cutting edge at its tip is rotated with pressure applied. Drill bits must be chosen to match the hole size, the material being drilled and the tool used to rotate it. There are various manual and power tools for holding and turning drill bits.

Figure 4.77 Flooring saw

Hand drill

This works by rotating the handle on the wheel. A larger version – the breast drill – has a support at the end of the handle that allows you to apply pressure using your body weight.

Power drills

The two most common are the hand-held electric/battery drill and the pillar/bench-mounted electric drill. Both need to be handled carefully and safely to avoid accidents.

Figure 4.78 Hand drill

Spanners

These are used for tightening nuts, bolts and setscrews. Spanners are usually made for one size of nut, marked on the spanner. (One exception is BSF spanners, which are designed to fit a nut one size larger than marked.)

Double-ended spanners usually fit two different sizes. The head of a ring or box spanner may be square, hexagonal (with six points) or bi-hexagonal (with 12 points). These spanners grip all sides of the nut, reducing the possibility of damaging it. However, they have to be placed over the nut and so cannot be used when you cannot get to the end of its bolt or rod.

Figure 4.79 Hand-held electric drill

Box spanners, which are cylindrical, can be used on deeply recessed nuts that are out of reach of normal spanners.

The jaws of open-ended spanners close on four sides of a hexagonal nut (or three sides of a square one), giving easier access to nuts on a long bolt or rod, but providing a less secure grip.

Figure 4.80 Adjustable spanner

Figure 4.81 Footprint wrench

Adjustable spanner

Adjustable spanners or wrenches allow you to change the jaw opening to the size needed by adjusting the screw wheel. They will fit a range of sizes of nut, but do not give as good a grip as a spanner and can slip, damaging the workpiece. Use them only when you do not have the proper-sized spanner, or need to grip something round.

Figure 4.82 Vice grip

Footprint wrench

This has serrated jaws that adjust to the size of the nut.

Vice grip

This is adjusted with a screw to a preset jaw width and released by a clip. Sometimes called a 'mole grip', the vice grip is good for holding work in progress securely.

Stillson wrench

This is similar to the footprint wrench, but is larger and more powerful.

Figure 4.83 Stillson wrench

Files

Files have a rough face of hardened metal that is pushed across the surface of a workpiece to remove particles of material. They can be used to make an object smoother or smaller, or to change its shape.

Files are classified according to length, cut, grade and shape. The main shapes are flat, square, half-round and round. The main cuts are bastard (intermediate), fine and medium. Files of this type must be used with a firmly fitted handle.

Figure 4.84 Files

Surform files have a perforated blade (like a cheese grater) held in a frame. They can be used on wood, plastic and mild steel, and are good for removing material quickly without clogging.

Chisels

As with most tools, there is a wide range available, with many used for special purposes.

Figures 4.85 and 4.86 show the most common ones.

Figure 4.85 Cold chisels

Figure 4.86 Wood chisel

Crimping tools

Many electrical cables are terminated using metal lugs. These can be fixed to conductors by crimping, where part of the lug slides over the conductor and is squeezed tightly onto the conductor using a crimping tool.

Crimping tools can be operated by hand or, for larger conductor sizes, by hydraulics.

Figure 4.87 Crimping tool

Adhesives

You may think that an electrician will have no need for adhesives. However, as an example, you may need to repair the fabric of a building after a small electrical repair or modification.

There are many types of adhesive suited to different tasks. If you are replacing timber, such as a skirting board or door frame, you can use fast-grip wood adhesive instead of nails or screws. Two-part epoxy resins or 'super glues' are also good for repairing plastics.

> **Safety tip**
>
> Take great care when using Super Glue® as it can bond skin together in a matter of seconds.

Figure 4.88 Steel ruler

Figure 4.89 Steel tape measure

Figure 4.90 Spirit level

Measuring and marking tools

During the course of a normal working day, an electrician will have to measure something, whether it be to check that something is level or to measure lengths and distances. Measuring tools range from a simple rule to a laser-levelling device. Although measuring tools are robust, you should handle them with care to maintain their accuracy.

Here are some of the most common ones.

Steel rulers are used for general measurement of trunking, tray plate or enclosures.

Steel tape measures are used for longer measurements, such as room sizes, trunking and timber.

Spirit levels are used to check that an object is vertical or horizontal; a bubble in a glass tube sits between two lines to show when this is so.

Plumb bobs and chalk lines are used to find or check verticals. A plumb bob is simply a weight on a piece of cord; when the weight stops moving, the string is a true vertical. With a chalk line, once it is straight, you snap the string against the wall, to leave a vertical chalk mark.

Laser levellers are the electronic equivalent of the spirit level and can be controlled remotely. They can be used to establish a set level around the perimeter of a room or to establish a true vertical line. Take care never to look directly at the laser light source.

Laser rangefinders are normally hand-held devices used for accurate measurements over longer distances, such as a corridor.

Figure 4.91 Laser level

Figure 4.92 Laser rangefinder

Fixing tools

There are many different types of fixing tool, some of which you have already looked at. With the exception of the cartridge hammer gun and cable stapler, all of the tools listed below are a powered version of their hand-tool equivalent.

- Timber nail gun
- Electric impact wrench
- Cable stapler
- Cartridge hammer gun
- Cordless riveter
- Cordless screwdriver

Safety awareness

Cartridge guns and compressed airlines are examples of particularly dangerous tools and equipment and should never be abused. You must be taught how to use them safely.

Fixings

The electrical industry uses a large variety of fixing and fastening methods, and there can be confusion over terminology. This section looks at the various types of fixings and fastenings and where they are used.

Screws and bolts

Wood screws

These are usually used when fixing items to wood. In the electrical field, they are more commonly employed in conjunction with rawl plugs where fastenings to masonry are required. When ordering wood screws, it is important to give the correct description, including four key pieces of information:

- size of screw
- length of screw
- type of head
- type of metal finish used.

Size of screw

Wood screws are measured across the shank. This diameter is equal all the way down the screw until the end, which tapers to a point for ease of starting.

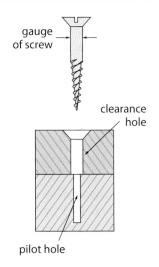

Figure 4.93 Screwing two materials together

The sizes or gauges of wood screws are given numbers, from No. 2 (the smallest) to No. 24 (the largest). Wood screws are available in all the intermediate sizes between 2 and 24, although the even sizes (that is 2, 4, 6, 8, 10 etc.) are the most commonly used.

In the electrical field, sizes 6, 8 and 10 are usually used for cable fixing and for mounting boxes and accessories. Larger sizes may be needed when mounting distribution boards or panels.

Figure 4.93 shows the difference between a clearance hole and a threaded hole. A clearance hole is designed to take the screw without the screw gripping its sides, while the pilot hole is made to give the thread of the screw a start. A pilot hole may be necessary if the screw is going to be used in thick or particularly hard wood. These holes are made either with a drill of the correct size or with a bradawl.

Length of screw

The length of a screw is measured in sizes ranging from ½ inch to 6 inches. Although they are now produced in metric sizes, they are generally still referred to by their old imperial size. You should be familiar with both. You can get screws in ⅛ inch steps for the smaller size screws, ¼ inch steps in the medium screw range, and ½ inch steps for the larger screw lengths.

The most commonly used screw sizes in the electrical field are in the ½ inch to 2½ inch range.

Type of head

A range of screw heads is available to suit different applications. Table 4.08 shows the most common types.

Screw type	Usage
Countersunk (flat-head)	• For general woodwork, fitting miscellaneous hardware, spacer bar saddles, etc. • Must be driven until the head is flush with the work surface or slightly below the surface
Countersunk (Pozidriv® head)	• Used with special screwdrivers that will not slip from the cross-slots • Can be carried into confined spaces on the end of the screwdriver • Twinfast type • For use in low-density chipboards, blockboard and softwood • Can be driven home in half the time of conventional screws

Table 4.08 Types of screw

Screw type	Usage
Raised head	• Used to fix door-handle plates and decorative hardware • Must be countersunk to the rim • Usually nickel- or chrome plated
Round head	• Used to fix surface work, fittings, accessory boxes, etc. when countersunk screws are not required
Dome head	• A concealed screw for fixing mirrors, bath panels and splash-backs • Screw head is covered by a chrome cap that is either screwed into the end of the screw or push-fitted onto it
Coach bolt	• Provides strong fixing in heavy construction and framework • Turned into the wood with a spanner

Table 4.08 Types of screw (cont.)

Type of metal used

The two main materials used are steel and brass.

Steel screws may be supplied with a bare finish or with a black, enamel-like paint known as black japanned, which helps prevent rust. They can also be cadmium coated for rust prevention. Brass screws are either left bare or are chrome plated.

The choice of material may be dictated by strength requirement – if a load-bearing capability is required, you must use steel screws.

The finish of screw is often just a matter of aesthetics.

Self-tapping hardened steel screws

Self-tapping screws are used primarily with sheet steel. A hole, slightly smaller than the screw to be used, is drilled through the steel. When the screw is driven into the steel, it will cut its own thread and become fast. This is particularly useful when joining two pieces of steel together.

Self-tapping screws and their head type are ordered as for wood. You will see them used in particular on cookers and heaters and on steel boxes.

Safety tip

Take care when using brass screws as the shaft of the screw can snap if over-tightened.

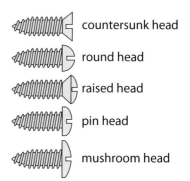

countersunk head

round head

raised head

pin head

mushroom head

Figure 4.94 Self-tapping screws

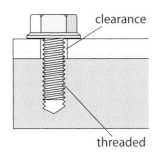

Figure 4.95 Machine-screw fixing

Machine screw

Unlike a bolt, a machine screw is threaded along its whole length. This avoids the need for long bolts when the parts to be joined are very thick, but requires the hole in one of the parts to be threaded, and the other hole to be a clearance hole.

Machine-screw heads vary in shape depending on their application. Some have a slot for a screwdriver, others a socket for an Allen key. The heads may stand proud or may be sunk below the surface for neater appearance.

The machine screws that are particularly favoured in the electrical trade have in the past been the BA thread ranges. Metric conduit boxes now use 4 mm screws, and socket outlet and switch covers use 3.5 mm metric screws, which are also often found on panels and equipment.

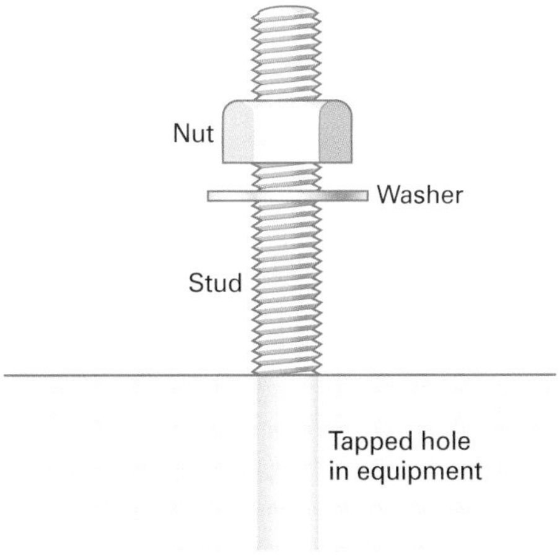

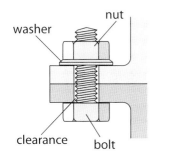

Figure 4.96 Stud, nut and washer

Stud, nut and washer

If a machine screw is frequently removed and replaced, the threads can wear and strip. This is overcome by using a stud, which is tightly fastened into a tapped hole in the equipment and remains in this position whenever the nut is removed.

Bolts

The bolt, in conjunction with a nut and washer, is used widely in all branches of engineering. The bolt passes through a clearance hole in the parts to be joined. The clearance for general work is 1.5 mm, so for a 16 mm bolt the hole would be drilled 17.5 mm.

Locking devices

Mechanical fastenings like bolts and studs are used so that parts may be removed for overhaul or replacement, or to allow access to other parts. When these fastenings are subject to vibrations, such as on machines and engines, they tend to work loose – and this could result in serious damage. Locking devices prevent this. Table 4.09 shows some of the more common types.

Figure 4.97 Bolt fixing showing clearance

Spring washer	• Similar to a coil spring • When the nut is tightened, the washer is compressed, and because the ends of the washer are chisel edged they dig into the nut and the component, preventing the nut from turning loose • Spring washers may have either a single or a double coil • Depending on the condition, spring washers are only used once
Locknut	• The bottom nut is tightened with a spanner • The top nut is then tightened, and friction in the threads and between the nut faces prevents them from rotating • Locknuts are always bevelled at the corners to ensure good setting of the faces
Split pin	• This can be used with ordinary nuts or castle nuts (see below) • When used with ordinary nuts, ensure that the split pin is in contact with the nut when tightened • The split pin is opened out after insertion to prevent it falling out • Split pins can only be used once. • The bolt is left 2 to 3 threads longer for drilling.
Castle nut	• The castle nut has a cylindrical extension with grooves • The nut is tightened, the stud is drilled opposite a groove, then the split pin is passed through the nut, and the stud prevents the nut from turning • The split pin is opened out after insertion to prevent it falling out • Split pins can only be used once
Simmonds locknut	• The Simmonds nut has a nylon insert • When the nut is screwed down, the threads on the end of the stud bite into the nylon • Friction keeps the nut tightened • This nut can only be used once
Serrated washer	• When the nut is tightened, the serration is flattened out, causing increased friction between the faces, preventing rotation • This type of washer can only be used once
Tab washer	• This is a more positive type of locking device • When the nut is tightened, one tab is bent up onto the flat side of the nut and the other tab is bent over the edge of the component • Tab washers can only be used once – they tend to fracture when straightened out and re-bent

Table 4.09 Locking devices

Fixing devices

Whenever it is necessary to fix a piece of apparatus to a wall, ceiling or partition, a fixing device is required to ensure a good hold without causing damage. The drill, screw and device should all be of the correct size with respect to each other. The choice of size will depend on the material from which the fixing surface is made and on the weight to be supported.

You should make fixing holes with an appropriate drill: use a plain drill bit for timber, but a tungsten carbide-tipped masonry drill for brick/block work. Select a slow speed for drilling masonry with an electric drill. Hammer or percussion drills are recommended for concrete.

Light fixing devices

Light fixing devices are used for relatively light fixing jobs and for partition walls and thin sheet materials. Table 4.10 shows the more common ones.

Fibre plugs	• General purpose • Size numbers match the screw numbers, so a number 12 screw should be used with a number 12 plug • Provide good holding power but may weaken with age • Supplied by the hundred, either of the same size or mixed
Plastic plugs	• More popular than fibre plugs • Should not be used when fixing a heating appliance, such as a storage radiator, as the heat will cause the plastic to soften and the appliance could become insecure • Come in strips of 10 or 20 in boxes of 100 • Colour-coded to denote size and usage, although colours can vary between manufacturers.
Plastic filler type plugs	• Use loose, powdery substance tamped into the hole • Some are mixed with water first • Holding strength is not equal to fibre or plastic • Have the advantage of fitting any hole size
Gravity toggles	• Only suitable for vertical surfaces • Intended for use in hollow partition walls (plasterboard), partition thickness 10 mm minimum • When inserted horizontally through the hole the long end falls to a vertical position

Table 4.10 Light fixing devices

Spring toggles	• Used with partition walls and ceilings (plasterboard) • Wings are spring-activated and automatically open out when inserted vertically through the hole
Rawl nuts	• Gives a secure fixing in thin, thick, solid or hollow material • Vibration-proof and waterproof
Expansion toggles	• Designed to make permanent fixings in thin sheet materials such as plywood and hardwood
Self-drilling plastic plugs	• For light to medium weight use in plasterboard • Drill a pilot hole first, then use a Pozidrive screwdriver to drive the plug home

Table 4.10 Light fixing devices (cont.)

Heavy fixing devices

These are used for heavier jobs such as fixing a large fuseboard or securing a motor to a concrete plinth. Because of the dangers associated with these heavier fixings, those shown in Table 4.11 should only be seen as a rough guide to different methods: you should get more information before using them.

Rawlbolt	• Used for fixing materials to walls, floors, etc.
	• Two types:
	  ○ bolt end protruding from the body onto which the washer and nut are placed
	  ○ bolt threaded separately

Self-drill anchor	• Expensive but faster to use • Self-drilling bolt, which is fastened in the chuck of the drill • The bolt is then removed and a tapered plug inserted • The bolt is then reinserted and, with the drill set to hammer, knocked into place • The end of the bolt is then snapped off, leaving an inserted shaft ready to accept a bolt
Ragbolts	• Bolts with a fluted end for use in floors • A hole is drilled in the ground larger than the bolt and the whole thing is cemented in • It is then left to dry before fixing the piece of equipment

Table 4.11 Heavy fixing devices

Other fixing devices

There are many types of fixing available. Your choice will depend on both the application and the aesthetics. Here are some that you may come across.

- **Roundhead nail** – used for general woodwork
- **Oval nail** – used for general woodwork; prevents splitting of timber, especially thin or heavily grained timber
- **Brad** – used as floorboard fixing; difficult to remove
- **Galvanised clout nail** – handy for fixing channelling over cables before plastering
- **Panel pin** – small pin for fastening hardboard or wood sheets; used with buckle clips
- **Masonry nail** – hard nail for use with plastic clips (PVC-sheathed cable)
- **Rivet** – a device for joining together two or more pieces of metal. They should be of the same material as the metal being joined; if this is not possible the rivets should be of a softer metal than the sheets being joined
- **Unistrut** – most commercial installations use this as a fixing method for supporting tray or trunking. It can also be used to make frames for mounting distribution boards and switchgear.

The gripple hanger

The gripple hanger is a system for supporting false ceilings, cable basket or other similar loads. It uses a principle of mini-tirfor jacks and can be easily tensioned into the correct position.

The hanger can be released using the small key provided.

Figure 4.98 Gripple supporting cable tray

Progress check

1   Identify four light fixing devices.

2   List typical hand tools an electrician may carry.

K6. Understand practices and procedures for installing wiring systems, enclosures and equipment

Nearly all modern domestic wiring is recessed into the walls, as this leaves a flat surface and a neater appearance. However, most rewiring of houses involves buildings that are at least 25 years old.

Before 1956, wiring was usually installed in vulcanised rubber insulation (VRI) with a tough rubber sheath (TRS) or a lead sheath. This type of wiring is greatly affected by temperature changes and eventually the rubber becomes brittle; any interference with the cable usually results in the insulation breaking off. Modern PVC cables have a far greater lifespan.

Sometimes this wiring was installed in slip-gauge conduit (light-duty conduit with an open seam). This type of conduit is not to be used as an earth return under any circumstances.

Both of these types of systems are now due for rewiring.

This section covers some of the techniques required for the installation and rewiring of an existing building, and the regulations that apply for the protection of installed cables. Areas covered include:

- rewiring an existing building
- floorboards
- cables run into walls
- chasing
- wiring in partition walls
- ceiling fittings
- protection of cables
- protection against heat damage and spread of fire
- cable support and bends.

Rewiring an existing domestic property

Cables run under floorboards

It is often necessary to lift floorboards to install new cables. With some older buildings the boards are of the butt-type finish; lifting these boards is comparatively easy. However, more modern houses tend to have tongue and groove floorboards.

> **Remember**
>
> Most present-day domestic properties have high-density chipboard panels, which interlock with each other. These will require a different approach should you be required to install additional points in the future.

Floorboards are normally fitted starting from one wall: board one is placed in position, then board two is slotted onto board one, then board three is slotted onto board two, and so on, until the whole floor is covered. They are then nailed down.

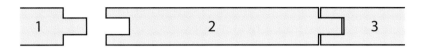

Figure 4.99 Tongue and groove

Lifting the boards

Lifting the boards with minimum damage entails lifting the last board laid and reversing the laying procedure. This is a time-consuming operation, especially when only one or two boards need to be lifted in the area you wish to work in. To lift a middle board, the two tongues holding the board need to be removed, ideally using a circular saw with a narrow blade or a flooring saw. A chisel can be used after cutting to lift the joint.

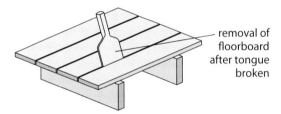

removal of floorboard after tongue broken

Figure 4.100 Lifting a middle floorboard

Before attempting to lift the floorboard, punch down the nails so that you can lift the board more easily. Take great care not to damage or cut any other service pipes, such as water or central heating pipes, while lifting the floorboards or chipboards.

Once the end of the board has been prised free, try to get the rest up without excess damage by using a fulcrum and gentle downward pressure.

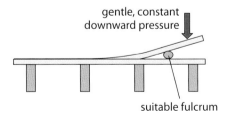

gentle, constant downward pressure

suitable fulcrum

Figure 4.101 Using a fulcrum to minimise damage

Sometimes you will need to cut a floorboard because it disappears under the skirting board or a built-in wardrobe. You should cut where the board crosses a joist; this is usually where the board is nailed. If you are unable to lift the floorboard enough to cut it, cut the board at the side of the joist, using a pad saw.

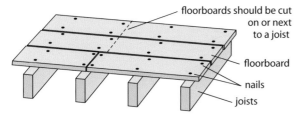

Figure 4.102 Refitting the floorboard

When refitting a board, all you need to do is screw a small piece of wood or fillet to the side of the joist.

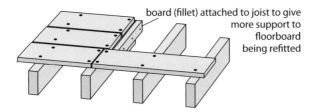

Figure 4.103 Attaching a fillet

The removed board can then be laid on it and nailed or screwed down as shown.

Cables run under floors or above ceilings

Regulation 522.6.100 states that a cable installed under a floor or above a ceiling should be run in such a position that it is not liable to be damaged by contact with the floor or the ceiling or their fixings. A cable passing through a joist within a floor or ceiling construction or through a ceiling support (for example, under floorboards), must do one of the following:

- be at least 50 mm measured vertically from the top/bottom as appropriate, of the joist or batten
- incorporate an earthed metallic armour or screen
- be enclosed in earthed conduit/trunking
- be mechanically protected against damage sufficient to prevent penetration of the cable by nails, screws and the like.

Take care to ensure that any holes or notches do not weaken the integrity of any load-bearing part of the structure.

Figure 4.104 demonstrates these requirements.

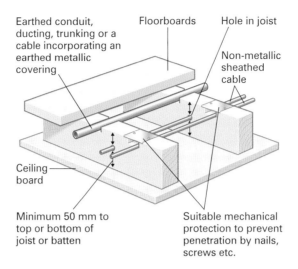

Earthed conduit, ducting, trunking or a cable incorporating an earthed metallic covering

Floorboards

Hole in joist

Non-metallic sheathed cable

Ceiling board

Minimum 50 mm to top or bottom of joist or batten

Suitable mechanical protection to prevent penetration by nails, screws etc.

Figure 4.104 Dealing with cables under floorboards

Cables run into walls

When cables have been installed in a wall at a depth less than 50 mm from any surface (for example, to feed a lighting switch or socket outlet), in accordance with Regulation 522.6.101 they must do one of the following conditions:

- incorporate an acceptable earthed metallic covering
- be enclosed in acceptable earthed conduit, trunking or ducting
- be mechanically protected to prevent penetration by nails or screws
- be installed in a zone as indicated in Figure 4.105. Note that a zone formed on one side of a wall of less than 100 mm thickness extends to the reverse side, but only if the location of that accessory can be determined from the reverse side.

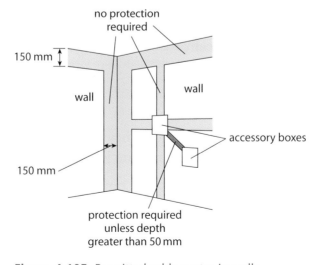

no protection required

150 mm

wall

wall

accessory boxes

150 mm

protection required unless depth greater than 50 mm

Figure 4.105 Permitted cable routes in walls

Regulation 522.6.102 also states that, where the finished installation is not intended to be under the supervision of a skilled or instructed person, cables that are not mechanically protected and are installed in a designated zone and in accordance with Regulation 522.6.101 must be provided with additional protection in the form of a suitable RCD.

Regulation 522.6.103 then states that, if the finished installation is not under the supervision of a skilled or instructed person, irrespective of the depth of the cable from the surface of the wall or partition, when that wall is constructed using metallic parts other than nails and screws, cables must do one of the following:

- incorporate an acceptable earthed metallic covering
- be enclosed in acceptable earthed conduit, trunking or ducting
- be mechanically protected to prevent penetration by nails or screws
- be provided with additional protection by means of an RCD.

Additionally, if the cable is installed at a depth of 50 mm or less from the wall or partition surface, the requirements of Regulation 522.6.101 will also apply.

Chasing

Chasing is the name given to cutting slots into a wall for conduit or cable. A chasing tool or a chasing tool attachment (available on most electric drills) makes this job easy; all that is required is a line on the wall as a guide to work to. Using a bolster chisel and hammer takes longer, but does the same job.

The back boxes for wall-mounted flush fittings must be mounted in a hole cut into the brickwork. The hole should be made slightly bigger than the box to allow plaster or filler to be applied around the edge and back of the hole, to make sure the box remains firm in the wall. The box can then be fitted in and secured. The front edge of the box must not protrude from the surface of the wall.

In new buildings, electrical wiring and back boxes are usually fitted before the plasterers render the walls – a process known as first fixing. Once all plastering and other decoration is complete, the accessories are then normally fitted – known as second fixing. Figure 4.106 shows a socket outlet fixed to an inner brick wall.

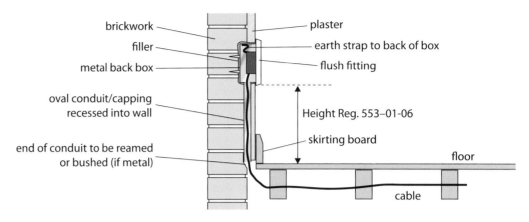

brickwork
filler
metal back box
oval conduit/capping recessed into wall
end of conduit to be reamed or bushed (if metal)

plaster
earth strap to back of box
flush fitting
Height Reg. 553–01-06
skirting board
floor
cable

Figure 4.106 Flush fitting fed through the wall via the floor

Figure 4.107 Dry lining box for wiring in partitions

Wiring in partitions

Wiring must be done before any lining (surface) is fixed in position. BS 7671 has the same requirements for partitions as for cables in walls (see pages 340–42).

The back boxes for fitting must be securely fixed, preferably to timber noggins in the structure of the partition. However, dry-lining boxes can be used if the partition lining is of sufficient strength and thickness.

Ceiling fittings

Boxes for ceiling outlets must be securely fixed to either a joist or to a noggin between the joists as shown in Figure 4.108.

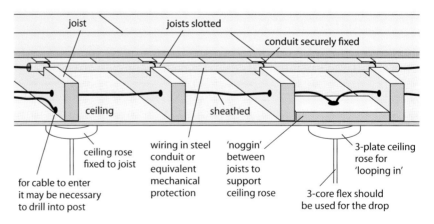

joist
joists slotted
conduit securely fixed
ceiling
sheathed
ceiling rose fixed to joist
for cable to enter it may be necessary to drill into post
wiring in steel conduit or equivalent mechanical protection
'noggin' between joists to support ceiling rose
3-plate ceiling rose for 'looping in'
3-core flex should be used for the drop

Figure 4.108 Cables through joists

Protection of cables

You have seen how cables should be run through floor joists or through walls. Regulation 522.6.1 states that precautions should be taken to prevent abrasion of any cable. When cables pass through metalwork, this can be achieved using an insulated grommet.

Where significant solar or ultraviolet radiation is experienced or expected, select or install a wiring system suitable for the conditions or provide adequate shielding. Special precautions may need to be taken to avoid the effects of solar or ionising radiation, such as from an X-ray set (Regulation 522.11.1 applies).

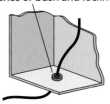

Insulated grommet or brass bushes or bush and locknut

Figure 4.109 Using insulated grommets

Protection against heat damage

To avoid the effects of heat from external sources, such as hot water systems, equipment, manufacturing processes or solar gain, one or more of the following should be used (Regulation 522.2.1 applies):

- shielding
- placing sufficiently far from the heat source
- selecting a wiring system with regard for the additional temperatures
- local reinforcement or substitution of insulating materials.

Regulation 522.2.100 states that 'parts of a cable within an accessory, appliance or luminaire shall be suitable for the temperatures likely to be encountered, as determined in accordance with Regulation 522.1.1, or shall be provided with additional insulation suitable for those temperatures'.

Protection against the spread of fire

You must first minimise the risk of fire spreading by selecting appropriate materials and equipment. However, Regulation 527.2.1 states that 'where a wiring system passes through elements of building construction such as floors, walls, roofs, ceilings, partitions or cavity barriers, the openings remaining after passage of the wiring system shall be sealed according to the degree of fire resistance (if any) prescribed for that element of the building's construction before penetration of the wiring system'.

Be aware that, during installation, temporary sealing arrangements might also be necessary (Regulation 527.2.1.1). Equally, any disturbances to sealing arrangements made during alterations must be repaired as soon as possible (Regulation 527.2.1.2).

Regulation 527.2.2 also states that, when a wiring system such as a conduit, cable ducting or cable trunking, busbar or busbar trunking penetrates elements of building construction having specific fire resistance, it should be internally sealed so as to maintain the degree of fire resistance of the respective element as well as being externally sealed (to comply with Regulation 527.2.1).

Cable/enclosure supports and bends

When a conductor or cable is not continuously supported, it should be supported by suitable means at appropriate intervals so that it doesn't suffer damage under its own weight (Regulation 522.8.4). Tables 4.12–4.16 show the support requirements and bending radii for wiring systems and enclosures.

Type of system	Maximum length of span (metres)	Minimum height of span above ground (metres)		
		At road crossings	Positions accessible to vehicular traffic, other than crossings	Positions inaccessible to vehicular traffic*
Cables sheathed with PVC or with an oil-resisting and flame-retardant or HOFR sheath, without intermediate support	3	5.8 for all types	5.8 for all types	3.5
Cables sheathed with PVC or with an oil-resisting and flame-retardant or HOFR sheath in heavy-gauge steel conduit of diameter not less than 20 mm and with no joint in its span	3	5.8	5.8	3
PVC-covered overhead lines on insulators without intermediate support	30	5.8	5.8	3.5
Bare overhead lines on insulators without intermediate support	30	5.8	5.8	5.2
Cables sheathed with PVC or with an oil-resisting and flame-retardant or HOFR sheath without intermediate support	No limit	5.8	5.8	3.5
Aerial cables incorporating a catenary wire	As specified by the manufacturer	5.8	5.8	3.5
Bare or PVC-covered overhead lines on insulators installed in accordance with the overhead line regulations	No limit	5.8	5.8	5.2

Table 4.12 Maximum length of span and minimum height above ground for overhead cables

Overall diameter of cable*	Maximum clip spacing							
	Non-armoured PVC, XLPE or lead-sheathed cable				Armoured cables		Mineral insulated with copper or aluminium sheath	
	Generally		In caravans					
	Horizontal mm	Vertical mm	Horizontal mm	Vertical mm	Horizontal mm	Vertical mm	Horizontal mm	Vertical mm
up to 9 mm	250	400	250	400	–	–	600	800
9 mm–15 mm	300	400	250	400	350	450	900	1200
15 mm–20 mm	350	450	250	400	400	550	1500	2000
20 mm–40 mm	400	550	250	400	450	600	–	–

* = For flat cables this is taken as the dimension of the major axis

Table 4.13 Spacing for supports for cables in accessible positions (Table 4A, On-site Guide)

Normal size of conduit (mm)	Maximum distance between supports (metres)					
	Rigid metal		*Rigid insulating*		*Pliable*	
	Horizontal	*Vertical*	*Horizontal*	*Vertical*	*Horizontal*	*Vertical*
Not exceeding 16	0.75	1	0.75	1	0.3	0.5
Exceeding 16 but not exceeding 25	1.75	2	1.5	1.75	0.4	0.6
Exceeding 25 but not exceeding 40	2	2.25	1.75	2	0.6	0.8
Exceeding 40	2.25	2.5	2	2	0.8	1

Table 4.14 Spacing of supports for conduits

| Cross-sectional area of trunking (mm²) | Maximum distance between supports (m) | | | |
| | Metal | | Insulating | |
	Horizontal	Vertical	Horizontal	Vertical
Exceeding 300 but not exceeding 700	0.75	1	0.5	0.5
Exceeding 700 but not exceeding 1500	1.25	1.5	0.5	0.5
Exceeding 1500 but not exceeding 2000	1.75	2	1.25	1.25
Exceeding 2000 but not exceeding 5000	3	3	1.5	2
Exceeding 5000	3	3	1.75	2

Table 4.15 Spacing of supports for trunking

Insulation	Finish	Overall diameter*	Factor to be applied to overall diameter to determine minimum internal radius of bend
XLPE, or PVC (with circular, or circular stranded copper of aluminium conductors)	Non-armoured	Not exceeding 10 mm Exceeding 10 mm but not exceeding 25 mm Exceeding 25 mm	3 (2) <> 4 (3) <> 6
	Armoured	Any	6
XLPE, or PVC (with solid aluminium or shaped copper conductors)	Armoured or non-armoured	Any	8
Mineral (with copper sheath)	With or without PVC covering	Any	6 #

* Denotes for flat cables that the diameter refers to the major axis

<> Figures in brackets relate to single-core circular conductors of stranded construction installed in conduit, ducting or trucking

\# Mineral-insulated cables may be bent to a radius not less than three times the cable diameter over the copper sheath, provided that the bend is not straightened and re-bent

Table 4.16 Minimum internal radii of bends in cable for fixed wiring

Miscellaneous

Other factors that affect the choice of a wiring system and its associated enclosures and equipment include:

- **ambient temperature** – this affects the insulation properties
- **moisture** – can the enclosures safely resist the ingress of water? **IP ratings** to be checked
- **corrosive substances** – chemicals can destroy insulation, enclosures and equipment
- **UV rays** – sunlight affects cable insulation
- **damage by animals** – by chewing cables or the corrosive effect of animal urine
- **mechanical stress and vibration** – use flexible conduits and helix loops at connections to motors to absorb vibration and make sure the cable is adequately supported throughout its length
- **aesthetic conditions** – what looks right and appropriate for the building and customer.

> **Key term**
>
> **IP rating** – International Protection Rating classifies and rates the degree of protection provided against the intrusion of solid objects, dust, accidental contact and water in mechanical casings with electrical enclosures

Remember that when you are installing cables, someone has to repair the building fabric once the job is finished. When carrying out electrical installations, you will have to work closely with other trades to co-ordinate the process, ensuring satisfactory completion of each stage and helping reduce damage to the building's structure and fabric.

Patching up afterwards will probably be the job of the main contractor (builder). However, on smaller jobs you may have this responsibility. The less damage is caused, the easier repairs are likely to be.

The repair needed will depend on the work done. In domestic installations, you may only need to use substances such as all-purpose filler to repair damage around ceiling outlets or switch boxes. However, in a domestic rewire, you should also think about flooring. Replace floorboards with care, screwing them back in place after you work to stop them squeaking.

On larger sites, you may have to think about replacing drop-in grid

Figure 4.110 Repairing a plastered wall

ceiling tiles or replacing fire barriers. Always clean up after yourself and dispose of waste materials in the correct manner.

Installing PVC/PVC cables

PVC/PVC cables are fixed using plastic clips that incorporate a masonry nail. The maximum spacing of clips for fixed wiring, when installed directly on a surface, plus the minimum bending radii were given in Tables 4.12–4.16.

Where PVC/PVC cables are installed on the surface, the cable should be run directly into the electrical accessory, ensuring that the outer sheathing of the cable is taken inside the accessory to a minimum of 10 mm. If the cable is to be concealed, a flush box is usually provided at each control or outlet position.

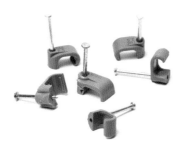

Figure 4.111 Cable clips

Installing and clipping PVC/PVC cable

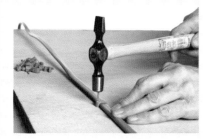

Step 1 Fix the first clip using a small cross-pein hammer.

Step 2 For a neat appearance, press PVC cable flat against the surface between the cable clips.

Step 3 Fit the end clip.

Step 4 Space other clips equally between the first and end clip.

Step 5 When a PVC cable is to be taken around a corner or changes direction, form the bend using your thumb and fingers, as shown.

Step 6 Fix the final clip after insertion into the enclosure.

Make sure that the bend does not cause damage to the cable or conductors, and that you space the cable clips at appropriate intervals. You can straighten the cable by running your thumb over it before clipping; use the palm of your hand for bigger cables.

Figure 4.112 shows what the finished product should look like.

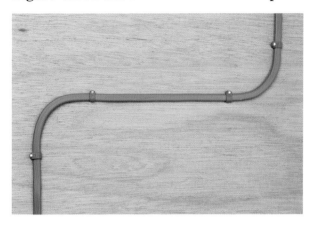

Figure 4.112 Properly executed bend in a cable

Cable suspension and catenary systems

Generally when installing cables between buildings they are laid in an excavated trench. However, cables can also be run between buildings by suspending them from an exposed catenary wire. This is seen in town centres every Christmas to support decorations.

In this context, a catenary wire is a length of steel wire fixed to the structure of both buildings by anchor plates and then tightened using a turnbuckle (sometimes referred to as a rigging screw) whose centre section rotates against a thread to draw the wire in and thus tension the wire. The cable is then fastened to the wire. To prevent rain water entering the building, a drip loop is formed at either end of the cable. Figure 4.113 shows one end of a catenary wire connected to a higher point on the other building.

Did you know?

Catenary wires are used to support lighting systems, most commonly in public spaces such as town squares or road networks, where the cables are supported between columns instead of buildings. The principle can also be found on 12 V decorative wire lighting systems often found in kitchens or public galleries.

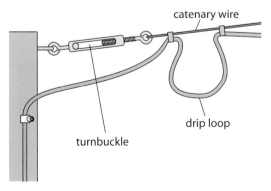

catenary wire

drip loop

turnbuckle

Figure 4.113 Cable suspension with drip loop

Cable runs

Cable runs should be planned to avoid cables having to cross one another, as this would result in an unsightly and unprofessional finish. When cables are to be installed in cement or plaster, they need protection from damage, so you should cover them with a metal or plastic channel or install them in oval PVC conduit.

When installing PVC cables, make sure that they are not allowed to come into contact with:

- gas pipes
- water pipes
- any other non-earthed metalwork.

PVC/PVC-sheathed cables should not come into contact with polystyrene insulation: a chemical reaction takes place between the PVC sheath and the polystyrene, resulting in the migration of polymers within the cable – a process known as 'marring'.

PVC/SWA/PVC cable

You are likely to come across three main variations of this cable. Two of them are PVC/ASA/PVC and PVC/AWA/PVC.

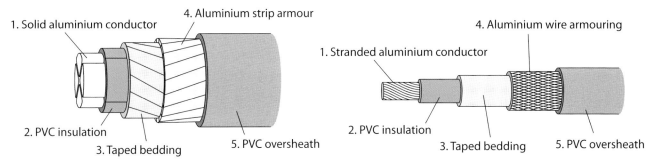

1. Solid aluminium conductor
4. Aluminium strip armour
2. PVC insulation
3. Taped bedding
5. PVC oversheath

4. Aluminium wire armouring
1. Stranded aluminium conductor
2. PVC insulation
3. Taped bedding
5. PVC oversheath

Figure 4.114 PVC/ASA/PVC four-core aluminium strip-armoured cable and PVC/AWA/PVC single-core sectoral cable

However, the most common version is PVC/SWA/PVC.

1. Shape stranded copper conductor
4. Galvanised steel wire armour
5. PVC oversheath
2. PVC insulation
3. Extruded bedding

Figure 4.115 PVC/SWA/PVC cable

Installing PVC/SWA/PVC cable

Cables can be laid directly in the ground or in ducts, or fixed directly onto walls using cable cleats and screws, plugs or other appropriate fixing devices.

If several cables are to follow the same route, they are best supported on cable trays or racks. On a cable tray, cable ties will hold cables in place, as Figure 4.116 shows.

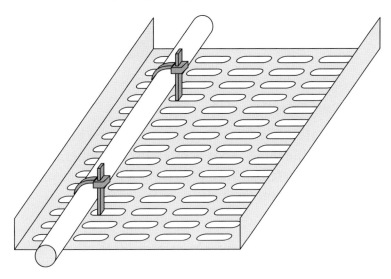

Figure 4.116 Cable tray and cleat

When several cables are installed in enclosed trenches, the current ratings will be reduced due to their disposition. The correction factors for cables run under these conditions are found in BS 7671 Appendix 4.

Installation is relatively easy for smaller size cables, for bigger sizes or multi-core cables, an installation team is needed.

Usually, one-hole nylon cable cleats are used. For bigger cables, cleats made of die-cast aluminium are used. These are often designed to slot into steel channels; once in place, the channels can accommodate multiple runs of cable.

See page 342 for information on support spacing and bending radii for this and other cables.

Figure 4.117 Cleats

PVC/GSWB/PVC cable

Installation techniques are virtually identical to SWA. The only difference lies in the type of gland required when terminating the cable.

MICC cable

Mineral-insulated copper-clad (MICC) cables have good fire-resisting properties: copper can withstand 1000°C and magnesium 2800°C. The limiting factor of the whole cable system is the seal; where a high working temperature is required, special seals must be used.

Figure 4.118 MICC gland and seal

Mineral-insulated (MI) cable is very robust and can be bent or twisted within reasonable limits, so it is often used in emergency lighting and fire-alarm systems. MI cables have a high current-carrying capacity and are non-ageing.

The cable itself is completely waterproof, although where it is to be run underground or in ducts, a PVC oversheath must be used. Bare copper unsheathed MI cables do not emit smoke or toxic gases in fires. The copper sheath can be used for earth continuity, saving the need for a separate protective conductor.

These cables come into their own in areas such as boiler houses, where the ambient temperature can become high and there is moisture present. They are ideal for buildings such as churches and castles, where they can be dressed into tight and awkward spaces, and will weather to blend aesthetically into timber and stonework. Fixings are made either with one-hole clips or two-hole saddles, and screws are usually round-headed brass.

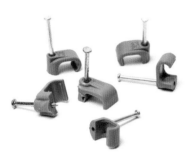

Figure 4.119 Cable clips with single nail

Installing MICC cable

One of the advantages of MI cables is that they can be run on the surface as well as under plaster. When they are run on the surface, for a neat appearance they must run straight. Lines should be run out before fixings are made and you should take care not to twist the cable unnecessarily before installation. Rollers can be used to straighten the cable.

Where a large number of cables are run together, you can use an adjustable saddle, which will take many cables. Often MI cables are run on cable tray. This provides easier fixing, a neat appearance and cleanliness. It also means that many cables can

Figure 4.120 Putting bent cable in roller; Closing roller; Running roller along cable

be run together. Where several cables leave or enter an enclosure – for example, at a distribution board – the cable should be distributed equally. If there are too many cables to fit into one row, drill two staggered rows of holes.

For cables buried in plaster, plaster-depth boxes are available. With this type of box no gland is required, but the earth clamp over the pot must be secured to give continuity. You can also get earth-tailed pots, which have an earth conductor manufactured as part of the pot, giving good earth continuity. If you are installing this cable directly into a motor, make a loop in the cable to prevent mechanical stress at the termination due to excessive vibration.

FP 200 and 400 cable

Often used for fire alarms and fire detection systems, there are various types of FP cable, including FP400, FP 200 Gold and FP 200 Flex. Some FP cables use stranded conductors and some solid.

In this section you will only look at FP 200 Gold as it is the most widely used.

FP 200 Gold

FP 200 Gold has solid copper conductors covered with a fire- and damage-resistant insulation (Insudite). An electrostatic screen is provided via a laminated aluminium tape screen bonded to the sheath of the cable, in full contact with a tinned annealed uninsulated cpc. The sheath is a robust thermoplastic low-smoke, zero-halogen sheath, which is an excellent moisture barrier.

Figure 4.121 FP 200 construction

The cable is used in fixed installations, in dry or damp premises, in walls, on boards, in channels or embedded in plaster for situations in which prolonged operation is required in the event of fire. It is primarily intended for use in fire-alarm and emergency-lighting circuits and has a low voltage rating of 300/500 volts. The cable is easy to handle, does not require any special tools and bends easily without the use of bending tools.

The sheath is available in red or white, although special colours are available on request. Core colours are:

- two core: brown and blue
- three core: brown, black and grey
- four core: brown, black, grey and blue.

Installing FP 200 Gold

When the cable is required to maintain circuit integrity during a fire, any clips or ties used to support the cable must also be able to withstand the fire. Clips should be copper, steel or copper coated; ties should be suitable for fixing a fire-rated cable. Standard plastic or nylon clips or ties should not be used to fix the cable. A new fixing system has now been introduced that uses coated stainless steel clips and gas nailing technology.

Depending on the installation, standard nylon compression glands can be used, though FP nylon or brass glands are available. The cable should be dressed by hand to prevent damage to the cable. The insulation strips easily from the conductors, leaving them in a bright, clean condition, eliminating the risk of any high-resistance terminations.

When bending this cable, the bending radius should not be less than six times the diameter of the cable.

Did you know?

Cable lacks mechanical strength and can be damaged if the bending radii are not observed.

Conduit

Annealed mild-steel tubing, known as conduit, is widely used as a commercial and industrial wiring system. PVC-insulated (non-sheathed) cables are run inside the steel tubing. Conduit can be bent without splitting, breaking or kinking, provided the correct methods are employed. Available with this system is an extensive range of accessories to enable the installer to carry out whole installations without terminating the conduit. It offers excellent mechanical protection to the wiring and, in certain conditions, may also provide the means of earth continuity.

The British Standard covering steel conduit and fittings is BS 4568. The two types of commonly used steel conduit are known as black enamel conduit, which is used indoors where there is no likelihood of dampness, and galvanised conduit, which is used in damp situations or outdoors.

In this section you will look at:

- screwed conduit
- bending machines
- types of bend
- bending methods
- conduit fixings
- conduit fitting
- running coupling
- termination of conduit
- use of non-inspection elbows and tees
- wiring conduit
- plastic conduit
- miscellaneous points
- conduit and cable capacities.

Screwed conduit

Screwed steel conduit can be either seam welded or solid drawn. Solid drawn is stronger but much more expensive. At one time solid-drawn conduit was the only conduit that could be used in hazardous areas. More recently, because of manufacturing improvements, seam-welded conduit can now also be used in hazardous areas.

The thread used on steel conduit is not used on any other pipe, so special conduit dies are required. Before 1970, conduit sizes were imperial, and of course many of these conduits are still around. Since 1970 metric sizes have been used, and where the two must be joined together adapters may have to be used. Adapters may have external imperial threads and internal metric threads or vice versa.

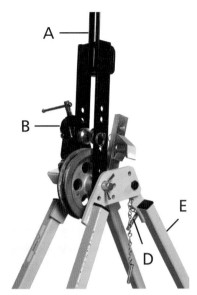

Figure 4.122 Correctly positioned bending machine

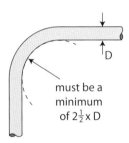

must be a minimum of $2\frac{1}{2} \times D$

Figure 4.124 Minimum-radius allowed

Bending machines

Bending machines used to be considered an expensive item but are available cheaply these days. They give consistent results every time and require only a small amount of practice.

To position the stand as shown in Figures 4.122 and 4.123, swing the rear leg (E) to its maximum. Place the safety pin (D) through the hole beneath where the pin hangs, locking the rear leg in place. The machine should now be standing with the swivel arm (A) hanging downwards. (B) illustrates the conduit guide and (C) is the adjusting arm for the conduit guide.

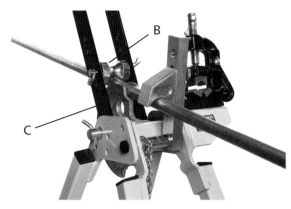

Figure 4.123 Bending machine with a piece of conduit inserted, which prevents the swivel arm from hanging downwards

Types of bend

Avoid sharp bends. The minimum radius of steel conduit is laid down in BS 7671 as two and a half times the outside diameter of the conduit, as shown in Figure 4.124.

Right-angled bend	This is used to go around a corner or change direction by 90°. When bending, measurements may be taken from the back, centre or front of the bend. Allowance should be made for the depth of the fixing saddle bases.	
Set	The set is used when surface levels change or when terminating into a box entry. Sets should be parallel and square, not too long and not too short so that the end cannot be threaded. Where there are numerous sets together all sets must be of the same length. The double set or saddle set is used when passing girders or obstacles as shown.	Set / Double set
Kick	The kick is used when a conduit run changes direction by less than 90°.	
Bubble set or crank set	The bubble set or crank set is used when passing obstructions, especially pipes or roof trusses etc. The centre of the obstruction should be central to the set.	

Table 4.17 Types of bend

Bending methods

This diagram shows the bend required to make a 90° bend from a fixed point.

Required set

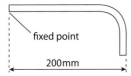

fixed point

200mm

Make a 90° bend from a fixed point

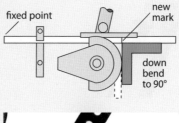

fixed point

mark 200mm from fixed point

new mark

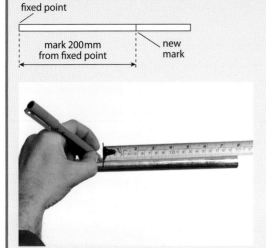

Step 1 Mark the conduit as shown, 200 mm from the fixed point. If the distance is given to the inside or centre of the tube, simply add on either the diameter or half the diameter respectively to give the back bend measurement and follow the same procedure as for outside measurement.

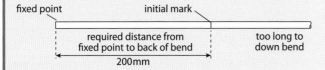

fixed point

initial mark

required distance from fixed point to back of bend

200mm

too long to down bend

Step 3 Where the remaining length of tube from the measured point is too long to down-bend and where it is not convenient or possible to up-bend using the method described, the problem can be overcome by using the following method.

fixed point

new mark

down bend to 90°

Step 2 Place the tube in the former, with the fixed point to the rear. Position the tube so that a square held against the tube at the fixed point touches and forms a tangent to the leading edge of the former.

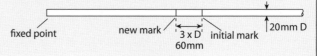

fixed point

new mark

3 x D 60mm

initial mark

20mm D

Step 4 Deduct three times the outside diameter of the tube from the initial mark.

Make a 90° bend from a fixed point (cont.)

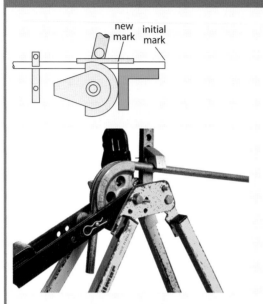

new mark   initial mark

Step 5 Place the tube in the former with a fixed point to the front with the mark at 90° to the edge of the former. This will give a 90° bend at the required distance from the fixed point to the back of the bend as shown.

Making double sets

This diagram shows the required double set.

50 mm

Double sets

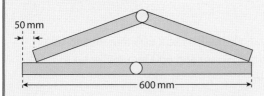

50 mm

600 mm

Step 1 To ensure re-entry of the bent tube into the bending machine to complete the return set, a determined angle of initial bend is required. Determine the distance of the set at 50 mm and deduct this distance from a 600 mm rule.

Step 2 The tube can be bent using the angled rule to indicate the angle of the first set.

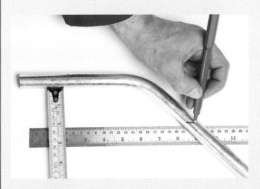

Step 3 Remove the tube from the bending machine and mark the tube for the return set, making sure to measure the height of the obstacle or accessory from the inside of the tube.

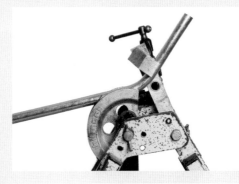

Step 4 Reposition the tube in the machine, ensuring that the mark on the tube forms a tangent to the edge of the former. The final set can be made parallel with the first.

Making bubble sets

Here is an example using a 50 mm tube.

Bubble sets

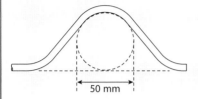

Step 1 To obtain the correct angle for the first set, multiply the external diameter of the obstacle by three. Here, for 50 mm tube, 50 × 3 = 150 mm.

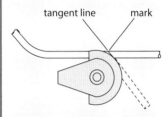

Step 3 Having marked the centre of the set on the tube, position the tube with the mark vertically above a mark on the former, which is determined by bisecting the angle of the rule when placed as shown.

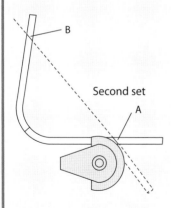

Step 5 Position the tube in the bending machine so that the mark A forms a tangent to the edge of the former. Bend down until the top edge of the tube is level and in line with mark B.

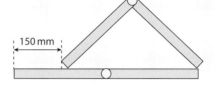

Step 2 Stagger the legs of a 600 mm folding rule between the 150 mm and 600 mm marks on a second rule.

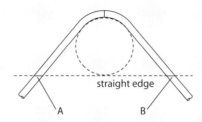

Step 4 Place the conduit over the obstacle; measure 50 mm from the inside of the first set to a straight edge and mark the tube at A and B as shown.

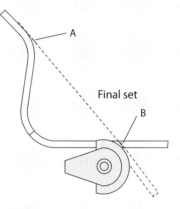

Step 6 Reverse the tube in the former and position as for mark A. Down-bend until the top edges of the tube are in line.

Conduit fixings

Conduits must be securely fixed in accordance with the following distances between supports.

Nominal size of conduit	Maximum distance between supports					
	Rigid metal		Rigid insulating		Pliable	
1 m	Horizontal 2 m	Vertical 3 m	Horizontal 4 m	Vertical 5 m	Horizontal 6 m	Vertical 7 m
Not exceeding 16	0.75	1	0.75	1	0.3	0.5
Exceeding 16 but not exceeding 25	1.75	2	1.5	1.75	0.4	0.6
Exceeding 25 but not exceeding 40	2	2.25	1.75	2	0.6	0.8
Exceeding 40	2.25	2.5	2	2	0.8	1

Table 4.18 Conduit-fixing parameters

Methods of supporting conduit

Figure 4.125 shows some common methods of supporting conduit.

Figure 4.125 Conduit fixings

- Strap saddle or half saddle – used for fixing conduit to cable tray or steel framework.
- Spacer bar saddle – used when fixing to an even surface; it gives a clearance of 2 mm.
- Distance saddle – used if the surface is uneven and where brick on concrete can give rise to heavy condensation.
- Hospital saddle – used where it is necessary to clean around the conduit fixing.
- Multiple saddle strip – used to fasten multiple runs of conduit together.
- Girder clamp will fix conduit to girders and I-beams without having to drill a hole in the girder.
- Pipe hook or crampet – used when conduits are secured to a wall or cast in concrete.

Conduit fitting

Cutting and screwing conduit

Step 1 Conduit should be cut with a hacksaw. The cut should be square and the full length of the blade should be used taking steady strokes. Hold the conduit in a pipe vice not a bench vice. The vice should be secured but not so tight that it cuts into the pipe.

Step 3 When the thread is finished the stock and die is removed and the inside of the conduit is cleaned and reamed. This removes all burrs and sharp edges, which would cut the cables (if not removed) when they are installed. Reaming can be carried out with a reamer or round file. The standard length of thread for a normal joint is half a coupling length.

Step 2 Before threading, the conduit should be chamfered with a file to help the die start. Screwing is carried out using stocks and dies. Another part of the stock and die is the guide, which ensures the screw cut is square. Stocks and dies should be kept clean and any lubricant or steel shavings should be removed after cutting. The cut is made by placing the stock and die on the conduit and then turning clockwise while applying forward pressure; sometimes a great deal of pressure may be required. Once the cut is started the stock and die are removed so that a cutting agent can be applied. Having applied the cutting agent the stock is placed on the conduit again and the threading begins. The stock and die is turned back every turn to clean out the cuttings.

Step 4 All couplings, bushes and conduit boxes must be fully tightened before installation. Where possible couplings, bushes and boxes should be tightened while the conduit is held firmly in a pipe vice.

Running coupling

Sometimes two conduits must be joined together and neither can be turned. This may be due to one conduit coming through a wall or ceiling or long runs combined with bends making turning impossible. In these cases a running coupling must be used. Running couplings are made by having one thread a normal half-coupling length and the other thread the length of a coupling plus locking ring.

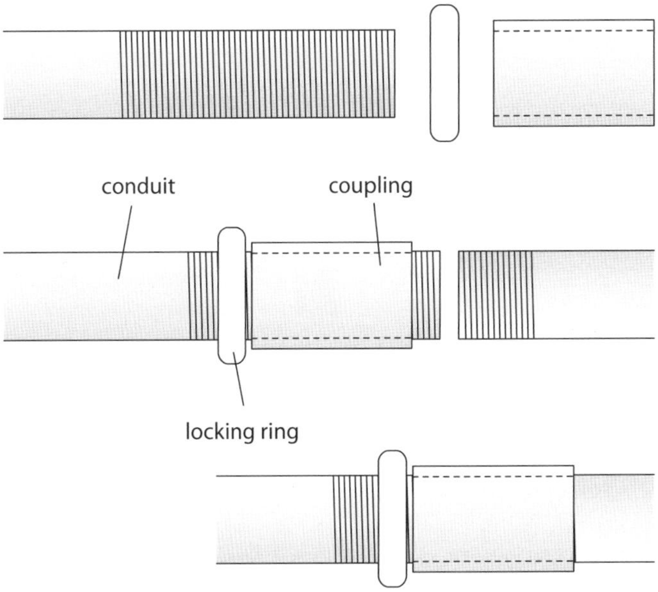

The coupling and locking ring are fixed on the long thread side and the two conduits are then butted together. The coupling is then removed from the long thread to the shorter thread and finally rests across the two sides. After tightening, the coupling is locked. A locking ring must be used because locknuts get caught on the ceiling in tight situations.

Figure 4.126 The running coupling

Conduit coupling with the locking ring

Because the coupling is traversing two threads simultaneously, the thread must be very clean and well cut. Reversing the dies and running them over the thread can help this. This is particularly important where the running coupling is in an awkward position (as it often is).

Termination of conduit

There are several methods available for terminating conduit, three of which are shown in Figures 4.127–129.

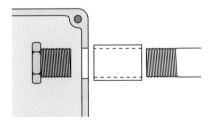

Figure 4.127 Terminating conduit at a box using a conduit coupling and brass male brush

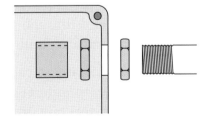

Figure 4.128 Terminating conduit at a box using locknuts and brass female brush

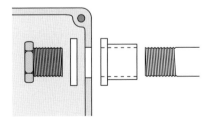

Figure 4.129 Flanged coupling washer and brass male method for use with PVC box

Figure 4.130 Fitting the brush

Figure 4.131 Tightening the brush

Use of non-inspection elbows and tees

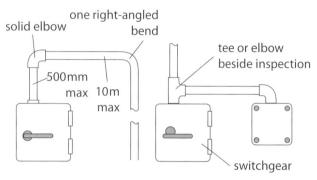

Figure 4.132 Non-inspection elbows and tees

The main consideration here is that damage to the cables does not occur during installation. Non-inspection elbows are only used adjacent to an outlet box or inspection-type fitting. One solid elbow may be used if positioned less than 500 mm from an accessible outlet, in a conduit run of less than 10 m that has other bends that are not more than the equivalent of one right angle.

Wiring conduit

Cables must not be drawn into a conduit system until the system is complete. When drawing in cables, you must first run off the reel or drum if there is no supporting mechanism such as a tube to support the reels or they must be allowed to freely revolve.

If a large number of cables are to be drawn into a conduit system at the same time the cable reels should be arranged on a stand or support so they can revolve freely.

In new buildings, cables should not be drawn in until the conduit is dry and free from moisture. If in any doubt, draw a draw tape with a swab at the end through the conduit to remove any moisture that may have accumulated.

It is usual to start drawing in cables from a midpoint in the conduit system so as to minimise the length of cable that has to be drawn in. Use a steel tape from one draw-in point to another. The draw tape should not be used for drawing in cables as it may become damaged. A steel tape should only be used to pull through a draw wire. The ends of the cables must be paired for a distance of approximately 75 mm and threaded through a loop in the draw wire.

Remember

If the cables are allowed to spiral off the reels they will become twisted and this would cause damage to the insulation.

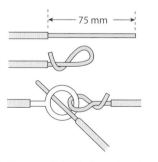

Figure 4.133 Drawing in cables

When drawing in a number of cables, feed them in carefully at the delivery end while someone pulls them at the receiving end. Take care to feed into the conduit in such a manner as to prevent any cables crossing. Always leave some slack cable in all draw-in boxes, and make sure that cables are fed into the conduit so as not to finish up with twisted cables at the draw-in point.

This operation requires care, and there must be synchronisation between the person who is feeding and the person who is pulling. If in sight of each other, you can do this by some pre-arranged signal; if within speaking distance, you the person feeding the cables can give instructions. If the two people are not within earshot or sight of each other the process is more difficult. A good plan is for the individual feeding the cables to give pre-arranged signals by tapping the conduit.

Drawing in cables

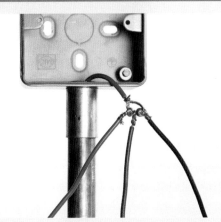

Cables attached to draw wire

Cable drum

Step 1 Pass draw tape through and between outlets.

Step 2 Fasten a draw wire securely to the draw tape.

Step 3 Feed draw wire into the conduit while withdrawing the draw tape. Ensure that the draw wire is long enough and strong enough for the job.

Step 4 Fasten cables to the draw wire. At least 75 mm of insulation should be stripped away and secured as illustrated. Fasten each cable separately.

Step 5 Where possible, cables should be drawn into the conduit directly off the cable drum.

Step 6 First make sure that there is sufficient cable for the job.

Step 7 Feed cables into the conduit, using the fingers of one hand to feed the cables and the other hand to keep cables straight.

Step 8 Ensure that no crossed or kinked cables enter the conduit.

Step 9 Keep hands close to the conduit entry and feed only short lengths of cable at a time.

Plastic conduit (PVC)

Plastic conduit is made from polyvinyl chloride (PVC), which is produced in both flexible and rigid forms. It is impervious to acids, alkalis, oil, aggressive soils, fungi and bacteria and is unaffected by sea, air and atmospheric conditions. It withstands all pests and does not attract rodents. PVC conduit is preferable for use in areas such as farm milking parlours. PVC conduit may be buried in lime, concrete or plaster without harmful effects.

Advantages	Disadvantages
• light in weight • easy to handle • easy to saw, cut and clean • simple to bend • does not require painting • minimum condensation due to low thermal conductivity in walls • quick to install	• care must be taken when applying glue to the joints to avoid forming a barrier across the inside of the conduit • if insufficient adhesive is used, the joints may not be waterproof • PVC expands around five times as much as steel and this expansion must be allowed for

Table 4.19 Advantages and disadvantages of plastic conduit

Working with PVC conduit

The techniques required for working with PVC conduits differ considerably from those in steel conduit installations.

PVC conduit is easily cut using a junior hacksaw. Any roughness of cut and burrs should be removed by simply wiping with a cloth.

The most common jointing procedure uses a PVC solvent adhesive. The adhesive should be applied to the female part of the joint and the conduit twisted into it to ensure a total coverage. Generally the joint is solid enough for use after two minutes, although complete adhesion takes several hours. In order to ensure a sound joint, the tube and fittings must be clean and free from dust and oil.

Where expansion is likely and adjustment is necessary, use a mastic adhesive. This is a flexible adhesive, which makes a weatherproof joint, ideal for surface installations and in conditions of wide temperature variation. It is also advisable to use mastic adhesive where there are straight runs on the surface exceeding 6 m in length.

PVC conduit expands considerably more than metal conduit with an increase in temperature. The expansion can be ignored where

Safety tip

Take care when using these adhesives as they are volatile liquids.

- Always replace the lid on the tin immediately after use.
- Take care when using in a confined space.
- Always read the manufacturer's instructions.

the conduit is buried in concrete or plaster. In surface work precautions must be taken to prevent such expansion from causing the conduit to bow. Usually where bends and sets are close together these take up any expansion. Where longer runs of conduit occur in conditions of varying temperatures, some provision for expansion must be made, using expansion couplers as shown here.

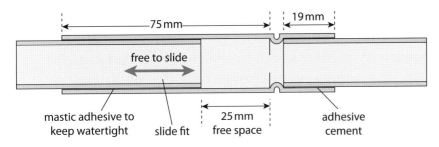

Figure 4.134 Expansion provision in conduits

PVC conduits not exceeding 25 mm diameter can be bent cold using a spring. The bend is then made by either the hands or across the knee. To achieve the angle required, the original bend should be made at twice the angle required and the tube allowed to return to the correct angle. Under no circumstances should you try to force the bend back with the spring inserted, as this can damage the spring. It is easier to withdraw the spring by twisting it in an anti-clockwise direction. This reduces the diameter of the spring, making it easier to withdraw.

In cold weather it may be necessary to warm the conduit slightly at the point where the bend is required. One of the simplest ways is to rub the conduit with your hand or a cloth. The PVC will retain the heat long enough for the bend to be made. To keep the bend at the correct angle, fasten the conduit to the surface with a saddle as soon as possible.

Miscellaneous points for steel and plastic conduits

- Ample capacity must be provided at junctions employed for cable connections.
- Where a steel conduit is used as a circuit-protective conductor, it must be tested in accordance with BS 7671.
- Where a steel conduit forms the protective conductor, the earthing terminal of each accessory (such as a socket outlet or switch grid) shall be connected by a separate protective conductor to an earthing terminal incorporated within the associated box or enclosure.

- Cable capacities should be calculated in accordance with the relevant tables found in Section 5 of the *Unite Guide to Good Electrical Practice*.

- Conduits run overhead should be run in accordance with the table on page 72 of the *Unite Guide to Good Electrical Practice*.

- Where conduits pass through walls the hole shall be made good during installation with a fire-resistant material.

Conduit capacities

The number of cables that can be drawn into or laid in any enclosure of a wiring system must be such that no damage can occur to the cables or the enclosure during installation.

You will need to calculate the size of conduit used to enclose the cables using a 'factor' system to compare the number of cables against the overall CSA (cross sectional area) of the conduit. The minimum conduit size will be one that has a higher factor than that of the cables.

Tables 4.20 and 4.21 are applicable to both plastic and metal conduit and use information found in Section 5 of the *Unite Guide to Good Electrical Practice*. These tables only give guidance to the maximum number of cables that should be drawn in and the electrical effects of grouping are not taken into account.

As the number of circuits increases, the current-carrying capacity of the cables will decrease so cable sizes would have to be increased, with a consequent increase in cost of cable and conduit. It may be more economical to divide the circuits concerned between two or more enclosures.

Tables 4.20 and 4.21 are for use with short, straight runs of up to 3 m.

Here is an example of how to use these tables:

Example

If 10 × 2.5 mm² cables with stranded copper conductors were to be installed in a 2 m straight length of conduit, their cable factor will be 10 × 43 = 430. You would need to select a conduit with a factor greater than 430.

From the conduit factor table, you can see that 16 mm conduit is too small (290) but that 20 mm conduit has a factor of 460, so this is what you would install.

This method is acceptable when using short, straight runs. However, when longer runs are involved, or the run has bends in it, you will need to use other tables.

Type of conductor	Conductor CSA mm²	Cable factor
Solid	1	22
	1.5	27
	2.5	39
Stranded	1.5	31
	2.5	43
	4	58
	6	88
	10	146
	16	202
	25	385

Table 4.20 Cable factors for use in short (up to 3 m) straight runs

16	290
20	460
25	800
32	1400
38	1900
50	3500
63	5600

Table 4.21 Conduit factors for use in short (up to 3 m) straight runs

Cable and conduit factors for runs over 3 m or with bends

Tables 4.22 and 4.23 apply to cable and conduit runs over 3 m or with bends. Conduit runs in excess of 10 m or with more than four bends should be divided into separate sections and then have the table values applied to each of those sections.

Tables with factors for conduit sizes larger than 32 mm are not available. Calculate these by multiplying the 32 mm factor as follows:

- for 38 mm conduit (1.4 x 32 mm term)
- for 50 mm conduit (2.6 x 32 mm term)
- for 63 mm conduit (4.2 x 32 mm term)

Solid or stranded cables	1	16
	1.5	22
	2.5	30
	4	43
	6	58
	10	105
	16	145
	25	217

Table 4.22 Cable factors for use in long (over 3 m) straight runs, or runs of any length with bends

Length of run (m)	Conduit diameter (mm)																			
	Straight				One bend				Two bends				Three bends				Four bends			
	16	20	25	32	16	20	25	32	16	20	25	32	16	20	25	32	16	20	25	32
1	Covered by short straight run tables				188	303	543	947	177	286	514	900	158	256	463	818	130	213	388	692
1.5					182	294	528	923	167	270	487	857	143	233	422	750	111	182	333	600
2					177	286	514	900	158	256	463	818	130	213	388	692	97	159	292	529
2.5					171	278	500	878	150	244	442	783	120	196	358	643	86	141	260	474
3					167	270	487	857	143	233	422	750	111	182	333	600	Divide into 2 or more parts with draw in boxes			
3.5	179	290	521	911	162	263	475	837	136	222	404	720	103	169	311	563				
4	177	286	514	900	158	256	463	818	130	213	388	692	97	159	292	529				
4.5	174	282	507	889	154	250	452	800	125	204	373	667	91	149	275	500				
5	171	278	500	878	150	244	442	783	120	196	358	643	86	141	260	474				
6	167	270	487	857	143	233	422	750	111	182	333	600	Divide into 2 or more parts with draw in boxes							
7	162	263	475	837	136	222	404	720	103	169	311	563								
8	158	256	463	818	130	213	388	692	97	159	292	529								
9	154	250	452	800	125	204	373	667	91	149	275	500								
10	150	244	442	783	120	196	358	643	86	141	260	474								

Table 4.23 Conduit factors for use in long (over 3 m) straight runs, or runs of any length with bends

Here is an example.

Example

A lighting circuit in a school requires 12 × 1.5 mm² cables inside a conduit of 8 m with two right angled bends.

From the cable factor table, you can see the cable has a factor of 264 (12 × 22).

From the conduit factor table, you can see that 20 mm is too small (159) to handle this, but a 25 mm conduit will be acceptable, with a factor of 292.

Steel and PVC trunking

Trunking is a fabricated casing for cables. To facilitate connections, terminations and so on, and in order to run the trunking with the contours of the building in which it is installed, a wide range of trunking fittings and accessories is available.

Trunking may be wall-mounted, using round head screws (to avoid damage to cables) or suspended using appropriate brackets or Unistrut.

Prefabricated bends and sets will usually be used to install trunking systems because they are quicker and cheaper. However, there will be times when the prefabricated bends and sets are either not suitable or are not available. In this situation you will need to be able to fabricate your own.

While steel trunking accessories use nuts and bolts to join pieces together, PVC ones usually just snap on to fit. PVC trunking can easily be cut at angles to form bends. Steel trunking is more difficult.

Here you will look at how to fabricate the following:

- right-angled internal bend
- right-angled vertical bend
- trunking sets
- tee junction.

Figure 4.135 Steel trunking

Safety tip

Remove any rough or jagged edges with a medium-cut file in order to prevent damage to the cables.

Right-angled internal bend

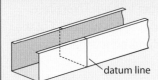

Step 1 Select a short section of trunking between 900 and 1000 mm in length. Using a soft pencil and a reliable set square, draw a line (called a datum line) around the outside (periphery) of the trunking. This should be done at the mid point positions as shown.

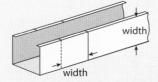

Step 2 Check the width of the workpiece and transfer this measurement to either the top left- or right-hand side of the central datum line as shown.

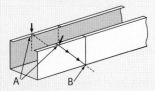

Step 3 Using an adjustable set square as a guide, draw a pencil line from the marked trunking to the bottom of the centre datum line as shown. Repeat this guideline on the opposite vertical side.

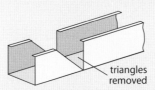

Step 4 At this stage there will be a right-angled triangle drawn on each outer side of the trunking. Remove the two triangles using a hacksaw with a blade fitted with 25–30 teeth per 25 mm of blade. Once removed, file smooth all rough or jagged edges as these may damage the cables.

Right-angled internal bend (cont.)

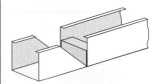

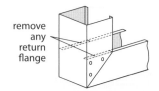

remove
any
return
flange

Step 5 Cut a wooden block with a good square edge on one side able to be fitted comfortably across the internal width of the trunking. Place the wooden block to the vertically cut side as shown in the diagram.

Step 6 Hold the block firmly in place and with the other hand push up the side of the trunking adjoining the angled cut. Allow the vertical sides to be sandwiched between the angled trunking sides. The wooden block will help provide a sharper edge at the bending point. Once completed, dress the bend with a hammer and remove the wood. Check for squareness and strengthen with pop rivets. Nut and bolt, or spot-weld.

Right-angled vertical bend

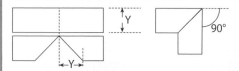

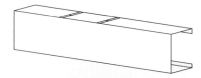

Step 1 Mark out the position of the bend on all sides of the trunking.

Step 2 Drill small holes in the corners at the point of the bend to stop the metal from folding. Then place wooden blocks inside the trunking for support. Cut the sides of the trunking with an appropriate hacksaw.

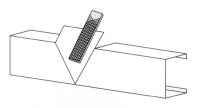

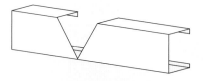

Step 3 The edge of the trunking can be cut with a file (as shown in (c)) and the waste broken off.

Step 4 Cut away the back of the trunking using a suitable hacksaw. Then file all the rough and jagged edges and bend the trunking to shape as shown.

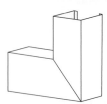

Step 5 Make a fishplate out of some scrap trunking and drill in some fixing holes.

Step 6 Finally mark out the trunking from the holes in the fishplate and drill.

Right-angled vertical bend (cont.)

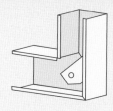

Step 7 Secure the assembly with nuts and bolts or pop rivets. Alternatively the joint may be spot-welded.

Making a right-angled vertical bend (alternative method)

Step 1 Mark the trunking for cutting.

Step 2 Place a wooden block inside trunking to secure vice.

Step 3 Use a hacksaw to cut the trunking.

Step 4 Bend the trunking into a right-angled bend.

Step 5 Secure with rivets.

Step 6 Here is the final result.

Trunking sets

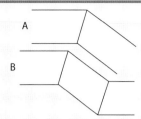

A trunking set and a return set are both constructed in a similar fashion when making a right-angled bend. The diagram illustrates a trunking set (A) and a trunking return set (B).

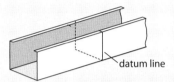

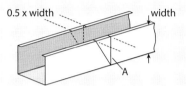

Step 1 Select a section of trunking that can be worked comfortably yet is long enough to accommodate either a set or return set. Draw a datum line using a soft pencil and set square as a guide.

Step 2 Measure the width of the trunking and transfer half of this measurement to either the top left or top right-hand side of the datum line. Draw a line from this mark to the base of the datum line A.

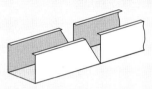

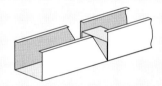

Step 3 Cut out both triangular shapes from each side of the trunking.

Step 4 Cut a section of a wooden block that will fit comfortably across the internal diameter of the trunking. Place the piece of wood as shown in the diagram so that it is at the side adjacent to the vertically cut datum line.

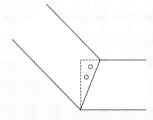

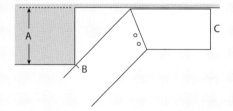

Step 5 Secure the wooden block with one hand while gently bending up the remaining half of the workpiece until the set is formed. Check that the required angle is correct and secure using pop rivets, nut or bolts or spot-welding.

Step 6 If a return set is involved then extra work is required. Draw a line of reference on a suitable flat surface and offer the shortest section of the set to the line as shown. Measure the depth of the required set (A) and mark on the trunking at (B). (C) represents the shortest leg.

Trunking sets (cont.)

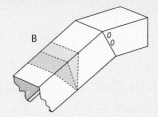

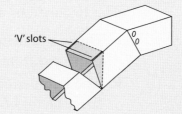
'V' slots

Step 7 Prepare the trunking as shown in the diagram. Cut out the complete left-hand side of the centre line comprising two side triangles bridged by a rectangular base section. To provide electrical continuity the trunking lip must remain unbroken.

Step 8 A 2 mm diameter V shape should be cut along both sides of the bottom edge of the trunking from the vertical centre line to the angled dotted line, as shown. This will act as a supplementary lip and help stabilise the return set when it is assembled.

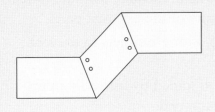

Step 9 Gently bend the workpiece and unite both sections of the return set, allowing the vertically cut centre edges to be sandwiched between the angled sides of the trunking as shown in the diagram. Check that the set has been worked to meet the required measurement. Secure using pop rivets, nuts and bolts or spot-welding. Dress the supplementary base lip to accommodate the changed angle and secure as necessary.

Tee junction

Mark out the position of the tee using a second piece of trunking to gauge the width. Cut out a space for the tee as shown. Use blocks of wood to support the sections being cut. File all rough edges to protect the cables.

Cut away the section to leave two lugs. These can be bent in a vice using a hammer to give a clean edge. File all edges smooth and check the fit as necessary.

Mark out the holes, drill and secure with nuts and bolts or pop rivet, or alternatively spot-weld.

Regulations concerning trunking

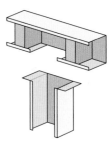

Figure 4.136 Tee junction

- **Regulation 521.5.1** – every conductor or cable shall have adequate strength and be so installed as to withstand the electromechanical forces that may be caused by any current (including fault current) it may have to carry in service.

- **Regulation 521.5.2** – the conductors of an a.c. circuit installed in ferromagnetic enclosure shall be arranged so that all line conductors, the neutral conductor (if any) and the appropriate protective conductor are contained in the same enclosure.

- **Regulation 521.6** – cable trunking or ducting shall comply with the appropriate part of the BS EN 50085 series.

The Regulations also lay down the following guidelines:

- Ducts and metallic trunking must be securely fixed and protected from mechanical damage.
- The number of cables installed in trunking shall be such that a space factor of 45 per cent must not be exceeded.
- During installation where conduit, ducts or trunking passes through walls, floors, ceilings or partitions, the surrounding hole must be made good with a non-combustible material, and internal fire barriers are to be installed within trunking at these positions.
- Copper links are used across joints in metallic conduit systems in order to maintain the continuity of the exposed parts.
- Cable entries must be protected with grommets to prevent damage or abrasions to cables.
- Straight runs of trunking are best joined by using a joining section and the nuts and bolts supplied.
- When terminating metal trunking at a distribution board it is essential to ensure that the junction between trunking and distribution board will not cause abrasion to the cables.
- Grommet strips should be fitted over the edges of any holes drilled in trunking to prevent damage to cables.
- To prevent the effects of eddy currents (electromagnetic effects), cables of an a.c. circuit should pass through a single hole.

Essentially, for PVC trunking you use the same range of accessories and installation techniques as for steel. Various styles exist for both materials, some of which are shown in Figures 4.137–4.139.

Figure 4.137 Dado trunking

Figure 4.138 Multi-compartment trunking

Figure 4.139 Self-adhesive mini trunking

Trunking capacities

The number of cables that can be drawn into or laid in any enclosure of a wiring system must be such that no damage can occur to the cables or the enclosure during installation.

Therefore, the size of trunking used to enclose the cables needs to be calculated and this is done by using a 'factor' system in a similar way to conduit.

Tables 4.24 and 4.25 are applicable to both plastic and metal trunking and use information found in Section 5 of the *Unite Guide to Good Electrical Practice*. These tables only give guidance to the maximum number of cables that should be drawn in.

Here is an example.

Example

The following XLPE insulated, stranded copper conductor cables are to be installed in trunking: 25 x 1.5 mm², 20 x 2.5 mm², 6 x 4.0 mm², 2 x 10 mm² and 2 x 16 mm².

This gives you factor values of 240 (25 x 9.6), 278 (20 x 13.9), 108.6 (6 x 18.1), 72.6 (2 x 36.3) and 100.6 (2 x 50.3), which gives you a total 'term' value of 799.8.

If you now look at the trunking factors table, you can see that the best option is a 50 mm x 50 mm trunking, which has a term capacity of 1037.

Type of conductor	Conductor CSA mm²	Thermoplastic cable factor	Thermosetting cable factor
SOLID	1.5	8.0	8.6
	2.5	11.9	11.9
STRANDED	1.5	8.6	9.6
	2.5	12.6	13.9
	4	16.6	18.1
	6	21.2	22.9
	10	35.3	36.3
	16	47.8	50.3
	25	73.9	75.4

Table 4.24 Cable factors for trunking

Dimensions of trunking mm × mm	Term	Dimensions of trunking mm × mm	Term
50 x 38	767	200 x 100	8572
50 x 50	1037	200 x 150	13001
75 x 25	738	200 x 200	17429
75 x 38	1146	225 x 38	3474
75 x 50	1555	225 x 50	4671
75 x 75	2371	225 x 75	7167
100 x 25	993	225 x 100	9662
100 x 38	1542	225 x 150	14652
100 x 50	2091	225 x 200	19643
100 x 75	3189	225 x 225	22138
100 x 100	4252	300 x 38	4648
150 x 38	2999	300 x 50	6251
150 x 50	3091	300 x 75	9590
150 x 75	4743	300 x 100	12929
150 x 100	6394	300 x 150	19607
150 x 150	9697	300 x 200	26285
200 x 38	3082	300 x 225	29624
200 x 50	4145	300 x 300	39428
200 x 75	6359		

Table 4.25 Trunking factors

Cable tray

A wide range of designs of cable tray and accessories is available to match any cabling requirement, from lightweight instrumentation cable through to the heaviest multi-core power cable. In situations where heavy multi-core cables are required to cross long, unsupported spans, cable ladders should be used.

A wide variety of factory-made accessories is available to suit both standard and return flange trays of various sizes.

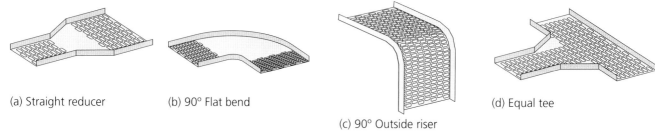

(a) Straight reducer (b) 90° Flat bend (c) 90° Outside riser (d) Equal tee

Figure 4.140 Accessories for cable trays

Installing cable tray

When installing cable tray it is essential that it is well supported and that all supports are secured. It is usually possible to complete the installation by making use of the wide range of accessories and fittings generally available, although it may sometimes be necessary to fabricate joints, bends or fittings to meet particular requirements.

Cable trays can be joined in a number of ways. Different manufacturers will supply a variety of patent couplings and fasteners. Links or fish plates are commonly used, and some cable trays are designed with socket joints. In some circumstances a welded joint may be required. If this method is used care must be taken to restore the finish around the weld to prevent corrosion.

Most methods of jointing cable tray involve the use of nuts and bolts. A round-headed or mushroom-headed bolt (roofing bolt) or screw should be used, and this should be installed with the head inside the tray. This reduces the risk of damage to the cables being drawn along the tray.

Fabricating on site (the tray bending machine)

It is sometimes necessary to fabricate joints, bends and fittings to meet particular requirements or where factory-made accessories are not available. Careful measurement and marking out is required. Cable trays can be cut quite easily with a hacksaw but care should be taken to remove any sharp edges or burrs.

Bends can be formed by hand after a number of cuts have been made in the flange to accommodate the bend, although a far better job can be made by using a crimping tool or a tray-bending machine (Figure 4.141). Bending machines are available from various manufacturers. Where a lot of cable tray work is to be installed, machine bending is quicker and more practical. The machines are made to accommodate the various widths and gauges of cable tray. They may also be used to bend and form flat strips of metal, and have a vice to hold the length of cable tray being worked upon. Cable-tray chain vices are also available for this purpose.

'Making good' cannot normally match the protection qualities of the original factory-applied finish, but the absence of any protection at all can seriously reduce the effectiveness of the original finish through corrosion spreading from this point. Primed tray should have a proper finishing paint applied over the primer as soon as possible. The purpose of the finish is to protect the cable tray from corrosion.

Remember

When installing a cable tray run, consider:

- ease of installation
- economy of time and materials
- facility for extending the system to take additional cables.

Remember

Whenever the tray has been cut the steel edges must be protected as far as possible by sealing with either zinc paint for galvanised tray or an appropriate primer and topcoat for painted tray, or a liquid plastic solution for plastic-coated tray.

(a) Cable tray bending machine

(b) Sharp-radius being formed

(c) Large upward-radius bend being formed

Figure 4.141

Bending cable tray by hand

Light-duty cable tray may be bent by hand with the aid of a crimping tool. This can be made from a piece of 6 mm mild steel bar.

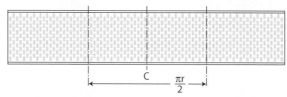

Step 1 To make an inside bend, first determine the radius of the bend from the diameter of the largest cable to be installed. Mark the points at which the bend will begin and end, and the centre of the bend on the piece of tray to be bent.

Step 2 Using the crimping tool, crimp evenly on each side of the tray. Work slowly, checking the form of the bend.

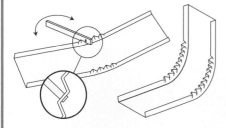

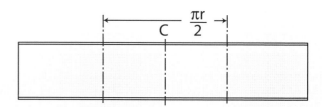

Step 3 To make an 'outside' bend, saw through the flange and bend the tray in a vice. Mark out the bend, as before, taking care that the radius of the bend accepts the largest size cable as stated in Table 4E from the *On-Site Guide*.

Step 4 Mark off along both flanges on the cable tray a series of equal distance points. Make appropriate hacksaw cuts that are equal to the depth of the tray flange on either side of the centre line as shown.

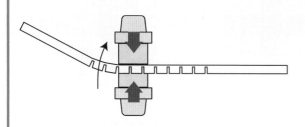

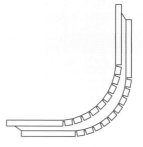

Step 5 Grip the tray with wooden blocks in a vice and bend gradually, moving the tray along and checking the bend for evenness.

Step 6 The completed bend.

Fabricating a flat 90° bend

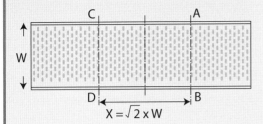

$$X = \sqrt{2} \times W$$

Step 1 Measure and mark the mid-point of the bend. Mark off X when X = W, where W = width of tray.

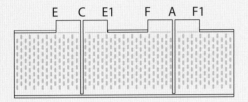

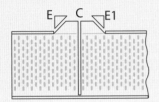

Step 3 Mark the flange at the points of overlap E, F and E1, F1. Cut through the flanges at these points and bend these flanges flat. Cut away the tongues at both slots A and C as shown.

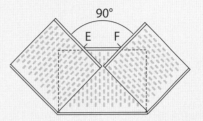

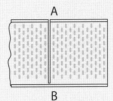

Step 2 Cut through the flange with a hacksaw at point A and along line A–B but do not cut through the opposite flange. Make a similar cut at point C and along line C–D. Bend the two outer sections of the tray together to form a 90° angle.

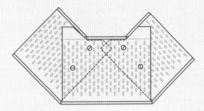

Step 4 Remove all sharp edges and burrs with a file. Make up the assembly as shown and secure with round-headed bolts. Ensure that the bolt heads are uppermost.

Fabricating a flat 90° bend from three pieces of cable tray

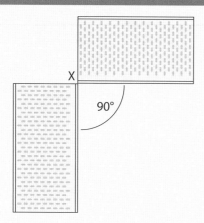

Step 1 Place two pieces of tray together to form a 90° angle at the point of contact marked X.

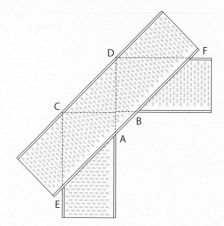

Step 2 Place a third piece of tray over these and mark points of contact A, B, C, D, E, F.

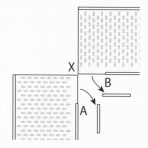

Step 3 Cut away flanges XA and XB as shown.

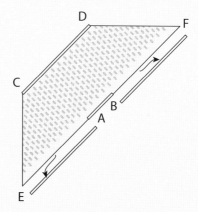

Step 4 Cut along CE and DF. Cut away inner flange AE, BF.

Step 5 Remove all sharp edges and burrs with a file. Assemble as shown, using round-headed bolts.

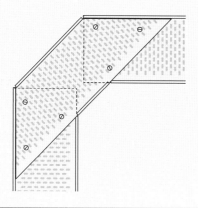

Forming a cable-tray reduction

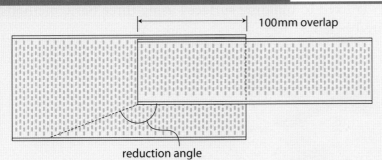

Step 1 Overlap the two segments of cable tray by at least 100 mm. Mark the tray width and reduction angle as shown.

100mm overlap

reduction angle

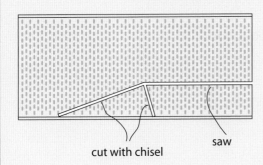

Step 2 Cut with a hacksaw and cold chisel but do not cut through the flange.

cut with chisel

saw

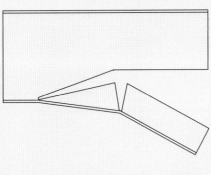

Step 3 Bend the cable tray into shape. Remove all sharp edges and burrs with a file.

Step 4 Assemble the joint using round-headed bolts.

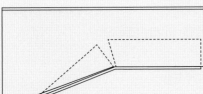

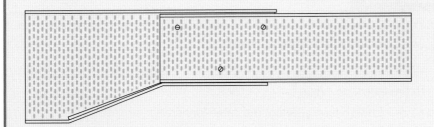

Step 5 Ensure that the bolt heads are uppermost to prevent any damage to the cables.

Fixing cable tray

Cable tray is fixed to building surfaces using three basic methods, as illustrated below.

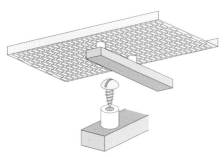

(a) Using spacers and round-headed screws

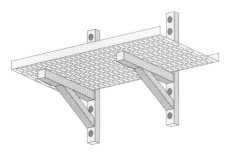

(b) Bolting to brackets

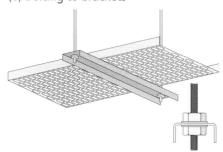

(c) Support channels are fixed to the underside of the tray when suspended from ceilings

Figure 4.142 Three methods for fixing cable tray

Progress check

1   When a cable is to be installed in a wall less than 50 mm from the surface what protection should be applied?

2   What are the maximum horizontal distances supports should be applied to the following cables?
   • A 12 mm PVC non-armoured cable (generally)
   • A 16 mm MICC cable

3   If 10 x 1.5 mm² cables with stranded copper conductors were to be installed in a 4 m straight length of conduit, what would be the minimum size of conduit required?

K7. Understand the regulatory requirements that apply to the installation of wiring systems, enclosures and equipment

This section covers:

- plan and style of the Regulations
- the Parts of the Regulations
- Appendices of the Regulations
- the Electricity At Work Regulations 1989.

When the first electrical regulations were laid down in 1882, hardly anyone had electricity in their homes or workplaces; now everyone has access to electricity. Over the years the Regulations have taken account of the new types of electrical equipment available and its usage, from lighting to computers, and they now affect all aspects of the installation process.

The Regulations apply to the design, erection and verification of electrical installations and are designed to protect people, property, and livestock from electric shock, fire and burns and injury from mechanical movement of electrically operated equipment. They are not designed to instruct untrained people, take the place of a detailed specification or to provide information for every circumstance.

The use of other British Standards is needed to supplement the information contained in the Regulations, such as BS 5266 that deals with emergency lighting.

In 1992 the 16th Edition of the IET Wiring Regulations became a British Standard, BS 7671. The Regulations are not a statutory document – they are not legislated by an Act of Parliament. However, compliance with the Regulations within this book will allow compliance with such legislation as Electricity at Work Regulations 1989, Health and Safety at Work Act 1974 and Electricity Supply Regulations 1988, which are all statutory documents and are enforceable by law. The Regulations have since been revised to become the 17th edition.

Many other British Standards are referred to throughout these Regulations. In some cases these other standards have a BS EN number that refers to European harmonisation standards, whereby all such standards will become applicable throughout Europe. The harmonised standards are co-ordinated by representatives from all the countries in the European Union via an organisation known as CENELEC.

In addition to BS 7671, the IET has published a series of Guidance Notes, which seek to clarify or simplify the requirements in BS 7671.

How the Regulations are laid out

There are seven parts to the Regulations and appendices. These are based on an internationally agreed arrangement that follows the pattern laid out within IEC 60364.

In this numbering system:

- the first digit signifies a part of the Regulations
- the second digit combines with the first to signify a chapter
- the third digit combines with the first two to signify a section
- any further digits signify sub-sections and the specific Regulation number.

With a copy of the Regulations in front of you, look at Regulation 412.1.3.

412.1.3	First digit	Part 4 of the Regulations (Protection for Safety)
4**1**2.1.3	Second digit	First chapter within Part 4 (Chapter 41: Protection against electric shock)
41**2**.1.3	Third digit	Second section of Chapter 41 (Protective measure: Double or reinforced insulation)
412.**1**.3	Fourth digit	First sub-section of section 2 (General)
412.1.**3**	Fifth digit	Specific regulation within sub-section 1 ('Where this measure – used as the sole…')

A 2010 amendment to the 17th edition introduces a slight change to the numbering system. To accommodate future IEC changes, it has been decided to have a 100 numbering system to represent UK-only regulations. This means that each UK regulation remains in the same order within the Regulations, but affected regulations within any section begin at 100.

Here is an example showing how what was regulation 422.3.14 (indicated in red), has become 423.3.100 and its effect on other regulations, for example 423.3.15 becomes 23.3.101.

422.3.12	A PEN conductor shall not be used. This requirement does not apply to a circuit traversing the location.
422.3.13	Except as permitted by Regulation 537.1.2, every circuit shall be provided with a means of isolation from all live supply conductors by a linked switch or a linked circuit-breaker.
422.3.100 (was 422.3.14)	Flexible cables and flexible cords shall be of the following construction: (i) Heavy duty type having a voltage rating of not less than 450/750 V, or (ii) Suitably protected against mechanical damage.
422.3.101	A heating appliance shall be fixed.
422.3.102	A heat storage appliance shall be of a type which prevents the ignition of combustible dusts or fibres by the heat storing core.
422.4	Combustible constructional materials

(Reverts to original sequence)

The same will occur for each UK-only Regulation. If you keep moving through section, 422.4 you see:

422.4.2	Except for equipment for which an appropriate product ...
422.4.100 (was 422.4.3)	Electrical equipment, e.g. installation boxes and distribution boards ...
422.4.101	Electrical equipment that does not comply with Regulation 422.4.100 ...
422.4.102	Cables and cords shall comply ...
422.4.103	Conduit and trunking systems shall be ...
422.5	Fire propagating structures

(Reverts to original sequence)

The Parts of the Regulations

Part 1: Scope, object and fundamental principles

There are three chapters within this part and numerous sections within each chapter. The first chapter (Chapter 11) deals with what the Regulations actually cover and what they do not.

The Regulations apply to the design, erection and verification of electrical installations such as:

- residential premises
- commercial premises
- public premises
- industrial premises
- agricultural and horticultural premises
- prefabricated buildings
- caravans, caravan parks and similar sites
- construction sites, exhibitions, shows, fairgrounds and other installations for temporary purposes including professional stage and broadcast applications
- marinas
- external lighting and similar installations
- mobile or transportable units
- photovoltaic systems
- low voltage generating sets
- highway equipment and street furniture
- medical locations
- operating and maintenance gangways.

However, they do not cover the following:

- systems for the distribution of electricity to the public
- railway traction equipment, rolling stock and signalling equipment
- equipment of motor vehicles except to those to which the requirements of the regulations concerning caravans or mobile units are acceptable
- equipment on board ships covered by BS 8450
- equipment for mobile and fixed offshore installations
- equipment of aircraft
- those aspects of mines and quarries covered by statutory regulations
- radio interference suppression equipment (unless it affects installation safety)
- lightning protection systems covered by BS EN 62305
- aspects of lift installations covered by BS 5655 and BS EN 81-1
- electrical equipment of machines covered by BS EN 60204
- electric fences covered by BS EN 60335.

Remember

Note that 'premises' covers the land and all facilities including buildings belonging to it.

The Regulations also include requirements for:

- circuits supplied at nominal voltages up to and including 1000 V a.c. or 1500 V d.c.
- all consumer installations external to buildings
- fixed wiring for information and communication technology.

Chapter 12 (Objects and effects) explains that the Regulations contain the rules for the design and erection of electrical installations so as to provide for safe and proper functioning for the intended use.

Chapter 13 (Fundamental principles) (131.1 General) states that the requirements of this chapter 'are to provide for the safety of persons, livestock and property against danger and damage that may arise in the reasonable use of electrical installations. The requirements to provide for the safety of livestock are applicable in locations intended for them.'

In electrical installations, risk of injury may arise from:

- shock currents
- excessive temperatures likely to cause burns and fires
- ignition of potentially explosive atmospheres
- undervoltages, overvoltages and electromagnetic influences
- mechanical movement of electrically actuated equipment
- power supply interruption and interruption of safety services
- interruption of safety services
- arcing or burning likely to cause blinding effects, excessive pressure and/or toxic gases.

Sections 131.2 to 131.8 clarify these points.

You should note an important change of terminology here between the 16th and 17th editions.

Regulation 131.2.1 introduced the phrase 'basic protection' to replace 'protection against direct contact', and Regulation 131.2.2 introduced the phrase 'fault protection' to replace 'protection against indirect contact'.

Part 2: Definitions

In this section of BS 7671 the terms used throughout the Regulations are given a specific meaning, so that when one person talks about a circuit breaker, for example, everyone knows what they mean. In a way, this section is a dictionary relating to electrical words.

Part 3: Assessment of general characteristics

Part 3 has seven chapters:

- Assessment of general characteristics (Chapter 30)
- The purpose of the installation and its supplies (Chapter 31)
- The external influences it will be exposed to (Chapter 32)
- The compatibility of its equipment (Chapter 33)
- Its maintainability (Chapter 34)
- Recognised safety services (Chapter 35)
- Assessment for continuity of service (Chapter 36).

Part 4: Protection for safety

Introducing more changes in terminology, this part of the Regulations has four chapters.

Chapter 41 – Protection against electric shock

BS EN 61140 states the fundamental rule of protection against electric shock as being 'that hazardous live parts shall not be accessible and that accessible conductive parts shall not be hazardous-live, either in normal use without a fault or under fault conditions'.

According to BS EN 61140, protection under normal conditions is met by 'basic protection' provisions, and protection under fault conditions is met by 'fault protection' provision. These provisions are referred to as 'protective measures', where Chapter 41 states that a protective measure shall consist of:

- an appropriate combination of a provision for basic protection and an independent provision for fault protection OR
- an enhanced protective provision (such as reinforced insulation) that provides both basic and fault protection.

Sections within Chapter 41 then deal with the following protective measures:

- Section 411 – Automatic disconnection of supply
- Section 412 – Double or reinforced insulation
- Section 413 – Electrical separation
- Section 414 – Extra low voltage (SELV and PELV).

Chapter 42 – Protection against thermal effects

This chapter deals with protection against the effects of heat or thermal radiation, the ignition, combustion of materials, flames and smoke from an electrical installation and against fire services being cut off by the failure of electrical equipment.

Chapter 43 – Protection against overcurrent

This chapter deals with protection of live conductors from the effects of overcurrent.

Chapter 44 – Protection against voltage disturbances and electromagnetic disturbances

This chapter deals with protection against the likes of temporary overvoltages due to earth faults in the H.V. system, overvoltages of atmospheric origin or due to switching and where a reduction in voltage could cause danger.

Part 5: Selection and erection of equipment

Revised under the 17th edition, this part provides common rules for compliance with measures of protection for safety, requirements for proper functioning of intended use and requirements pertinent to the external influences. This information is given over the following six chapters. As it is impossible to fully cover the content, here are just one or two key points within each one.

Common rules (Chapter 51)

512.1.2 Current
Every item of equipment shall be suitable for:

(i) the design current, taking into account any capacitive and inductive effects, and

(ii) the current likely to flow in abnormal conditions for such periods of time as are determined by the characteristics of the protective devices concerned.

Switchgear, protective devices, accessories and other types of equipment shall not be connected to conductors intended to operate at a temperature exceeding 70°C at the equipment in normal service, unless the equipment manufacturer has confirmed that the equipment is suitable for such conditions.

512.1.5 Compatibility
Every item of equipment shall be selected and erected so that it will neither cause harmful effects to other equipment nor impair the supply during normal service including switching operations.

Switchgear, protective devices, accessories and other types of equipment shall not be connected to conductors intended to operate at a temperature exceeding 70°C at the equipment in normal service, unless the equipment manufacturer has confirmed that the equipment is suitable for such conditions.

Where cables that are permitted to run at a temperature exceeding 70°C (such as thermosetting insulated cables) are connected to equipment or accessories designed to operate at a temperature not exceeding 70°C, the conductor size shall be chosen based on the current ratings for 70°C cables of a similar construction.

514.2.1 General

As far as is reasonably practicable, wiring shall be so arranged or marked that it can be identified for inspection, testing, repair or alteration of the installation. (This removes the use of orange as an identifying colour for conduit.)

514.9 Warning notice: isolation
514.9.1

A notice of durable material in accordance with Regulation 537.2.1.3, shall be fixed in each position where there are live parts which are not capable of being isolated by a single device. The location of each disconnector (isolator) shall be indicated unless there is no possibility of confusion.

515.2

Where equipment carrying current of different types or at different voltages is grouped in a common assembly (such as a switchboard, a cubicle or a control desk or box), all the equipment belonging to any one type of current or any one voltage shall be effectively segregated wherever necessary to avoid mutual detrimental influence.

The immunity levels of equipment shall be chosen taking into account the electromagnetic influences that can occur when connected and erected as for normal use, and taking into account the intended level of continuity of service necessary for the application.

Selection and erection of wiring systems (Chapter 52)

521.5.2

Single-core cables armoured with steel wire or steel tape shall not be used for an a.c. circuit.
NOTE: The steel wire or steel tape armour of a single-core cable is regarded as a ferromagnetic enclosure. For single-core armoured cables, the use of aluminium armour may be considered.

521.10 Installation of cables

521.10.1

Non-sheathed cables for fixed wiring shall be enclosed in conduit, ducting or trunking. This requirement does not apply to a protective conductor complying with Section 543.

Non-sheathed cables are permitted if the cable trunking system provides at least the degree of protection IP4X or IPXXD, or if the cover can only be removed by means of a tool or a deliberate action.

522.8.1

A wiring system shall be selected and erected to avoid during installation, use or maintenance, damage to the sheath or insulation of cables and their terminations. The use of lubricants containing silicone oil for drawing-in of cables is not permitted.

TABLE 52.3 – Minimum cross-sectional area of conductors

The minimum copper conductor size for a circuit wired in sheathed/non-sheathed cables is now 1.0 mm^2 for lighting circuits and 1.5 mm^2 for power circuits.

527.2.4

No wiring system shall penetrate an element of building construction which is intended to be load bearing unless the integrity of the load bearing element can be assured after such penetration.

527.2.1

Where a wiring system passes through elements of building construction such as floors, walls, roofs, ceilings, partitions or cavity barriers, the openings remaining after passage of the wiring system shall be sealed according to the degree of fire-resistance (if any) prescribed for the respective element of building construction before penetration.

Protection, isolation, switching, control and monitoring (Chapter 53)

537.1.3

Each installation shall have provision for disconnection from the supply.

Where the distributor provides switchgear complying with Chapter 53 at the origin of the installation and agrees that it may be used as the means of isolation for the part of the installation between the origin and the main linked switch or circuit-breaker required by Regulation 537.1.4, the requirement for isolation and switching of that part of the installation is satisfied.

Table 53.4 then gives guidance on the selection of protective, isolation and switching devices.

Earthing arrangements and protective conductors (Chapter 54)

542.1.4
The earthing arrangements shall be such that:

(i)   the value of impedance from the consumer's main earthing terminal to the earthed point of the supply for TN systems, or to earth for TT and IT systems, is in accordance with the protective and functional requirements of the installation, and considered to be continuously effective, and

(ii)   earth fault currents and protective conductor currents which may occur are carried without danger, particularly from thermal, thermomechanical and electromechanical stresses, and

(iii) they are adequately robust or have additional mechanical protection appropriate to the assessed conditions of external influence.

Other equipment (Chapter 55)

559.3 Outdoor lighting installation
An outdoor lighting installation comprises one or more luminaires, a wiring system and accessories.

The following are included:

(i)   lighting installations such as those for roads, parks, car parks, gardens, places open to the public, sporting areas, illumination of monuments and floodlighting

(ii)   other lighting arrangements in places such as telephone kiosks, bus shelters, advertising panels and town plans

(iii) road signs and road traffic signal systems.

The following are excluded:

(iv)   equipment of the owner or operator of a system for distribution of electricity to the public

(v)   temporary festoon lighting

(vi)   luminaires fixed to the outside of a building and supplied directly from the internal wiring of that building

(vii) road traffic signal systems.

Safety services (Chapter 56)

560.1 SCOPE

This chapter covers general requirements for safety services, selection and erection of electrical supply systems for safety services and electrical safety sources. Standby electrical supply systems are outside the scope of this chapter. This chapter does not apply to installations in hazardous areas (BE3), for which requirements are given in BS EN 60079-14.

Examples of safety services include:

- emergency (escape) lighting
- fire detection and alarm systems
- CO (carbon monoxide) detection and alarm systems
- fire evacuation systems
- fire services communication systems
- essential medical systems
- industrial safety systems.

Part 6: Inspection and testing

Every installation shall during erection and upon completion, before being put into service, be inspected and tested to verify that the requirements of the Regulations have been met.

During the inspection and testing process precautions shall be taken to prevent danger to persons, livestock and to protect property and installed equipment from damage.

Part 6 has three chapters:

- Initial verification (Chapter 61)
- Periodic inspection and testing (Chapter 62)
- Certification and reporting (Chapter 63).

Note that the Period Inspection and Testing Report has now been replaced by an Electrical Installation Condition Report.

This part of the Regulations does not give any detail on how to carry out any of the tests required, this information normally being found in the IET Guidance Note 3 or the *Unite Guide to Good Electrical Practice*. However, you must refer to the Regulations to ensure that any results obtained are within acceptable limits in this respect.

Table 4.26 shows the minimum values of insulation resistance.

Circuit normal voltage	d.c. test voltage (V)	Minimum insulation resistance (MΩ)
SELV and PELV	250	0.5
Up to and including 500 V with the exception of the above	500	1.0
Above 500 V	1000	1.0

Table 4.26 Minimum values of insulation resistance

Part 7: Special installations or locations

This part of the Regulations now comprises the 15 sections listed below, each of which either supplements or modifies the general requirements of other parts of the Regulations.

- Locations containing a bath or shower (701)
- Swimming pools and other basins (702)
- Rooms and cabins containing hot air saunas (703)
- Construction and demolition site installations (704)
- Agricultural and horticultural premises (705)
- Conducting locations with restricted movement (706)
- Caravan/camping parks and similar locations (708)
- Marinas and similar locations (709)
- Medical locations (710)
- Exhibitions, shows and stands (711)
- Solar photovoltaic (pv) power systems (712)
- Mobile or transportable units (717)
- Electrical installations in caravans and motor caravans (721)
- Operating and maintenance gangways
- Temporary installations for structures, amusement devices and booths at fairgrounds, amusement parks and circuses (740)
- Floor and ceiling heating systems (753)

Here are details for some of the more relevant sections.

Section 701: Locations containing a bath or shower

- All circuits supplying the bathroom (irrespective of the points they are serving) have to be protected by 30 mA RCDs.
- Supplementary equipotential bonding is not required when all of the following are met:

- All circuits for the bathroom meet the requirements for ADS.
- All final circuits have 30 mA RCD protection.
- All extraneous conductive parts of the bathroom are connected to the protective equipotential bonding of the installation (411.3.1.2).
- SELV socket outlets and shaver sockets are permitted outside Zone 1. 230 V socket outlets are permitted provided they are more than 3 m horizontally from Zone 1.

Figure 4.143 (a) to (f) show the zone arrangements and dimensions in plan view.

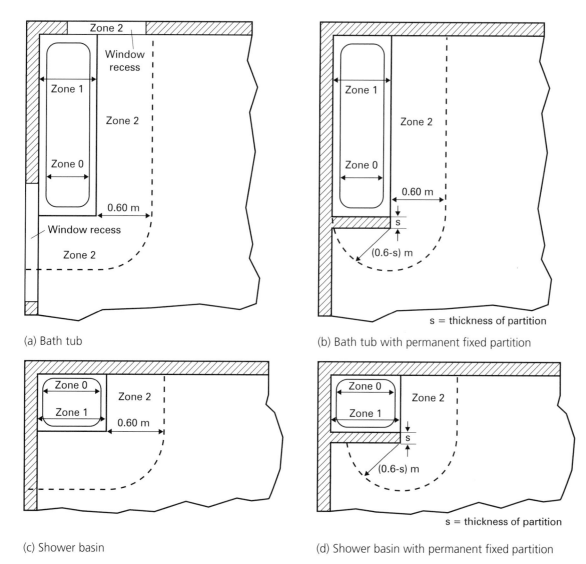

(a) Bath tub

(b) Bath tub with permanent fixed partition

(c) Shower basin

(d) Shower basin with permanent fixed partition

Figure 4.143 Zone arrangements and dimensions

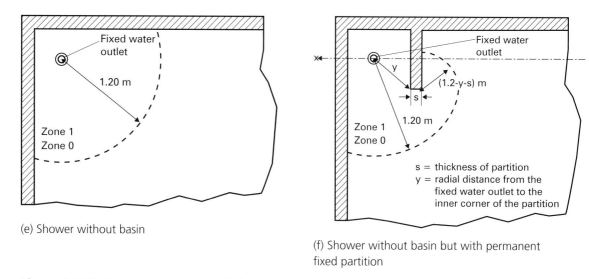

(e) Shower without basin

(f) Shower without basin but with permanent fixed partition

Figure 4.143 Zone arrangements and dimensions (cont.)

Figure 4.144 (a) – (c) give the same information but in elevation view.

Please note that ★ denotes a Zone 1 if the space is accessible without the use of a tool. Spaces under the bath that are only accessible with the use of a tool are outside the zones.

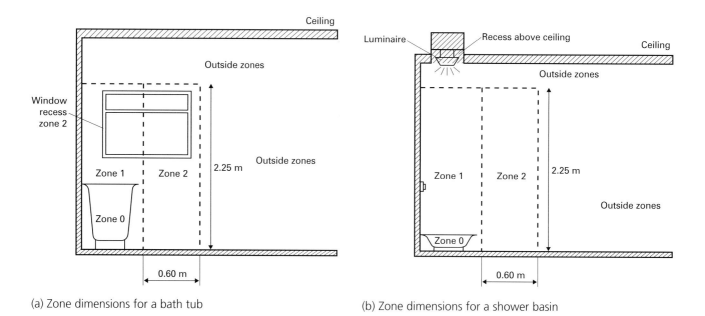

(a) Zone dimensions for a bath tub

(b) Zone dimensions for a shower basin

Figure 4.144 Zone arrangements and dimensions in elevation view

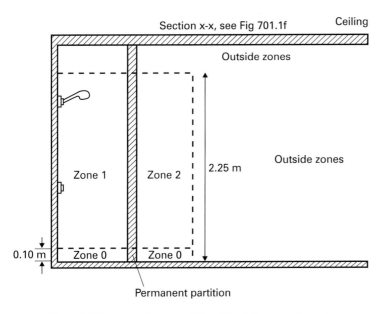

Zone 1 if the space is accessible without the use of a tool.
Spaces under the bath accessible only with the use of a tool are
outside the zones.

(c) Zone dimensions for a shower basin with no permanent fixed partition

Figure 4.144 Zone arrangements and dimensions in elevation view (cont.)

Section 702: Swimming pools and other basins

In this section the addition of 'other basins' reflects the fact it
refers to the basins of swimming pools, paddling pools and
fountains and, only where designated as swimming pools, natural
waters, such as the sea and lakes.

The zone arrangements are shown in Figures 4.145–4.147.

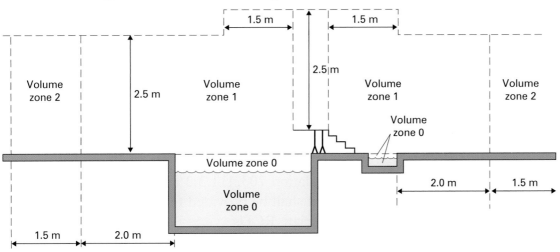

Figure 4.145 Zone dimensions for swimming pools and paddling pools

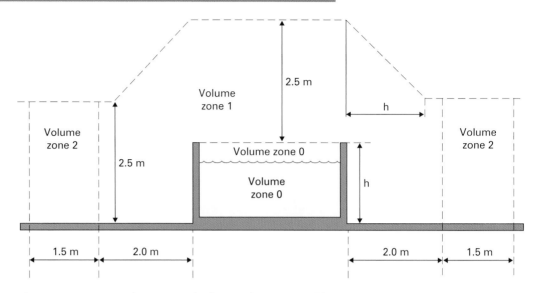

Figure 4.146 Zone dimensions for basin above ground level

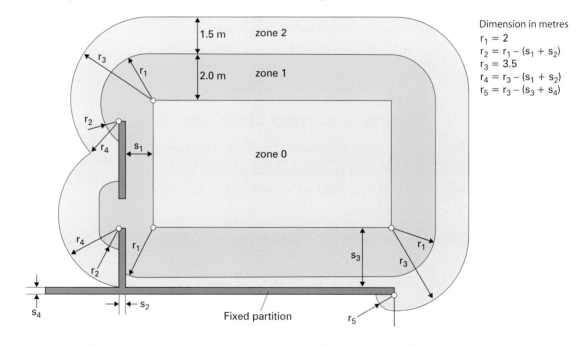

Figure 4.147 Zone dimensions with fixed partitions of height at least 2.5 m

Dimension in metres
$$r_1 = 2$$
$$r_2 = r_1 - (s_1 + s_2)$$
$$r_3 = 3.5$$
$$r_4 = r_3 - (s_1 + s_2)$$
$$r_5 = r_3 - (s_3 + s_4)$$

Section 703: Rooms containing saunas

In this section, Regulation 703.411.3.3 now specifies that additional protection shall be provided for all circuits of the sauna by the use of appropriate RCDs.

However, RCD protection is not required for the sauna heater unless recommended by the manufacturer. The zone arrangements are shown in Figure 4.148.

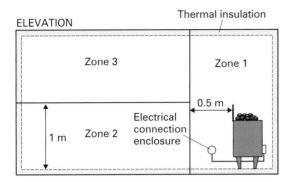

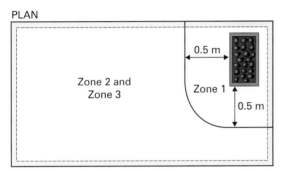

Figure 4.148 Rooms containing saunas

Section 705: Agricultural and horticultural premises

Regulation 705.411.1 now specifies that, irrespective of the earthing arrangements, the following disconnection device shall be provided:

- for final circuits supplying socket outlets not exceeding 32 A, an RCD not exceeding 30 mA
- for final circuits supplying socket outlets exceeding 32 A, an RCD not exceeding 100 mA
- in all other circuits, RCDs not exceeding 300 mA.

In livestock locations and where accessible to the livestock, all exposed conductive parts and extraneous conductive parts must be connected together with supplementary equipotential bonding. Figure 4.149 gives an example (see page 398).

Section 708 (Caravan/camping parks and similar locations)

The note to regulation 708.521 states the preferred method of supply is by underground distribution cables buried to a depth of at least 0.6 m and, unless having additional mechanical protection, placed away from caravan pitches or areas where tent pegs or similar could be present.

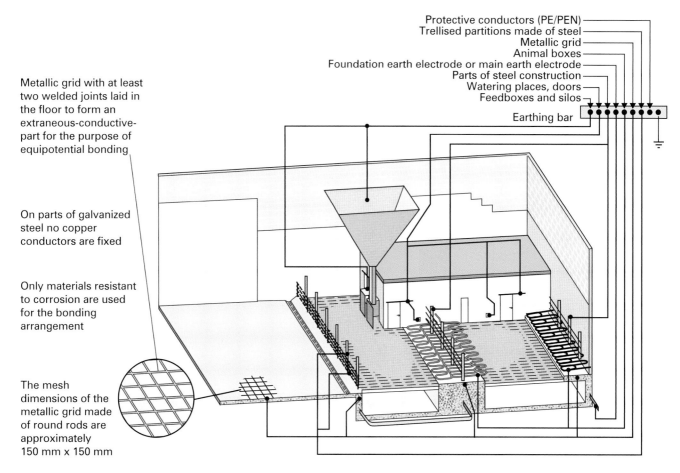

Protective conductors (PE/PEN)
Trellised partitions made of steel
Metallic grid
Animal boxes
Foundation earth electrode or main earth electrode
Parts of steel construction
Watering places, doors
Feedboxes and silos
Earthing bar

Metallic grid with at least two welded joints laid in the floor to form an extraneous-conductive-part for the purpose of equipotential bonding

On parts of galvanized steel no copper conductors are fixed

Only materials resistant to corrosion are used for the bonding arrangement

The mesh dimensions of the metallic grid made of round rods are approximately 150 mm x 150 mm

Figure 4.149 Supplementary equipotential bonding within a cattle shed

Overhead supplies are permitted, but must be no less than 6 m above ground where there is vehicular movement and 3.5 m in other areas.

For the actual caravan pitch:

- the socket outlet and its enclosure that forms part of the caravan pitch shall be to IP44 and mounted at a height of between 0.5 m and 1.5 m
- at least one socket outlet shall be provided for each pitch
- the socket outlet rating shall be no less than 16 A
- each socket outlet will be protected individually by overcurrent protection and a 30 mA RCD.

Section 710 (Medical locations)

This section applies to electrical installations in medical locations to ensure safety of patients and medical staff. These requirements, in the main, refer to hospitals, private clinics, medical and dental practices, healthcare centres and dedicated medical rooms in the workplace.

This section exists because patients may have their skin resistance broken while under treatment or have their body's defence system reduced or stopped while under anaesthetic. Also life-support equipment should not be affected by any loss of supply.

The requirements of this section do not apply to medical electrical equipment.

Section 711 (Exhibitions, shows and stands)

This section only applies to temporary exhibitions and shows. It does not apply to permanent shows or to the fixed supplies and installations of exhibition halls that house temporary exhibitions.

Care must be taken over the assessment of the event, which could be anything from an art exhibition to an agricultural show. As a result:

- all supply cables shall be protected by a 30 mA RCD except for emergency lighting
- the protective measures of obstacles and placing out of reach are not permitted
- cable size shall be a minimum of $1.5\,mm^2$
- the installation shall be inspected and tested on site, in accordance with Chapter 61 after each assembly on site.

Section 712 (Solar photovoltaic [PV] power systems)

As covered in the Unit ELTK 02, this system is based on inter-connected PV cells that turn daylight into electrical energy. As these systems are normally installed outside on a roof, the electrical installation must be suitable for the environment and therefore have a suitable IP rating. The equipment must conform to the relevant BS for this type of system.

Equally, to allow maintenance of the PV inverter, means of isolating it from both the a.c. and d.c. sides shall be provided.

Section 753 (Floor and ceiling heating systems)

Such systems are most commonly installed in domestic premises. For protective measure ADS, 30 mA RCDs shall be used.

To avoid overheating of the system, within the zone of its installation the unit shall be limited to a temperature of 80°C. To protect against burns to the skin, the temperature of the floor shall be limited to 35°C in areas where skin contact is possible.

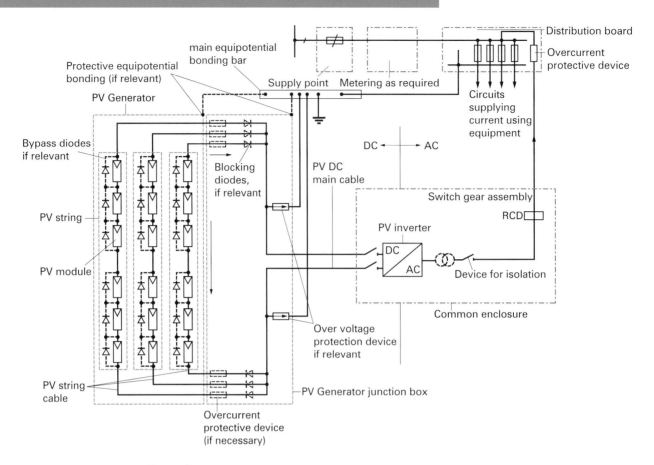

Figure 4.150 PV installation for one array

Appendices

BS 7671 contains the following appendices:

- Appendix 1 lists those British and European Standards referred to in the Regulations

- Appendix 2 lists relevant UK statutory legislation passed by Parliament

- Appendix 3 gives the time/current characteristics of overcurrent protective devices

- Appendix 4 gives the current carrying capacity and voltage drop of different types of cables along with the various installation methods and correction factors

- Appendix 5 gives classification of external influences that effect the selection and erection of equipment

- Appendix 6 gives model examples of the documents required for certification and reporting when carrying out inspection and testing of installations

- Appendix 7 explains the harmonised cable-colouring system introduced in 2004

- Appendix 8 gives the current-carrying capacity and voltage drop for busbar trunking and Powertrack systems
- Appendix 9 gives definitions of earthing arrangements relative to multiple source, d.c. and other systems
- Appendix 10 relates to the protection of conductors in parallel against overcurrent
- Appendix 11 is deleted by Amendment 1
- Appendix 12 is deleted by Amendment 1
- Appendix 13 gives methods for measuring the insulation resistance/impedance value of floors/walls to earth or to the protective conductor system
- Appendix 14 gives consideration to the increase of resistance in conductors with increase in temperature, relative to the measurement of earth fault loop impedance
- Appendix 15 gives details of ring and radial final circuit arrangements.
- Appendix 16 gives details of devices for protection against overvoltage.

The Electricity at Work Regulations 1989 (EAWR)

The Electricity at Work Regulations came into force in 1990. Their purpose is to require precautions to be taken against the risk of death or personal injury from electricity in work activities.

The Regulations were made under the Health and Safety at Work Act 1974, which imposes duties on employers, employees and the self-employed. These Regulations (EAWR) are more specific and concentrate on work activities at or near electrical equipment. They make one person primarily responsible to ensure compliance in respect of systems, electrical equipment and conductors; this person is referred to as the 'duty holder'.

The EAWR were also designed to include all the systems that BS 7671 does not cover, such as voltages above 1000 volts a.c.

There are 33 Regulations and three appendices within EAW 1989 but not all of them apply to all situations. This chapter contains an overall look at the relevant Regulations and what they mean, to give you an appreciation of what they are about. For detailed information read *The memorandum of guidance on the Electricity at Work Regulations 1989.*

What particular Regulations cover

Regulation 1 only states that these Regulations came into force on 1 April 1990.

Regulation 2 contains definitions of certain words or phrases.

Regulation 3 deals with duty holders and the requirements imposed on them by these Regulations. There are three categories of duty holder:

- employers
- employees
- self-employed people.

The duty holder is the person who has a duty to comply with these Regulations because they are relevant to circumstances within their control. Such a person must be competent.

Whenever a Regulation does not use the phrase 'so far as is reasonably practicable', it is an absolute duty: it must be done regardless of cost or any other consideration.

Whenever a Regulation does use the phrase 'as far as is reasonably practicable', the duty holder must assess the magnitude of the risks against the costs in terms of physical difficulty, time, trouble and expense involved in minimising the risks. The onus is on the duty holder to be able to prove in a court of law that he or she took all steps, as far as is reasonably practicable.

Regulation 4 has four parts.

- The first part deals with the construction of the electrical systems (all parts) in that the equipment should be suitable for its intended use so that it does not give rise to danger 'as far as is reasonably practicable'.
- The second part deals with the maintenance of systems, whereby all systems should be maintained to prevent danger (this includes portable appliances) 'as far as is reasonably practicable'. Records of maintenance including test results should be kept.
- The third part deals with ensuring safe work activities near a system, including operation, use and maintenance 'as far as is reasonably practicable'. This could include non-electrical activities such as excavation near underground cables, and erecting scaffolding near overhead lines. Safe work activities include things such as:

- company health and safety policy
- Permit to Work systems
- clear communication
- use of competent people
- personnel attitudes.

- The fourth part deals with provision of protective equipment such as insulated tools, test probes, insulating gloves and rubber mats which must be suitable for use, maintained in that condition and properly used. This is an absolute duty.

Regulation 5 has four parts. This Regulation states that 'no electrical equipment must be used where its strength and capability may be exceeded and give rise to danger': for example, switchgear should be capable of handling fault currents as well as normal load currents, correct size cable, and so on.

An example is the installation of a socket near a back door of a house. There may be nothing plugged into it at present, but it is reasonable to assume that it could be used in the future for plugging in an appliance to be used outside, such as a lawnmower, which would therefore necessitate protection by an RCD. This is an absolute duty.

Regulation 6 deals with the siting and/or selection of electrical equipment and whether it would be exposed to, or you could foreseeably predict it being exposed to, adverse or hazardous environments.

The following list provides what needs to be considered when siting and/or selecting electrical equipment, not just for the present situation but for what you could reasonably expect could be the situation in the future 'as far as is reasonably practicable':

- protection against mechanical damage
- effects of weather, natural hazards, temperature or pressure
- effects of wet, dirty, dusty or corrosive conditions
- any flammable or explosive substances, including dusts, vapours or gases.

Regulation 7 is concerned with conductors in a system and whether they present a danger to persons. All conductors must either be suitably covered with insulating material and protected or, if not insulated (such as overhead power lines), placed out of reach 'as far as is reasonably practicable'. The definition of 'placed out of reach' is in Part 2 of BS 7671.

Regulation 8 deals with the requirements for earthing or other such suitable precautions that are needed to reduce the risk of electric shock when a conductor (other than a circuit conductor) becomes live under fault conditions. This consists of such things as earthing the outer conductive parts of electrical equipment that can be touched and other conductive metalwork in the vicinity such as water and gas pipes. Other methods could be reduced voltage systems, double-insulated equipment and RCDs. This is an absolute duty.

Regulation 9 is about maintaining the integrity of referenced conductors, which in simple terms means that the neutral conductor must not have a fuse or switch placed in it. The only exception is that a switch may be placed in the neutral conductor if that switch is interlocked to break the phase conductor(s) at the same time. This is an absolute duty.

Regulation 10 requires that all joints and connections in a system must be mechanically and electrically suitable for its use. For example, things like taped joints on extension leads are not allowed. This is an absolute duty.

Regulation 11 states that every part of a system must be protected from excess current that may give rise to danger. This means that suitably rated fuses, circuit breakers and so on must be installed so that, in a fault situation, they will interrupt the supply and prevent a dangerous situation happening. This is an absolute duty.

Regulation 12 deals with the need for switching off and isolating electrical equipment and, where appropriate, identifying circuits. Isolation means cutting off from every source of electrical energy in such a way that it cannot be switched back on accidentally, in other words a means of 'locking off' the switch securely. This is an absolute duty.

Regulation 13 refers to the precautions required when work is taking place at or near electrical equipment that is to be worked on, whether it be electrical or non-electrical work that is taking place. The electrical equipment must be isolated, locked off and tested for absence of live parts before the work takes place and must remain so until all persons have completed their work. A safe isolation procedure or written Permit to Work scheme should be used. This is an absolute duty.

Regulation 14 refers to the precautions needed when working on or near live conductors. An absolute duty is imposed that the conductors must be isolated unless certain conditions are met. Precautions considered appropriate are:

- properly trained and competent staff
- provision of adequate information regarding nature of the work and system
- use of appropriate insulated tools, equipment, instruments, test probes and protective clothing
- use of insulated barriers
- accompaniment of another person
- effective control of the work area.

Regulation 15 concerns matters relating to work being carried out at or near electrical equipment whereby, to prevent danger, adequate working space, adequate means of access and adequate lighting must be provided. This is an absolute duty.

Regulation 16 deals with the competency of people working on electrical equipment to prevent danger or injury. To comply with this Regulation, a person should conform to the following or be under such a degree of supervision as appropriate given the type of work to be carried out:

- adequate understanding and practical experience of the system to be worked on
- an understanding of the hazards that may arise
- the ability to recognise at all times whether it is safe to continue.

The Regulation tries to ensure that no one places themselves or anyone else at risk due to their lack of technical knowledge or practical experience. This is an absolute duty.

Regulations 17 to 28 apply to mines and quarries only.

Regulation 29 is what is known as the 'Defence' Regulation. If an offence is committed by the duty holder under these Regulations (the absolute ones) and criminal proceedings are brought by the HSE, if the duty holder can prove that they took all reasonable steps and exercised due diligence to avoid committing that offence, they will not be found guilty.

Regulation 30 says that a duty holder can apply to the HSE for exemption from these Regulations for:

- any person
- any premises
- any electrical equipment

- any electrical system
- any electrical process
- any activity.

Exemptions will only be granted by the HSE if they do not prejudice the health and safety of any persons.

Regulation 31 deals with work activities and premises outside Great Britain. If the activity or premises is covered by sections 1 to 59 and sections 80 to 82 of the Health and Safety at Work Act 1974, these regulations apply.

Regulation 32 details what these regulations do not apply to:

- sea-going ships
- aircraft or hovercraft moving under their own power.

Regulation 33 deals with changes and modifications to these Regulations since they were brought in.

Appendices

Appendix 1 lists the HSE guidance publications available for help in understanding and applying the regulations.

Appendix 2 lists various other codes of practice and British Standards that could help in the understanding and application of these regulations.

Appendix 3 deals with the legislation concerning working space and access regulations.

Installations in potentially explosive areas

Within **hazardous areas** there exists the risk of explosions and/or fires occurring due to electrical equipment 'igniting' the gas, dust or flammable liquid. These areas are not included in BS 7671 but are covered by IEC Standard BS EN 60079 as follows:

- BS EN 60079 Part 10 – Classification of hazardous areas
- BS EN 60079 Part 14 – Electrical apparatus for explosive gas atmospheres
- BS EN 60079 Part 17 – Inspection/maintenance of electrical installations in hazardous areas.

The BS EN 60079 has been in place since 1988, replacing the old BS 5345. However, many installations obviously still exist that were completed in accordance with BS 5345 and new

Key term

Hazardous area – according to the Regulations, 'an area in which explosive gas/air mixtures are, or may be expected to be, present in quantities such as to require special precautions for the construction and use of electrical apparatus'

European Directives (ATEX) address safety where there is a danger from potentially explosive atmospheres.

Other statutory regulations such as the Petroleum Regulation Acts 1928 and 1936 and local licensing laws govern storage of petroleum.

Zoning

Hazardous areas are defined in the Dangerous Substances and Explosive Atmospheres Regulations 2002 (DSEAR) as 'any place in which an explosive atmosphere may occur in quantities such as to require special precautions to protect the safety of workers'. In this context, 'special precautions' is best taken as relating to the construction, installation and use of apparatus, as given in BS EN 60079 Part 10.

Area classification is a method of analysing and classifying the environment where explosive gas atmospheres may occur. The main purpose is to facilitate the proper selection and installation of apparatus to be used safely in that environment, taking into account the properties of the flammable materials that will be present. DSEAR specifically extends the original scope of this analysis, to take into account non-electrical sources of ignition, and mobile equipment that creates an ignition risk.

Hazardous areas are classified into zones based on an assessment of the frequency of the occurrence and duration of an explosive gas atmosphere, as follows:

- Zone 0 – an area in which an explosive gas atmosphere is present continuously or for long periods
- Zone 1 – an area in which an explosive gas atmosphere is likely to occur in normal operation
- Zone 2 – an area in which an explosive gas atmosphere is not likely to occur in normal operation and, if it occurs, will only exist for a short time.

Various sources have tried to place time limits on to these zones, but none have been officially adopted. The most common values used are:

- Zone 0 – explosive atmosphere for more than 1000h/yr
- Zone 1 – explosive atmosphere for more than 10, but less than 1000h/yr
- Zone 2 – explosive atmosphere for less than 10h/yr, but still sufficiently likely as to require controls over ignition sources.

Key terms

Here are some of the key terms used in BS EN 60079.

Explosive limits – the upper and lower percentages of a gas in a given volume of gas/air mixture at normal atmospheric temperature and pressure that will burn if ignited

Lower explosive limit (LEL) – the concentration below which the gas atmosphere is not explosive

Upper explosive limit (UEL) – the concentration of gas above which the gas atmosphere is not explosive.

Ignition energy – the spark energy that will ignite the most easily ignited gas/air mixture of the test gas at atmospheric pressure. Hydrogen ignites very easily, whereas butane or methane require about ten times the energy

Flash point – the minimum temperature at which a material gives off sufficient vapour to form an explosive atmosphere

Ignition temperature or **auto-ignition temperature** – the minimum temperature at which a material will ignite and sustain combustion when mixed with air at normal pressure, without the ignition being caused by any spark or flame. Note: This is not the same as flash point, so don't confuse them.

Where people wish to quantify the zone definitions, these values are the most appropriate, but for the majority of situations a purely qualitative approach is adequate. When the hazardous areas of a plant have been classified, the remainder will be defined as non-hazardous, sometimes referred to as 'safe areas'.

Selection of equipment

DSEAR sets out the link between a zone and the equipment that may be installed in that zone. This applies to new or newly modified installations. The equipment categories are defined by the ATEX equipment directive, set out in UK law as the Equipment and Protective Systems for Use in Potentially Explosive Atmospheres Regulations 1996.

Standards set out different protection concepts, with further subdivisions for some types of equipment according to gas group and temperature classification. Most of the electrical standards have been developed over many years and are now set at international level, while standards for non-electrical equipment are only just becoming available from CEN.

The DSEAR ACOP describes the provisions concerning existing equipment.

There are different technical means (protection concepts) of building equipment to the different categories. These, the standard current and the letter giving the type of protection are listed below.

Correct selection of electrical equipment for hazardous areas requires the following information:

- temperature class or ignition temperature of the gas or vapour involved according to Table 4.27:

Temperature classification	Maximum surface temperature °C	Ignition temperature of gas or vapour °C
T1	450	>450
T2	300	>300
T3	200	>200
T4	135	>135
T5	100	>100
T6	85	>85

Table 4.27 Temperature class or ignition temperature

- classification of the hazardous area, as in zones shown in Table 4.28:

Zone 0	Zone 1	Zone 2
Category 1	Category 2	Category 3
'ia' intrinsically safe EN 50020, 2002	'd' flameproof enclosure EN 50018 2000	Electrical Type 'n' EN 50021 1999 Non electrical EN 13463-1, 2001
Ex s – Special protection if specially certified for Zone 0	'p' pressurised EN 50016 2002	
	'q' powder filling EN 50017, 1998	
	'o' oil immersion EN 50017, 1998-	
	'e' increased safety EN 50019, 2000 'ib' Intrinsic safety EN 50020, 2002	
	'm' encapsulation EN 50028, 1987	
	's' special protection	

Table 4.28 Classification of hazardous areas

If several different flammable materials may be present within a particular area, the material that gives the highest classification dictates the overall area classification. The IP code considers specifically the issue of hydrogen containing process streams as commonly found on refinery plants.

Consideration should be shown for flammable material that may be generated due to interaction between chemical species.

Ignition sources – identification and control

Ignition sources may be:

- flames
- direct fired space and process heating
- use of cigarettes/matches etc.
- cutting and welding flames
- hot surfaces

- heated process vessels such as dryers and furnaces
- hot process vessels
- space heating equipment
- mechanical machinery
- electrical equipment and lights
- spontaneous heating
- friction heating or sparks
- impact sparks
- sparks from electrical equipment
- stray currents from electrical equipment
- electrostatic discharge sparks
- lightning strikes
- electromagnetic radiation of different wavelengths
- vehicles (unless specially designed or modified are likely to contain a range of potential ignition sources).

Sources of ignition should be effectively controlled in all hazardous areas by a combination of design measures, and systems of work:

- using electrical equipment and instrumentation classified for the zone in which it is located. New mechanical equipment will need to be selected in the same way (see above)
- earthing of all plant/equipment (see Technical Measures Document on Earthing)
- elimination of surfaces above auto-ignition temperatures of flammable materials being handled/stored (see above)
- provision of lightning protection
- correct selection of vehicles/internal combustion engines that have to work in the zoned areas (see Technical Measures Document on Permit to Work Systems)
- correct selection of equipment to avoid high intensity electromagnetic radiation sources, e.g. limitations on the power input to fibre optic systems, avoidance of high intensity lasers or sources of infrared radiation
- prohibition of smoking/use of matches/lighters
- controls over the use of normal vehicles
- controls over activities that create intermittent hazardous areas, such as tanker loading/unloading
- control of maintenance activities that may cause sparks/hot surfaces/naked flames through a Permit to Work System.

Petrol-filling stations

The primary legislation controlling the storage and use of petrol is the Petroleum (Consolidation) Act 1928 (PCA). This requires anyone who keeps petrol to obtain a licence from the local Petroleum Licensing Authority (PLA).

Figure 4.151 Petrol filling station

The licence may be, and usually is, issued subject to a number of licence conditions. The PLA sets the licence conditions, but they must be related to the safe keeping of petrol.

The Local Authority Co-ordinating Body on Food and Trading Standards (LACOTS) has issued a set of standard licence conditions. Most, if not all, PLAs apply these to their sites.

Installations within petrol filling stations are effectively also covered by BS EN 60079 Parts 10, 14 and 17.

There is also industry-developed guidance for this sector in the form of the electrical section of IP/APEA's Guidance for the Design, Construction, Modification and Maintenance of Petrol Filling Stations (Institute of Petroleum and the Association for Petroleum and Explosives Administration).

This guidance replaced most of HS(G).41 and was published in 1999 by IP/APEA with input from HSE.

Progress check

1   When planning an electrical installation the general characteristics need to be considered. What areas will this cover?

2   What is the main purpose of inspecting and testing an installation before it is put into service as required by Part 6 of the IET Wiring Regulations?

3   BS 7671 has a number of special installations and locations including bathrooms. Summarise the main requirements.

Getting ready for assessment

This unit introduces you to many of the craft and practical skills that you will use throughout your career as an electrician. These skills will become some of the cornerstones of your professional careers, and it is vital that you make sure you are practised at, and familiar with, their procedures.

For this unit you will need to be familiar with:

- procedures, practices, statutory and non-statutory requirements relative to preparing and installing
- procedures for checking the work location before starting work activities
- requirements for safe isolation of circuits and complete systems
- types, application and limitations of wiring systems and associated equipment
- procedures for selecting and using tools, equipment and fixings for wiring systems, enclosures and equipment
- practices and procedures for installing wiring systems, enclosures and equipment
- regulatory requirements that apply to the installation of wiring systems, enclosures and equipment

For each Learning Outcome, there are several skills you will need to acquire, so you must make sure you are familiar with the assessment criteria for each outcome. For example for Learning Outcome 4 you will need to be able to state the constructional features, applications, advantages and limitations for a range of different types of cable. You will also need to be able to state the constructional features, applications, advantages and limitations of a range of cable and conductor containment systems. You will need to be able to describe the key environmental factors that can affect the selection of wiring systems, equipment and enclosures and state the types of wiring system and associated equipment used for a selection of installations and systems.

It is important to read each question carefully and take your time. Try and complete both progress checks and multiple choice questions, without assistance, to see how much you have understood. Refer to the relevant pages in the book for subsequent checks. Always use correct terminology as used in BS 7671. There are some simple tips to follow when writing answers to exam questions:

- **Explain briefly** – usually a sentence or two to cover the topic. The word to note is 'briefly' meaning do not ramble on. Keep to the point.
- **Identify** – refer to reference material, showing which the correct answers are.
- **List** – a simple bullet list is all that is required. An example could include, listing the installation tests required in the correct order.
- **Describe** – a reasonably detailed explanation to cover the subject in the question.

This unit has a large number of practical skills in it, and you will need to be sure that you have carried out sufficient practice and that you feel you are capable of passing any practical assessment. It is best to have a plan of action and a method statement to help you work. It is also wise to check your work at regular intervals. This will help to ensure that you are working correctly and help you to avoid any problems developing as you work. Remember, don't rush the job as speed will come with practice and it is important that you get the quality of workmanship right.

Good luck!

CHECK YOUR KNOWLEDGE

1. When selecting the correct type of cable for an electrical installation, which of the following is **not** a consideration?
   a) Environmental conditions
   b) Conductor size
   c) Insulation
   d) Cost

2. The only striped conductor used in fixed electrical installations is?
   a) Phase
   b) Neutral
   c) CPC
   d) Functional earth

3. Which of the following cable types is most suitable for burying direct in the ground?
   a) PVC insulated
   b) PVC insulated and sheathed
   c) PVC Steel Wire Armoured
   d) PVC insulated bell wire

4. Which of these containment systems is not suitable for single insulated cables?
   a) Trunking
   b) Ducting
   c) PVC conduit
   d) Steel conduit

5. Where lighting circuits are utilised on long corridors, they normally incorporate two-way and intermediate switching arrangements. How many intermediate switches may be employed?
   a) 1
   b) 2
   c) 3
   d) unlimited

6. A radial circuit feeding sockets outlets and protected by a 30/32A protective device may cover a floor area of?
   a) 20 m²
   b) 50 m²
   c) 75 m²
   d) 100 m²

7. Emergency lighting systems must follow the detailed guidance in which document?
   a) BS 5266
   b) BS 5839 Part I
   c) BS 5839 Part 6
   d) BS 7671

8. Fire alarm systems are classified into general types, which one of the following represents protection for life?
   a) Type S
   b) Type P
   c) Type M
   d) Type L

9. In security systems passive infrared detectors are activated by?
   a) Pressure
   b) Vibration
   c) Body heat
   d) Sound waves

10. Final connections to water heating systems must be made in?
    a) Single-insulated cable
    b) PVC-insulated and sheathed cable
    c) Heat-resistant flexible cord
    d) PVC-insulated cord

11. When selecting hand tools an electrician would not consider?
    a) Pliers
    b) Screwdrivers
    c) Cutters
    d) Pipe bender

12. An **unsuitable** fixing method to fasten an electrical accessory to a wall would be?
    a) Nails
    b) Screw and plastic plug
    c) Screw and fibre plug
    d) Spring toggle

CHECK YOUR KNOWLEDGE

13. Which of the following would not be considered when selecting a wiring system?
    a) Moisture
    b) Cost
    c) Mechanical stress
    d) Ambient temperature

14. Part 4 of BS 7671 covers the following topic.
    a) Definitions
    b) Selection and erection of equipment
    c) Protection for safety
    d) Special installations and locations

15. Where an insulation resistance test has been carried out on a 230 V lighting circuit the minimum acceptable value should be?
    a) 0.5 MΩ
    b) 1.0 MΩ
    c) 0.5 Ω
    d) 1.0 Ω

16. Which of the following is **not** regarded in BS 7671 as a special installation or location?
    a) Bathroom
    b) Swimming pool
    c) Kitchen
    d) Construction site

17. The temperature class or ignition temperature of the gas or vapour involved would **not** be classified as?
    a) T1
    b) T2
    c) T4
    d) T8

UNIT ELTK 05

Understanding the principles, practices and legislation for the termination and connection of conductors, cables and cords in electrical systems

You looked at the topic of isolation in Units ELTK 01 and ELTK 04. This unit extends your knowledge, looking at the procedures, requirements and methods you should use for safe isolation in more detail. As you read through, it may be helpful to check back with the information in the earlier units.

This unit will cover the following learning outcomes:

- understand the principles, regulatory requirements and procedures for completing the safe isolation of circuits and complete electrical installations

- understand the regulatory requirements and procedures for terminating and connecting conductors, cables and flexible cords in electrical wiring systems and equipment

- understand the procedures and applications of different methods of terminating and connecting conductors, cables and flexible cords in electrical wiring systems and equipment.

K1. Understand the principles, regulatory requirements and procedures for completing the safe isolation of circuits and complete electrical installations

Working life

You are working in a tower block of flats and your supervisor asks you to go and connect up that fluorescent fitting in the caretaker's large store. It's all switched off and locked off; the distribution room is locked and the keys are locked away in the site cabin. The cables have been pulled through. It's just on a one-way switch and the fitting is propped up against the back wall of the store ready for you.

Do you believe everything you've just been told?

Safe working practices rely on clearly thought-out systems of work, carried through by trained, competent people who are aware of their own limitations. Your supervisor may believe that what they told you is correct, but with the best will and planning in the world, circumstances can change. You must always be monitoring such changes and be prepared to deal with them.

Look back at the scenario again. What would you do?

K2 and K3. Understand regulatory requirements, procedures and applications for terminating and connecting conductors and cables

Sources of relevant information

As you read in ELTK 01, both employer and employee have legal duties to establish and maintain a safe working environment. Most construction sites have similar hazards and risks.

Once you understand what the requirements are, you should undertake risk assessments (using the HSE 'Five steps to risk assessment' model) to make sure that:

- you have safe access and egress to, from and around the work area
- you eliminate or control all work area hazards to property, personnel and livestock.

Once risk assessments have been completed, you must ensure that their findings are implemented before any work begins. If you discover a variation to the installation or any problem, you must report it to the relevant person, usually, in the first instance, the site supervisor.

In relation to terminating and connecting, you will use statutory documents (HASAWA, EAWR etc.), site drawings, wiring diagrams, the specification and relevant technical data such as manufacturer's instructions. These topics were covered in greater detail in Unit ELTK 03 of this book (see pages 159–73).

You will also use relevant British Standards, perhaps the most important of which is BS 7671, Requirements for Electrical Installations, more commonly referred to as the IET Wiring Regulations. The requirements for electrical connections are given in Part 5, Chapter 52, Section 526. Here are the main points.

- Regulation 526.1 requires that any connection between conductors or equipment must provide durable electrical continuity as well as adequate mechanical strength and protection.

- Regulation 526.2 requires the means of connection to take into account the conductor material and CSA, its insulation, the number of conductors to be connected and the terminal temperature in normal service.
- Regulation 526.3 generally requires every connection to be accessible for inspection, testing and maintenance.

Connections

What Regulation 526.2 calls a 'means of connection' includes the terminal of a wiring accessory, such as a socket, a junction box, a distribution board or compression lugs. It is the entry of the cable end into an accessory that is known as a termination.

In the case of a stranded conductor, the strands should be twisted together with pliers before terminating. Take care not to damage the wires. BS 7671 Chapter 52, Section 526 requires that a cable termination of any kind should securely anchor all the wires of the conductor that may impose any appreciable mechanical stress on the terminal or socket, as a termination under mechanical stress is liable to disconnection.

When current is flowing in a conductor, a certain amount of heat is developed and the consequent expansion and contraction may be enough to allow a conductor under stress – particularly tension – to be pulled out of the terminal or socket.

One or more strands or wires left out of the terminal or socket will reduce the effective cross-sectional area of the conductor at that point. This may result in increased resistance and probably overheating. When terminating flexible cords into a conduit box, the flex should be gripped with a flex clamp. Any cord grips that are used to secure flex or cable should be clamped down onto the protective outer sheathing.

Table 5.01 shows most common terminal types.

The joints in non-flexible cables shall be made by soldering, brazing or welding mechanical clamps, or be of a compression type and be insulated. The devices used should relate to the size of cable and be insulated to the voltage of the system being used. Connectors used must be of the appropriate British Standard, and the temperature of the environment must be considered when choosing the connector. Additionally the EAWR state that every joint and connection in a system shall be installed so as to prevent corrosion.

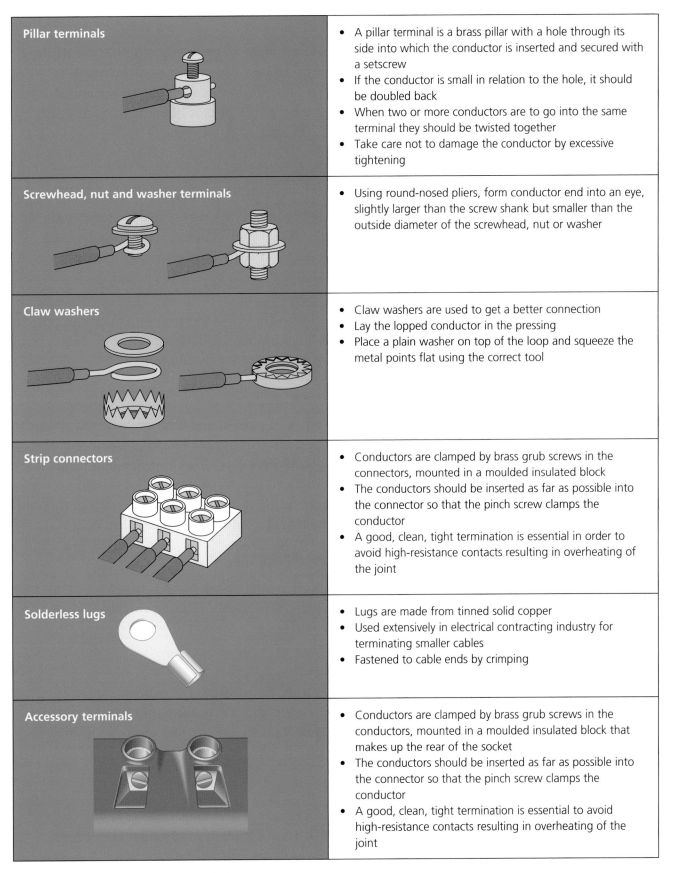

Pillar terminals	• A pillar terminal is a brass pillar with a hole through its side into which the conductor is inserted and secured with a setscrew • If the conductor is small in relation to the hole, it should be doubled back • When two or more conductors are to go into the same terminal they should be twisted together • Take care not to damage the conductor by excessive tightening
Screwhead, nut and washer terminals	• Using round-nosed pliers, form conductor end into an eye, slightly larger than the screw shank but smaller than the outside diameter of the screwhead, nut or washer
Claw washers	• Claw washers are used to get a better connection • Lay the lopped conductor in the pressing • Place a plain washer on top of the loop and squeeze the metal points flat using the correct tool
Strip connectors	• Conductors are clamped by brass grub screws in the connectors, mounted in a moulded insulated block • The conductors should be inserted as far as possible into the connector so that the pinch screw clamps the conductor • A good, clean, tight termination is essential in order to avoid high-resistance contacts resulting in overheating of the joint
Solderless lugs	• Lugs are made from tinned solid copper • Used extensively in electrical contracting industry for terminating smaller cables • Fastened to cable ends by crimping
Accessory terminals	• Conductors are clamped by brass grub screws in the conductors, mounted in a moulded insulated block that makes up the rear of the socket • The conductors should be inserted as far as possible into the connector so that the pinch screw clamps the conductor • A good, clean, tight termination is essential to avoid high-resistance contacts resulting in overheating of the joint

Table 5.01 The most common terminal types

Plastic connector

Among the most common types of connector, plastic connectors often come in a block of 10 or 12, sometimes nicknamed a 'chocolate block'. The size required is very important and should relate to the current rating of the circuit being used. They are available in 5, 15, 20, 30 and 50 A ratings.

Porcelain connectors

These are used where a high temperature may be expected. They are often found inside appliances such as water heaters, space heaters and luminaries. They should also be used where fixed wiring has to be connected into a totally enclosed fitting.

Screwits

These are porcelain connectors with internal porcelain threads that twist onto the cable conductors. They are now obsolete, but you may come across them in older installations.

Compression joints

This includes many types of connectors that are fastened onto the conductors, usually by a crimping tool. The connectors may be used for straight-through joints or special end configurations. If the conductors are not clean when making the joint with a crimping tool, this may result in a high-resistance joint that could cause a build-up of heat and eventually lead to a fire risk.

Uninsulated connectors

These are often required inside wiring panels, fuse boards, and so on and are used to connect earth cables and protective conductors.

Junction boxes

The two important factors to consider when choosing junction boxes are the current rating and the number of terminals. Junction boxes are usually either for lighting or socket outlet circuits.

Soldered joints

Although soldering of armoured and larger cables is still common, the soldering of small-circuit cables is rarely carried out for through joints these days. This joint was often referred to as a marriage joint, and involved stripping back the insulation,

Figure 5.01 A four-terminal junction box

Figure 5.02 A modern RB4 junction box

carefully twisting together conductors, soldering and then finally taping up.

Bear in mind that when a conductor end is formed into an 'eye', the conductor should be placed over the post or terminal such that the motion of the screw or nut when tightened draws the eye with it. If it is placed on in the reverse position, tightening the nut or screw will cause the eye of the conductor to unwind and result in a less than perfect connection.

Proving terminations and connections are electrically and mechanically sound

Look at Figure 5.03, showing two pieces of metal joined together fitting over an insulated post terminal and held by a nut.

If you take an insulation resistance tester acting as a low-resistance ohmmeter, and apply a lead to each piece of metal, you will find that the meter shows the effect of a 'dead short': in other words, you have zero resistance. There is nothing to impede the meter signal from input at one side, through the metals to the lead on the other. It is the perfect joint.

However, if the two metals are not tightly connected, a resistance will be present and the signal will struggle to get from one side to the other. The opposite situation of no resistance is a resistance of infinite value – infinity. This means that the gap between the two metals is so great that the signal from one lead can never reach the other lead, as shown in Figure 5.04.

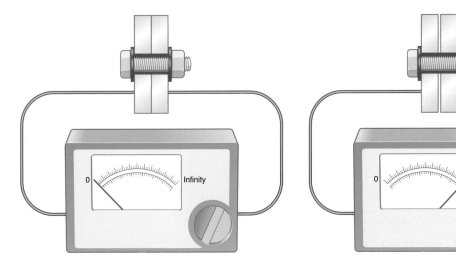

Figure 5.03 The perfect joint   **Figure 5.04** Joint not tight enough

Resistance, whether caused by loose connections or the effects of corrosion, will be a value between zero and infinity. However, none the less this is called a high-resistance joint.

You can use a low-resistance ohmmeter as part of the process of testing to see whether a joint or termination is of acceptable quality. Other, more basic methods include visual inspection and to tug the cable to see if it is secure.

Remember that, when current flows through a resistance, heat is developed. At a terminal, the consequent expansion and contraction due to heating and cooling can loosen the conductor in the terminal. If a large current is passing through a high-resistance joint, a lot of heat can be generated and this can lead to arcing, damage to the equipment or cable and pose a serious fire risk.

Terminating cables and conductors

Before you look at terminating the various types of cable, it would be wise to remind yourself of the colours of conductors, both before and after April 2004.

Insulation colours

Throughout the EU and in the UK since April 2004, all new installations in the UK had to use cables whose conductors complied with Table 51 of BS 7671.

You have already looked at this in Unit ELTK 04. Turn back now to page 250 and remind yourself what the insulation colours are.

Cable termination techniques, methods and procedure

In this section you will look at terminating:

- PVC/PVC flat-profile cable
- mineral-insulated cable
- SWA and SWB cable
- data cables
- fibre-optic cable
- fire-resistant cables.

Safety tip

If you are asked to extend an existing installation you may well have a situation where cables of both colour systems exist. Great care must be taken as black and blue coloured installation is used in both systems.

PVC insulated PVC sheathed flat profile cable with integral CPC (6241Y, 6242Y, 6243Y)

Terminating PVC/PVC cables

Step 1 Nick the cable at the end with your knife and pull apart.

Step 2 When the required length has been stripped, cut off the surplus sheathing with the knife.

Step 3 The insulation can be stripped from the conductors with the knife or with a pair of purpose-made strippers. Examine the conductor insulation for damage.

Once the cable ends are prepared, the conductor should be 'doubled over' before connecting to the equipment. This ensures that the equipment grub screw has a large surface area to grip.

Don't forget to sleeve the CPC with green and yellow sleeving before connecting to the equipment and back box.

Mineral-insulated cable

Mineral-insulated cables must be sealed at each end, otherwise the magnesium oxide insulator will absorb moisture, resulting in a low insulation resistance reading. A complete termination comprises two sub-assemblies – the gland and the seal – each performing a different function.

- The gland is used to connect or anchor the cable to a piece of equipment and consists of three brass components (gland body, compression ring and gland nut).
- The purpose of the seal is to exclude moisture from the cable. The seal consists of a brass pot that is screwed onto the copper sheath of the cable and filled with a special compound. A plastic disc is then slotted over the conductors and compressed into the 'mouth' of the pot, and the neoprene rubber sleeves are fitted to insulate the conductor tails.

The first thing that you must do when terminating MI cables is to strip off the copper sheath. If the cable has a PVC oversheath, you must remove this before the copper sheath. Some electricians recommend that the cable should be indented with a ringing tool before sheath stripping. It may, however, be better to ring nearer the end of the stripping process. You can strip off the copper sheath in several ways, although you will need to practise before you can carry out the task in a reasonable time.

Stripping

Figure 5.05 Using side cutters

Using side cutters

When using side cutters or pliers it is essential that the points of the blades are in good condition, otherwise difficulties can arise.

First make a small tear in the sheath and then peel off the sheath with the side cutters, working around the cable in a clockwise direction, at the same time keeping the side cutters at an angle of approximately 45 degrees to the cable.

Using stripping bar (fork-ended stripper)

Figure 5.06 Using a stripping bar

Having started to break the copper sheath away with the side cutting pliers, flatten the torn-off sheath portion and insert it into the slot of the stripping bar.

Rotate the stripping bar while keeping the bar at 45 degrees to the cable.

Using rotary stripping tools

Rotary stripping tools are available in many shapes and forms. They are much easier to use than either side cutters or the stripping bar, but can be difficult to set up. This is not a problem if you are using the same size cables all the time.

Rotary strippers usually require a good, square end to the cable. The stripper is then pushed over the cable, a little pressure is applied and then it is rotated.

If the sheath is to be stripped for a longer distance, then the sheath being removed should be cut away at intervals to avoid fouling the tool. Another advantage of the rotary stripping tool is that a ringing tool is not required. When the stripping tool has reached a required position, apply pliers to the cable sheath and turn the stripper further.

Figure 5.07 Rotary stripping tool

Ringing cable sheath

Before the pot can be fixed the sheath must have a good, clean circular end. To achieve this you use a ringing tool, which makes an indent in the sheath as shown.

This is required when the sheath is stripped with either side cutters or the stripping bar, but is not necessary (as previously mentioned) when rotary strippers are used.

Figure 5.08 Ringing cable sheath

Fitting and sealing the pot

Actions prior to fitting the pot

Before fitting the pot, check that the gland nut and olive are in position.

If the cables have the PVC oversheath, fit a shroud now!

Fitting the pot

The following steps describe the fitting of a screw-on pot. Pots are best fitted using a pot wrench as shown in Figure 5.09, which will ensure that the pot goes on square.

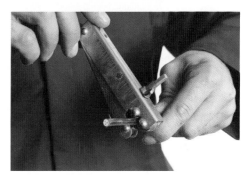

Figure 5.09 Fitting the pot

The pot has an internal thread, which screws onto the copper sheath. Screw the pot onto the copper sheath until it just lines up in the inside of the pot.

Finally tap the pot to empty it of any filings or loose powder. Do not be tempted to blow into the pot: the moisture in your breath could reduce the insulation resistance of the magnesium oxide, causing a short circuit at a later stage.

Actions prior to sealing

Before sealing, wipe the conductors with a clean, dry rag to remove any loose powder. If the cables have become twisted, straighten them pulling them firmly with a pair of pliers.

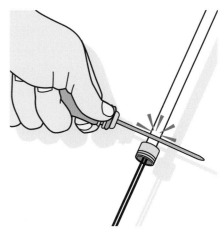

Figure 5.10 Tapping the pot empty of powder

Figure 5.11 MICC pot and seal

Now prepare the cable insulation and seal disc. The insulation is usually neoprene rubber sleeving, which you should cut to the required length then fit to the conductors and disc. The disc ensures the conductors are kept apart while the compound is inserted into the pot.

Sealing the pot

Now fill the pot with sealing compound as shown. It is important that the pot is filled with compound from one side only, directed towards and between the conductors. This prevents air locks, which could lead to condensation problems and subsequent failure of the termination. Make sure your hands are clean when using the compound and keep the compound covered to stop dirt getting into it.

Figure 5.12 Sealing the pot

Crimping

Before applying the crimping tool, remove any excess compound. Apply the crimping tool and gradually tighten it, while keeping it straight. Stop at intervals to allow the compound to seep out.

Finished seal and identification

When crimping is completed, test the seal to ensure a high value of insulation resistance between cores and from cores to the copper sheath. Wring the cable out and seal the other end. Then test the whole cable and give the conductors a continuity test to indicate which conductor is which before marking them for identification.

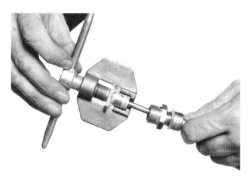

Figure 5.13 Crimping the pot

Gland connection

Once you have tested the cable and identified the conductors, tighten the gland up and connect it into the equipment or accessory in which it is being terminated. After the conduit thread has been tightened into the accessory, tighten the back nut. This will compress the olive ring inside the gland, providing earth continuity. Once this has been tightened it is very difficult to remove.

Types of gland

Standard glands have their sizes stamped on them: for example, 2LI meaning 2 core, light duty, 1 mm² CSA (cross-sectional area), or 4H6 meaning 4 core, heavy-duty, 6 mm² CSA. The size of the conduit thread will depend on the overall size of the cable being used.

PVC/Steel Wire Armour/PVC cable (6942X, 3X, etc.) Single core PVC/Aluminium Wire Armour/PVC cable XLPE/Steel Wire Armour/PVC cable (6945XL7W TO 69448XL7W)

Although the photograph shows the construction of a PVC/SWA/PVC cable, the general principle of termination is the same for each of the cables listed above.

The only major difference would be that, when using any form of aluminium armouring, whether it be strip or wire, you must use an aluminium gland – brass and aluminium in contact with each other will create an electrolytic reaction.

Figure 5.14 PVC/SWA/PVC cable

In all versions, the armouring is used as a circuit protective conductor so special glands are employed to ensure good continuity between this and the metalwork of the equipment to which you are connecting.

An earth tag ('banjo') provides earth continuity between the armour of the cable and the box or panel. If PVC/SWA/PVC cable is used to connect directly to an electric motor mounted on slide rails, you should leave a loop (helix loop) in the cable next to the motor to permit necessary movement and absorb vibration.

These glands vary a little from one manufacturer to another and their design also depends on the environment in which they are to be used. However, in terms of UK manufacturers such as CMP, cable glands are currently manufactured in accordance with BS 6121, which categorises the different types of common gland as shown in Table 5.04.

Type	Definition
A2	Single seal to outer sheath cable gland for unarmoured cable
BW	No seal indoor use cable gland for SWA cable
CW	Single seal (outer sheath) for SWA cable
E1W	Double seal (outer sheath and inner sheath) cable gland for SWA cable
CX	Single seal (outer sheath) cable gland for braided cable
E1X	Double seal (outer sheath and inner sheath) cable gland for braided cable

Table 5.04 BS 6121 gland categories

The following diagrams are of an E1FW gland, the flameproof equivalent to the E1W listed in Table 5.04.

Figure 5.15 E1W gland

The basic differences between an industrial gland and its flameproof equivalent are the longer thread length and the fact that there must be a seal onto the inner (or outer, in the case of an A2 gland) sheath of the cable.

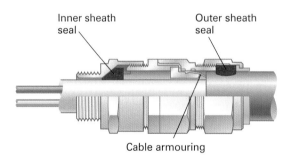

Inner sheath seal

Outer sheath seal

Cable armouring

Figure 5.16 Section through an E1W gland

The purpose of the flameproof gland is not to prevent gas entering an enclosure; instead, should a spark from the equipment in the enclosure ignite such gas, the gland and enclosure will prevent the resulting explosion leaving the enclosure and igniting the gas outside it.

The gland achieves this by a combination of the seal on the inner sheath and an increased thread length that will 'cool' the exploding gas as it travels around the thread: by the time it reaches the external gas, it is too cool to ignite it.

However, for the purpose of this section, we will use the termination of a BW gland onto an SWA cable.

Terminations are made by stripping back the PVC sheathing and steel-wire armouring, then fitting a compression gland, which is terminated into the switchgear or control gear housing.

Terminating SWA cables

Step 1 For correct termination, start by measuring the length of armouring required to fit over the cable gland armour clamp and make a note of this measurement. Then establish how long the conductors need to be to connect to your equipment and make a note of this. Add the two together and then mark that distance on the cable by measuring from the end of the cable (point A) to point B.

Step 2 Mark the length of armour required (point C) for the cable armour clamp by measuring back from point B. At this stage some people will strip off the PVC outer sheathing. However, this is best left on, as it will hold the steel-wire armouring in place for you. You must remember at this point to place the shroud over the cable. This is often forgotten and is a sign of poor workmanship.

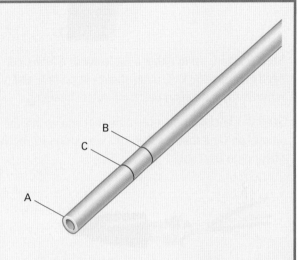

Step 3 Next, using a junior hacksaw, cut through the PVC outer sheathing and partly through the armouring at point B.

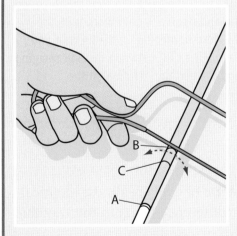

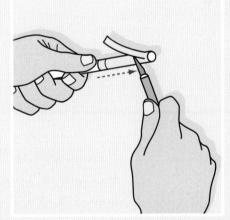

Step 4 The PVC outer sheath can now be cut away.

Step 5 The next step is to take each strand of the armouring in turn; snap them off at the point where they are partly cut through. In the photograph we have covered the section between point C and point B with yellow tape to make it easier to see.

Terminating SWA cables (cont.)

Step 6 Then, using either a hacksaw or a knife, cut neatly around the PVC outer sheath at point C and remove any remaining pieces of outer sheathing.

Step 7 Checking one more time that you've put the shroud on (and that it's the right way around!), fit the gland onto the cable.

Step 8 First, slide the backnut and compression rings (if any) onto the cable. Then, taking the gland body, slide this onto the cable, making sure that it fits under all the strands of armouring.

Step 9 Finally, slide up the backnut and screw it onto the gland body, clamping the armouring tightly. The inner PVC sheath can be stripped off like any other PVC cable. Now, the gland is ready.

Step 10 Before the termination can be inserted into the enclosure and connected to your equipment, if required you must slide the earth tag over the threaded part of the gland and clean any paintwork from the area of contact before tightening up the locknut and securing the gland. Fit a CPC between the bolt securing the earth tag and the earthing terminal of the equipment.

PVC/GSWB/PVC cable

The process of terminating this type of cable is almost identical to that of the SWA, the only major difference being in the way that the gland clamps the steel braid.

To illustrate, we are using SY cable, which is essentially a PVC/SWB/PVC-constructed cable usually used for flexible applications, such as between moving parts of plant on production lines.

Such cables are normally terminated using a BS 6121 CX gland where the braid is 'spread' over the cone of the gland in much the same way as an SWA gland. This method usually allows good

Figure 5.17 PVC/GSWB/PVC cable

EMC protection. However, another gland type uses a 'pigtail' arrangement, where the braid is tied into two pigtails that are then fed through slots cut in the length of the threaded part of the gland, then held in place between two washers and the locknut. Examples of both cable and CX gland are shown in Figures 5.17 and 5.18.

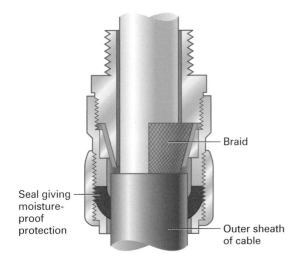

Figure 5.18 Cross section of a CX gland

Data cables

There are three basic types of cabling used in data systems: coaxial, fibre-optic and twisted pair.

Although Cat 8 cable is available, for many, Cat 5 (a twisted multi-pair cable) more than meets needs. Typically used for networking in domestic situations, it would be terminated using a specialist tool onto a connecting block within a plastic plate. The termination process is shown on page 432.

Before beginning establish the length of conductor required and carefully remove the outer sheaf. For Cat 6 cable, remove any sheath removal string and central plastic divider. Separate the twisted pairs, rotating the cable so that each pair lines up with its respective terminal.

Terminating a Cat 5 cable

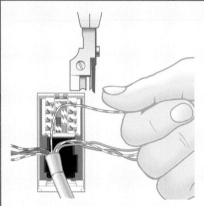

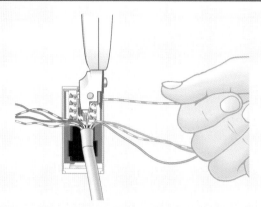

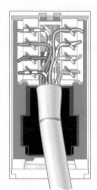

Step 1 Route each individual conductor to its specified terminal.

Step 2 Place the cable over the terminal slot and use a punchdown tool to make the connection.

Step 3 The punchdown tool normally removes any excess conductor. Repeat for remaining conductors.

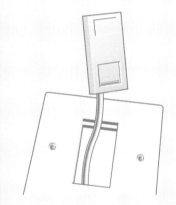

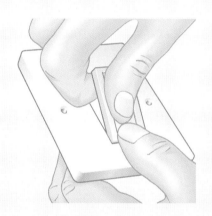

Step 4 Feed the face plate over the connected cable conductor.

Step 5 Clip the connector in place.

Step 6 Fix the assembled face plate to the accessory box or trunking.

Fire-resistant cables

FP 200 Gold

FP 200 does not possess the same mechanical strength as other cables, so it is important to terminate the cable according to the manufacturer's instructions. The following sequence explains a recognised method of termination utilising a nylon, A2 type compression gland.

Having measured the length of conductor required, score around the sheath with a knife or suitable cable-stripping tool, taking care not to cut right through to the aluminium tape.

Now flex the cable gently at the point of scoring until the sheath yields and then pull off the sheath, twisting gently to follow the lay of the cores.

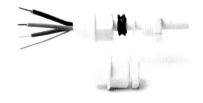

Figure 5.19 FP 200 Gold termination

Figure 5.20 Scoring the sheath

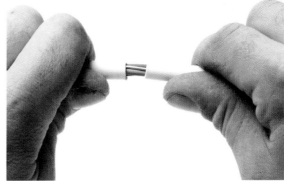

Figure 5.21 Pulling the sheath

As an A2 gland uses a compression seal onto the outer sheath of the cable, remove the compression backnut and seal and thread carefully over the outer sheath of the cable. At this point the termination may be carefully fed into the enclosure and secured with a lock nut.

FP 200 Gold does not require a ferrule; the cable may enter wiring accessories or fittings through a simple grommet. However, when installed in wet conditions or outdoors, a standard waterproof gland incorporating a PCP sealing ring must be used.

To prevent electrical faults occurring from phase to earth when terminating these cables, take care to avoid damage to the insulation by not bending the cores sharply over the end of the sheath.

Working life

At the start of this unit we gave a scenario. This was taken from a real and recent incident on a construction site, when an NVQ assessor visited an apprentice on site.

The assessor asked the apprentice to take him to the example of connecting that would be assessed and while they were walking the apprentice explained the activity given to him by his supervisor.

Once in the store, the apprentice located the light and then climbed the access equipment to connect it to the cables that were protruding from the conduit end box.

Before the apprentice could touch any cables, the assessor asked him how he knew that the circuit was dead and safe to work on. The assessor then asked the apprentice to forget what his supervisor had told him and instead demonstrate to his satisfaction that the circuit was indeed safe to work on.

Followed by the assessor, the apprentice followed the run of conduit and trunking down the corridor to the main switchgear cupboard, which was not locked, but instead wide open.

Inside the apprentice followed the cabling to the DB which was not isolated, but instead had been powered up by another electrician. Worse still the circuit feeding the store light had been connected into the DB and MCB and the MCB was switched on.

Because of the attending assessor, one potentially fatal accident was averted.

So...do you still believe everything you've just been told?

Getting ready for assessment

This unit will extend your knowledge of practical skills in relation to terminating and connecting cables, conductors and cords. This unit will help you to ensure that electrical connections are being made safely and correctly.

For this unit you will need to be familiar with:

- principles, regulatory requirements and procedures for completing the safe isolation of circuits and complete electrical installations
- regulatory requirements and procedures for terminating and connecting conductors, cables and flexible cords in electrical wiring systems and equipment
- procedures and applications of different methods of terminating and connecting conductors, cables and flexible cords in electrical wiring systems and equipment

For each Learning Outcome, there are several skills you will need to acquire, so you must make sure you are familiar with the assessment criteria for each outcome. For example for Learning Outcome 3 you will need to be able to explain the cable termination techniques for a range of different cables, and explain the advantages, limitations and application of the different connection methods. You will also need to describe the procedures for proving terminations and connections are electrically and mechanically sound and the consequences of this not being the case. You will also need to name the health and safety requirements for terminating and conducting cables, conductors and flexible cords.

It is important to read each question carefully and take your time. Try and complete both progress checks and multiple choice questions, without assistance, to see how much you have understood. Refer to the relevant pages in the book for subsequent checks. Always use correct terminology as used in BS7671. There are some simple tips to follow when writing answers to exam questions:

- **Explain briefly** – usually a sentence or two to cover the topic. The word to note is 'briefly' meaning do not ramble on. Keep to the point.
- **Identify** – refer to reference material, showing which the correct answers are.
- **List** – a simple bullet list is all that is required. An example could include, listing the installation tests required in the correct order.
- **Describe** – a reasonably detailed explanation to cover the subject in the question.

This unit has a large number of practical skills in it, and you will need to be sure that you have carried out sufficient practice and that you feel you are capable of passing any practical assessment. It is best to have a plan of action and a method statement to help you work. It is also wise to check your work at regular intervals. This will help to ensure that you are working correctly and help you to avoid any problems developing as you work. Remember, don't rush the job as speed will come with practice and it is important that you get the quality of workmanship right.

Good luck!

CHECK YOUR KNOWLEDGE

1. The reason BS 7671 requires all connections to be readily accessible is for?
   a) Ease of terminating
   b) Quicker to terminate
   c) Inspection, testing and maintenance
   d) Cost-effectiveness

2. BS 7671 requires all connections to be?
   a) Readily accessible
   b) Enclosed in junction boxes
   c) Enclosed in trunking
   d) Concealed under floors

3. Loose connections at terminals can result in?
   a) Excess current
   b) Excess heat
   c) Cooling of the terminal
   d) Operation of the protective device

4. Joints in non-flexible cables should **not** be made by?
   a) Soldering
   b) Brazing
   c) Terminal strips
   d) Compression

5. Which of the following results would indicate a good connection?
   a) 0.01 Ω
   b) 240 Ω
   c) 3 MΩ
   d) 100 MΩ

6. The cable that must be sealed at each end to avoid ingress of moisture is?
   a) PVC singles in conduit
   b) PVC multi-core cable
   c) PVC steel wire armoured cable
   d) Mineral-insulated cable

7. A BW gland used to terminate steel wire armoured cable has a?
   a) Single seal (outer sheath) for braided cable
   b) No seal
   c) Single seal to outer sheath
   d) Double seal

8. Safe isolation of an installation to be worked on is required when?
   a) Risk of electric shock is present
   b) Always before work commences
   c) Never
   d) Before client arrives

9. Which of the following cables is used in data wiring?
   a) MICC
   b) PVCSWA
   c) Bell wire
   d) Twisted pair

10. A suitable fire-resisting cable would be?
    a) FP200
    b) PVC insulated
    c) PVC insulated and sheathed
    d) PVC steel wire armoured

Glossary

This glossary is intended to support you throughout the whole qualification, and as a reference guide for you to use to quickly check the meaning of some key terminology you will encounter both during your study and in your professional life.

This glossary covers content from both this book and *Level 3 NVQ/SVQ Diploma Installing Electrotechnical Systems and Equipment Book B*. As such, some of these terms may not feature or be fully explained in this book, but are covered in depth in Book B.

a.c. – alternating current

addressable – digitally programmable

analogue device – represents measurement by position of, for example, a needle on a calibrated scale

apparent power – in an a.c. circuit the sum of the **true or active power** and the **reactive power**

arcing – a plasma discharge as the result of **current** flowing between two terminals through a normally non-conductive media (such as air), producing high light and heat

armature – moving part of electrical machine in which a **voltage** is induced by a magnetic field

average value – for **sine wave a.c.** this equals either **peak value** or **voltage** × 0.637

ballast (or **choke**) – **inductance** coil in fluorescent lights to limit variations of **a.c.** or alter its **phase**

bill of quantities – a list of all the materials required, their specification and the quantities needed; contractors who are tendering for the project use this information to prepare their estimates

capacitance – ability of circuit or device to store electric **charge** measured in farads – generally microfarads (μF), nanofarads (nF) or picofarads (pF); symbol C

capacitive – circuit containing components that can store electric **charge**, i.e. have **capacitance**

capacitive reactance – opposition to flow of **a.c.** produced by a capacitor, measured in ohms (Ω); symbol X_C

capacitor – component with ability to store required quantity of electric **charge** if **voltage** applied to it, which can be returned to a circuit

carbon footprint – the total amount of greenhouse gases produced by an organisation, event, product or person

catenary wire – a supporting wire fixed between two buildings from which cable is 'hung' with suitable supports/clips

charge – quantity of **electrons**, e.g. as stored in a **capacitor**, measured in **couloumbs**; symbol Q

child – according to the Education Act 1996, a person who is below Minimum School Leaving Age (MSLA). This will be 15 or 16 years old depending on when their birthday falls

circuit integrity – the ability of a circuit to keep operating

coercivity – due to **hysteresis**, reverse magnetic **field strength** that must be applied to reduce **flux density** to zero

combustion – rapid chemical combination of two or more substances accompanied by production of heat and light; commonly known as burning

commutator – device for connecting a rotating **current** carrying coil to a **d.c.** supply

conductor – any material allowing electrical charges to flow easily; includes metal pipework, metal structures of buildings, salt water or ionised gases

conduit – mild steel or PVC tubing through which electrical cabling can be run

contactor – widely used in electrical installations, using **solenoid** effect to make or break contacts

contract – a legally binding agreement between two or more parties

control panel – programmable 'brains' of any system to which all parts of the system are connected

CPC – circuit protective conductor

CSA – cross-sectional area, frequently used when describing **conductors**

current – flow of free **electrons** in a **conductor**; measured in amperes (A); symbol I; direction is opposite to actual electron flow

d.c. – direct current

dB – the abbreviation for decibel, the unit in which noise level is measured

data – factual information and statistics used as a basis for discussion, calculation or analysis

defect liability period – a contracted period of time and coverage, during which the installer is responsible for the repair of any defective items

delta connection – triangular arrangement of electrical three-phase windings

dermatitis – inflammation of the skin normally caused by contact with irritating substances

detailed visual inspection – a more detailed inspection than a **pre-check** for visible defects according to the manufacturer's instructions, the results of which should be recorded

digital device – represents measurement as numbers (liquid crystal or LED display) rather than a needle on a scale as in **analogue device**

diode – **semiconductor** device that allows current to flow in one direction only

directly proportional – increase in one property which results in an increase in another, e.g. the harder a coin is flicked along a smooth flat surface, the further it will travel (see also **indirectly proportional**)

disability – a physical or mental impairment that has a substantial and long-term effect on a person's ability to carry out normal everyday activities

discharge consent – written authorization that must be obtained from the Environment Regulator before discharging any sewage, effluent or contaminated surface water to surface waters or groundwater

discrimination – correct arrangement of fuses in sequence to protect parts of a circuit

dissipate – get rid of through dispersion

dual in-line ICs – the types of IC with the pins lined up down each side

duty holders – people who have duties under a particular piece of legislation, such as EaWR

ecological or **eco-** – linked with ecology, the branch of biology that looks at the relationship between organisms and their environments

efficiency – ratio of a system's output against the input, normally expressed as a percentage. In machines, degree to which friction and other factors reduce actual work output from the theoretical maximum of 100%

electromagnetic induction – production of an **e.m.f.** in a **conductor**, by moving it through a magnetic field across the lines of flux, which will, in a closed circuit, cause an electric **current** to flow in the conductor

electromotive force (e.m.f.) – total force from a source such as a battery that causes the motion of **electrons** due to potential difference between two points

electrons – particles in an atom that circle in orbit around the **nucleus**, said to possess a negative charge

electronic variable speed drive – device which provides variable **voltage** and frequency supply enabling motor speed and **torque** (**current**) to be precisely controlled. Used throughout industry for process automation, air handling, variable speed conveyor systems

embolism – an obstruction in a blood vessel due to a blood clot in the bloodstream

EPC – a certificate that gives buildings a rating for energy efficiency, with ratings from 'A' to 'G', 'A' being the most energy-efficient and 'G' the least. The average to date is 'D'

explosive limits – the upper and lower percentages of a gas in a given volume of gas/air mixture at normal atmospheric temperature and pressure that will burn if ignited (as per BS EN 60079)

fault current – **current** due to a fault in a circuit, including earth-fault current

fibrillate – make rapid twitching movements, such as your heart may make under electric shock

field strength – force field around a magnet or current carrying **conductor**, measured in teslas (T); symbol H

flash point – the minimum temperature at which a material gives off sufficient vapour to form an explosive atmosphere

'floating' neutral – where the star point of a transformer is not connected to the general mass of earth

flux density – the strength of a magnetic field at any point, calculated by counting the lines of magnetic flux at that point; symbol B

friction – force that opposes motion

fused spur – **spur** connected to a ring final circuit through a fused connection unit, the fuse incorporated being related to the **current**-carrying capacity of cable used for the spur but not exceeding 13 A

fusing factor – ratio of minimum current causing a fuse or circuit breaker to trip and stated current that either can sustain without blowing

Genuine Occupational Requirements – where an employer can demonstrate that there is a genuine identified need for someone of specific race or gender to the exclusion of others – for example a film company needs an Indian actor for a film set in India, or a modelling agency needs a woman to model female clothes

hazardous area – according to the Regulations, 'an area in which explosive gas/air mixtures are, or may be expected to be, present in quantities such as to require special precautions for the construction and use of electrical apparatus'

hygroscopic – tending to attract and absorb water

hysteresis – a generic term meaning a lag in the effect of a change of force, e.g. change in magnetism of a body which lags behind changes in magnetic field

ignition energy – the spark energy that will ignite the most easily ignited gas/air mixture of the test gas at atmospheric pressure. Hydrogen ignites very easily, whereas butane or methane require about ten times the energy (as per BS EN 60079)

ignition temperature or **auto-ignition temperature** – the minimum temperature at which a material will ignite and sustain combustion when mixed with air at normal pressure, without the ignition being caused by any spark or flame. Note: This is not the same as **flash point** so don't confuse them.

illuminance – measure of visible light reaching a surface, measured in lumens per square metre (lux); symbol lx

impedance – total opposition to **current** in a circuit, combining **resistance**, **inductance** and **capacitance**, measured in ohms (Ω); symbol Z

indirectly proportional – increase in one property which results in decrease of another: if a coin is flicked along a smooth angled surface, the greater the angle of the surface, the less distance the coin will travel

inductance – opposition created by a changing **current** in a magnetic field which induces a **voltage** to oppose change in current, either within a circuit (self-inductance) or a neighbouring circuit in the same magnetic field (mutual inductance), measured in henrys (H); symbol L

inductive circuit or **load** – containing components with windings, e.g. motor, generator or transformer, which have **inductance**

inductive reactance – opposition to flow of **a.c.** produced by an inductor, measured in ohms (Ω); symbol X_L

inductor – component introduced into a circuit to provide required amount of **inductance**

insulator – poor **conductor** of electric charges (i.e. a **current**) or heat

ions – electrically charged atoms formed by the loss or gain of one or more **electrons**, leaving them with a net positive or negative charge

IP Codes – International Protection, more commonly interpreted as Ingress Protection

IP rating – International Protection Rating classifies and rates the degree of protection

provided against the intrusion of solid objects, dust, accidental contact and water in mechanical casings with electrical enclosures

isolator – mechanical switching device which can cut off supply from all, or a section of, an installation by separating it from every source of electrical energy

kinetic energy – energy due to movement

Kirchhoff's law – the sum of the **voltage** drops around a closed loop in the network must equal zero

line current – **current** flowing in any one **phase** of a three-phase circuit between source and load

line voltage – **potential difference** between any two-phase **conductors** between source and load in a three-phase electrical circuit

loop in method – a method of wiring lighting circuits in multi-core cable without junction boxes. Three terminal (3 plate) ceiling roses or batten lampholders are employed. The additional terminal is often marked 'Loop'. Any additional points are 'looped' (connected) into the previous lamp position and so on to extend the circuit. This saves returning to the supply point each time a new lighting position is required

lower explosive limit (LEL) – the concentration below which a gas atmosphere is not explosive (as per BS EN 60079)

mass – amount of matter contained in an object regardless of volume or any forces acting on it. Not to be confused with weight

mechanical advantage – relationship between effort needed to lift something (input) and load (output). A machine which puts out more force than is put into it, is said to give good mechanical advantage

micro-controller – a microcomputer on a single chip

modules – sections that link together to form a whole

multi-core – having more than one core

mutual inductance – production of **e.m.f.** in a circuit by a change in **current** in an adjacent circuit

neutrons – sub-atomic particles within the **nucleus** of an atom, which have no charge and are said to be electrically neutral

noise pollution – excessive noise from any source, but particularly industrial sources, that spoils people's experience of the environment; examples could be noise from machinery, plant or power tools

non-fused spur – **spur** usually directly connected to a circuit at the terminal of socket outlets

nucleus – centre of an atom made up of **protons** and **neutrons**

Ohm's law – principle establishing that **current** flowing in a circuit is **directly proportional** to **voltage** and **indirectly proportional** to **resistance** in a constant temperature

out of phase – where **current** and **voltage** are alternating at different times

overcurrent – **current** exceeding rated value or current-carrying capacity at different times

overload current – **overcurrent** occurring in a circuit that is electrically sound

oxidation – a chemical process in which a substance combines with oxygen. During this process, energy is given off, usually in the form of heat

Ozone layer – a region of the upper atmosphere containing a relatively high concentration of ozone, which is what absorbs solar ultraviolet radiation and prevents the Earth overheating

packaging – 'all products, made of any materials of any nature, used for the containment, protection, handling, delivery and presentation of goods, from raw materials to processed goods, from the producer to the user or the consumer, including non-returnable items used for the same purposes' (official definition from the Regulations)

pattress – the container for the space behind electrical fittings, such as power outlet sockets and light switches

peak value – maximum value of **a.c.** or **d.c.** waveform. In a.c. this can refer to the negative half cycle

permit to work – official document giving authorisation to work within defined circumstances

phase current – **current** through any one component comprising a three-phase source or load

phase voltage – **voltage** measured across a single component in a three-phase source or load

phasor – straight line with length representing size of an **a.c.** quantity and direction representing relationship between **voltage** and **current**

pole pair – any magnetic system consisting of a north and south pole

potential difference – the amount of energy used up by one coloumb in its passage between any two points in a circuit

potential energy – the stored energy of an object

power factor – number less than 1.0 used to represent relationship between the **apparent** and **true power** of a circuit

pre-check – a basic visual check done for obvious defects

prospective fault current – The value of **overcurrent** that would flow at a given point in a circuit if a fault (a short-circuit or an earth fault) were to occur at that point. The prospective **fault current** tends to be higher than the fault current likely to occur in practice.

protons – particles within the **nucleus** of an atom said to possess a positive charge

pulping – where paper-based products are dropped into a huge tank and blitzed with water, which separates the paper from any impurities

reactive current – current in a **capacitive** or **inductive circuit**, with no **resistance** and no dissipation of energy

reactive power – power 'consumed' by a **capacitor** or **inductor** is not dissipated as heat but returned to source when **current** reverses

remanence – due to **hysteresis**, when an applied magnetic **field strength** reduces to zero some **flux density** remains

RCD (residual current device) – device which monitors the electrical **current** flowing both in the phase and neutral **conductors**

resistance – opposition to the flow of **electrons** (**current**)

resistivity – nature of a circuit containing pure **resistance** in which all power is dissipated by **resistor**(s). **Voltage** and **current** are in phase with each other

resistor – component in a circuit to give a desired amount of **resistance**

rotor – part of a motor or generator that rotates in magnetic field; it may carry conductors or the magnetic field system

scaled drawing – a drawing on which everything is drawn at a fixed ratio to the size of the actual object. This ratio is called the scale of the drawing, and this should be indicated on the drawing

semiconductor – material (commonly silison, germanium) that will conduct under certain conditions, allowing **electrons** (negative charge) or 'holes' (positive charge) to flow

server – a computer being used only to store information

short-circuit current – overcurrent resulting from a fault that creates negligible **impedance** between live conductors

sine wave – path traced on a graph by a pure **a.c. current** or **voltage** going from zero to positive maximum, to negative maximum and back to zero (one cycle)

slip – in a motor or generator, the difference between the speed of the **rotor** and the speed of the rotating magnetic field measured as a fraction or percentage

snagging – a list of omissions, normally prepared by the Consulting Engineer, that require correction before an installation can be classed as complete

solenoid – long, hollow cylinder round which is wound a uniform coil of wire. When a **current** is sent through the wire, a magnetic field is created inside the cylinder

spur – radial branch taken from a ring final circuit, perhaps to feed a new socket outlet

star connection – Y-shaped arrangement of three-phase electrical windings

star-delta – system of connection sometimes used in induction motors, whereby a **star connection** switches to a **delta connection**

stator – stationary part of a motor or generator which rotates in a magnetic field; it may carry **conductors** or the magnetic field system

statutory – requirement that is binding by law

stroboscopic – effect when the flicker rate in a light coincides with the rotation of an object and causes it to look as if it has slowed down or stopped

synchronous – used to describe motors in which the speed of the motor and/or the rotating magnetic field are the same

thermal comfort – an individual's preferred temperature. Environmental factors (such as humidity and sources of heat in the workplace) combine with personal factors (such as the clothing a worker is wearing and how physically demanding their work is) to influence this

thermistor – an electronic device whose resistance changes quickly in response to changes in temperature

thesaurus – a type of dictionary that lists words with similar meaning

thyristor – **semiconductor** device enabling high-speed switching, formerly known as a silicon-controlled rectifier (SCR)

tinnitus – a permanent ringing in the ears

torque – turning force, such as that produced by an electric motor on its rotating shaft; measured as force X distance from its point of rotation

torsional – to do with twisting, when 'torque' or turning force is applied

transformer – transfers electrical energy from one **a.c.** circuit to another by **mutual inductance** between two stationary coils wound into a former; output to input valves (e.g. **voltage**) are determined by the relative number of turns in each coil

transistor semiconductor – device that allows a larger **current** to be controlled by a smaller one, allowing **voltage** or **current** amplification or as a switch

true or active power – the rate at which energy is used

upper explosive limit (UEL) – the concentration of gas above which the gas atmosphere is not explosive (as per BS EN 60079)

valence electrons – the **electrons** in an atom's outermost orbit

VO-rated – a flame resistant plastic made from materials with flame protection above industry standards

voltage – difference in electrical charge between two points in a circuit, expressed in volts; force available to push **current** around a circuit; measured in volts; symbol V

voltage drop – loss of **voltage** (electrical pressure) due to **resistance** in **conductor** and components in a circuit

wind farm – a large number of wind turbines gathered in one location. There are currently on-shore and off-shore wind farms in the UK

windage – the air resistance of a moving object or the force of the wind on a stationary object

young people – according to the Management of Health and Safety at Work Regulations, people who have not reached the age of 18

Index